Periodensystem der Elemente mit Gmelin-Systemnummern

IA	IIA	IIIB	IVB	VB	VIB	VIIB	VIII			IB	IIB	IIIA	IVA	VA	VIA	VIIA	0
H 1,2																	He 2,1
Li 3,20	Be 4,26											B 5,13	C 6,14	N 7,4	O 8,3	F 9,5	Ne 10,1
Na 11,21	Mg 12,27											Al 13,35	Si 14,15	P 15,16	S 16,9	Cl 17,6	Ar 18,1
K 19,22 *	Ca 20,28	Sc 21,39	Ti 22,41	V 23,48	Cr 24,52	Mn 25,56	Fe 26,59	Co 27,58	Ni 28,57	Cu 29,60	Zn 30,32	Ga 31,36	Ge 32,45	As 33,17	Se 34,10	Br 35,7	Kr 36,1
Rb 37,24	Sr 38,29	Y 39,39	Zr 40,42	Nb 41,49	Mo 42,53	Tc 43,69	Ru 44,63	Rh 45,64	Pd 46,65	Ag 47,61	Cd 48,33	In 49,37	Sn 50,46	Sb 51,18	Te 52,11	J 53,8	Xe 54,1
Cs 55,25	Ba 56,30	La 57**,39	Hf 72,43	Ta 73,50	W 74,54	Re 75,70	Os 76,66	Ir 77,67	Pt 78,68	Au 79,62	Hg 80,34	Tl 81,38	Pb 82,47	Bi 83,19	Po 84,12	At 85,—	Rn 86,—
Fr 87,31	Ra 88,31	Ac 89***,40	104 71	105 71													

*NH₄ — 23: NH_4 *, 23

**Lanthanide 39:

Ce 58	Pr 59	Nd 60	Pm 61	Sm 62	Eu 63	Gd 64	Tb 65	Dy 66	Ho 67	Er 68	Tm 69	Yb 70	Lu 71

***Actinide:

Th 90,44	Pa 91,51	U 92,55	Np 93,71	Pu 94,71	Am 95,71	Cm 96,71	Bk 97,71	Cf 98,71	Es 99,71	Fm 100,71	Md 101,71	No(?) 102,71	Lr 103,71

Reihenfolge der Gmelin-Systemnummern siehe Innenseite des hinteren Deckels

Gmelin Handbuch der Anorganischen Chemie

Achte, völlig neu bearbeitete Auflage
8th Edition

Organometallic Compounds in the Gmelin Handbook

The following listing indicates in which volumes these compounds are discussed or are referred to:

Ag   Silber B 5 (1975)

Bi   Bismut-Organische Verbindungen (Erg.-Werk, Bd. 47, 1977)

Co   Kobalt-Organische Verbindungen 1 (Erg.-Werk, Bd. 5, 1973) und 2 (Erg.-Werk, Bd. 6, 1973) sowie Kobalt Erg.-Bd. A (1961), B 1 (1963) und B 2 (1964)

Cr   Chrom-Organische Verbindungen (Erg.-Werk, Bd. 3, 1971)

Fe   Eisen-Organische Verbindungen A 1 (Erg.-Werk, Bd. 14, 1974), A 2 (Erg.-Werk, Bd. 49, 1977), A 3 (Erg.-Werk, Bd. 50, 1978), A 6 (Erg.-Werk, Bd. 41, 1977), B 1° (Erg.-Werk, Bd. 36, 1976), B 2* (1978), B 3° (1979), B 4 (1978), B 5 (1978), C 1 (1979), C 2 (1979), C 3* (1980) (present volume) und Eisen B (1929–1932)

Hf   Organohafnium Compounds* (Erg.-Werk, Bd. 11, 1973)

Nb   Niob B 4 (1973)

Ni   Nickel-Organische Verbindungen 1 (Erg.-Werk, Bd. 16, 1975), 2 (Erg.-Werk, Bd. 17, 1974), Register (Erg.-Werk, Bd. 18, 1975) und Nickel B 3 (1966) und C (1968–1969)

Np, Pu   Transurane C° (Erg.-Werk, Bd. 4, 1972)

Pt   Platin C (1939) und D (1957)

Ru   Ruthenium Erg.-Bd. (1970)

Sn   Zinn-Organische Verbindungen 1 (Erg.-Werk, Bd. 26, 1975), 2 (Erg.-Werk, Bd. 29, 1975), 3 (Erg.-Werk, Bd. 30, 1976), 4 (Erg.-Werk, Bd. 35, 1976), 5 (1978) und 6 (1979)

Ta   Tantal B 2 (1971)

Ti   Titan-Organische Verbindungen 1 (Erg.-Werk, Bd. 40, 1977), 2 (1980)

V   Vanadium-Organische Verbindungen (Erg.-Werk, Bd. 2, 1971) und Vanadium B (1967)

Zr   Organozirconium Compounds°

* Completely or ° in part in English

Gmelin Handbuch
der Anorganischen Chemie

Achte völlig neu bearbeitete Auflage

BEGRUNDET VON

Leopold Gmelin

ACHTE AUFLAGE BEGONNEN

im Auftrag der Deutschen Chemischen Gesellschaft
von R.J. Meyer

FORTGEFUHRT VON

E.H.E. Pietsch und A. Kotowski
Margot Becke-Goehring

HERAUSGEGEBEN VOM

Gmelin-Institut für Anorganische Chemie
der Max-Planck-Gesellschaft zur Förderung der Wissen-
schaften
Direktor: Ekkehard Fluck

Springer-Verlag Berlin Heidelberg GmbH 1980

Gmelin Handbuch
der Anorganischen Chemie

Achte, völlig neu bearbeitete Auflage
8th Edition

Fe
Organoiron Compounds

Part C 3

Binuclear Compounds 3

With 42 illustrations

AUTHORS

Gerd Dettlaf (Hamburg), Ulrich Krüerke,
Norbert Kuhn (Mülheim), Marlis Mirbach (Aachen)

EDITOR

Ulrich Krüerke

FORMULA INDEX

Edgar Rudolph

CHIEF EDITOR

Ulrich Krüerke

Springer-Verlag Berlin Heidelberg GmbH 1980

LITERATURE CLOSING DATE: 1977

Die vierte bis siebente Auflage dieses Werkes erschien im Verlag von
Carl Winter's Universitätsbuchhandlung in Heidelberg

Library of Congress Catalog Card Number· Agr 25-1383

ISBN 978-3-662-06051-3   ISBN 978-3-662-06049-0 (eBook)
DOI 10.1007/978-3-662-06049-0

Gesamtherstellung Universitatsdruckerei H. Stürtz AG, Wurzburg

Preface

The present volume is a continuation of Series C on the polynuclear organoiron compounds. It covers the literature completely to the end of 1977 and includes occasional references to the 1978 and 1979 literature.

The first part of this volume, sections 2.4 to 2.4.3, pp. 1/81, describes all binuclear complexes with ligands bonded to iron through four carbon atoms (^{4}L ligands). The second part of the volume, sections 2.5 to 2.5.2.2.13, pp. 82/169, begins the description of binuclear complexes with ligands bonded to iron through five carbon atoms (^{5}L ligands). The large number of binuclear ^{5}L compounds will require one volume more in Series C (Volume C 4) which is planned for the beginning of 1981.

Formulas and symbols have been explained in the prefaces to "Kobalt-Organische Verbindungen" 1, New Suppl. Ser., Vol. 5, and "Nickel-Organische Verbindungen" 1, New Suppl. Ser., Vol. 16. It has not always been possible to indicate the compounds in a chapter exactly by general formulas in the chapter heading. In these cases the user is referred to the introductory remarks at the start of the chapter where the compounds are more precisely characterized.

Much of the data, particularly in the tables, is given in abbreviated form without dimensions; for explanations see p. 170. Additional remarks, if necessary, are given in the texts heading the tables. The location of substances in other organoiron volumes is given in the form "B 4, 1.1.6.1", i.e., Series B (mononuclear compounds), Volume 4, Chapter 1.1.6.1.

The volume contains an empirical formula index on page 171 and a ligand formula index on page 180.

Frankfurt am Main, February 1980   Ulrich Krüerke

Table of Contents

2.4   Compounds with Ligands Bonded by Four Carbon Atoms

2.4.1   Compounds with One ^{4}L Ligand

2.4.1.1   Carbonyl Complexes without Fe–Fe Bond

2.4.1.1.1   Compounds of the $(CO)_4Fe(\mu-^2L-^2L)Fe(CO)_4$ Type

The present chapter deals with compounds, see Table 1, in which two Fe(CO)$_4$ groups are separately coordinated to two olefinic double bonds of an organic molecule.

X-ray structure investigations [7, 13, 18, 19] have shown that the skeleton of the ligand molecule remains intact in all cases. The coordination around each Fe atom can be regarded as approximately trigonal bipyramidal with the olefinic double bond in an equatorial position.

Most of the complexes are prepared from Fe$_2$(CO)$_9$ (usually in excess) and the ligand molecule in an inert solvent at room temperature or slightly above. The conditions are indicated in Table 1 in the form solvent/temperature/time. The abbreviations "chromat." and "cryst." are used for chromatography and crystallization. Frequently, mononuclear compounds of the ^{2}LFe(CO)$_4$ and ^{4}LFe(CO)$_3$ type are also formed; they can be separated from the dinuclear products by chromatography or fractional crystallization. For other methods of preparation see compounds No. 11 and No. 13.

The complexes are quite air-stable in the crystalline state but readily decompose in solution and at higher temperatures. The mass spectra of compounds No. 3, 4, and 10 show the parent ion and the ions resulting from the successive loss of eight CO groups.

Explanations to Table 1: μ_D represents the electric dipole moment in Debye units.

Table 1
Compounds of the $(CO)_4Fe(\mu-^2L-^2L)Fe(CO)_4$ type.
Further information on compounds preceded by an asterisk is given at the end of the table.
For abbreviations and dimensions see p. 170.

No.	Ligand $^2L-^2L$	Preparative conditions (yield in %)	Properties and other remarks Explanations see above	Ref.
*1		hexane/40°/– cryst. from hexane or methanol (24)	orange yellow, dec. at 82 to 87° μ_D (C$_6$H$_6$): about 0.9 D ^{57}Fe-γ (77 K): $\delta=0.09$ (steel), $\Delta=1.80$ IR (C$_2$Cl$_4$): 1980, 2004, 2071	[2, 10]
*2		petroleum ether/40°/2 h, see further information (20)	yellow needles, m.p. 82.5 to 83° (dec.) μ_D (C$_6$H$_6$): about 2.4 D IR (heptane): 1978, 1982, 2003, 2011, 2074, 2085	[5]

References on p. 7

Table 1 [continued]

No.	Ligand ^{2}L–^{2}L	Preparative conditions (yield in %)	Properties and other remarks Explanations on p.1	Ref.
*3	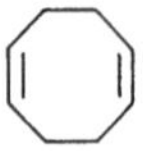	benzene/20°/−, with excess of the ligand in the dark, cryst. from pentane (30)	yellow, m.p. 85 to 88° (dec.) ^{1}H NMR (CH$_2$Cl$_2$): 1.8, 2.7, 3.7 (m's of equal intensity) IR (CH$_2$Cl$_2$): 1965, 1995, 2098 UV (pentane): shoulder at about 278	[11]
4	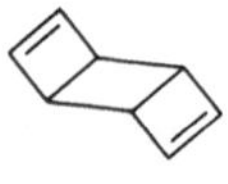	hexane/20°/9 h, separation by chromat. on SiO$_2$ (−)	bright yellow, brittle needles from pentane, benzene or acetone ^{1}H NMR (CD$_3$COCD$_3$): 2.91, 3.9 (s's of equal intensity) IR (hexane): 1988, 2015, 2085	[16]
*5	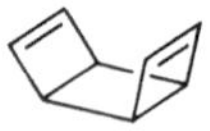	no details available	pale yellow crystals from heptane, no further information	[16, 18, 21]
*6		ether/20°/16 h (−)	yellow needles, m.p. 65 to 66° ^{1}H NMR (C$_6$D$_6$, 6°): 0.7 to 1.2 (m, 4 H), 2.96 (d, 2 H), 4.25 (d, 2 H) ^{13}C NMR (C$_6$D$_6$): 27.0 (CH$_2$), 41.0 (C-1), 70.1 and 74.3 (C-2 to C-5), 213.6 (CO) IR (CCl$_4$): 1971, 2002, 2071 UV (hexane): λ_{max}(lg ε) =249 (4.29), 305 (4.25)	[20]
*7	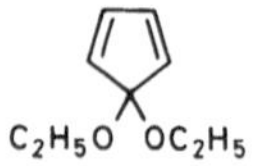 C$_2$H$_5$O OC$_2$H$_5$	pentane/reflux/2 h, chromat. on basic Al$_2$O$_3$ (−)	m.p. 86 to 88° (dec.) ^{1}H NMR (C$_6$D$_6$): 1.1 (t, CH$_3$), 4.45 (m, 4 H), 4.5 (m, CH$_2$) IR (hexane): 1990, 2011, 2022, 2080	[14]
*8	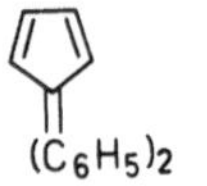 (C$_6$H$_5$)$_2$	petroleum ether/40°/3 h, chromat. on Al$_2$O$_3$ (49 to 66)	orange red, m.p. 106 to 108° (dec.) μ_D (C$_6$H$_6$) =1.73 ±0.2 D IR (KBr): 1965, 2004, 2075, 2088	[1, 4]
9	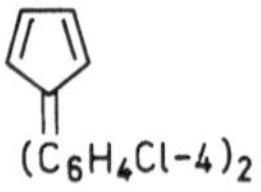 (C$_6$H$_4$Cl-4)$_2$	petroleum ether/40°/1 h, cryst. from petroleum ether (6.5)	orange, m.p. 95 to 125° (dec.) IR (KBr): 1965, 2004, 2083, 2088 soluble in benzene and acetone, slightly in petroleum ether; air-sensitive in solution	[4]
*10		hexane/20°/−, chromat. on Al$_2$O$_3$ and cryst. from hexane at −20° (−)	yellow, m.p. 88 to 90° (dec. with formation of semibullvalene) ^{1}H NMR (CD$_3$COCD$_3$): 1.55 (q, 1 H, J=5.5), 2.71 (d, 2 H, J=5.5), 3.81 (m, 5 H) IR (CS$_2$): 1970, 1995, 2076	[15]

References on p. 7

Table 1 [continued]

No.	Ligand $^2L-^2L$	Preparative conditions (yield in %)	Properties and other remarks Explanations on p. 1	Ref.
*11	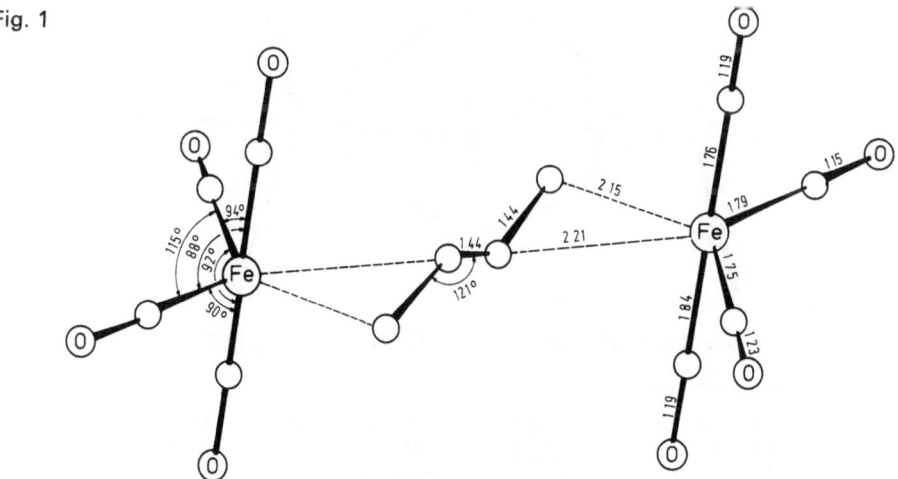	see further information	yellow, m.p. 95.5°	[6]
*12		no solvent/45 to 50°/ 16 h, chromat. on SiO_2, cryst. from light petroleum (3.5)	cream colored, m.p. 147° ^{19}F NMR ($CHCl_3$): 135.8 (F'), 155.6 (F) IR (C_6H_{12}): 2028, 2050, 2060, 2115	[9]
*13	(HOOC– CH=CH_2– CO–NH–)$_2$	$CH_3COOH/40°/5$ h, see further information (49)	bright yellow, m.p. 160 to 170° (dec.)	[17]
14		see further information for compound No. 13	crystals from CH_2Cl_2/C_6H_{14}, dec. at 145 to 155° IR ($CHCl_3$): 2027, 2049, 2118 v(C=O) at 1730	[17]

* Further information:

$C_4H_6Fe_2(CO)_8$ (Table 1, No. 1) crystallizes partially from the preparative solution on cooling to 0 °C; the main product of the reaction is $C_4H_6Fe(CO)_3$ (60% yield), which can be removed by distillation at reduced pressure [2].

Fig. 1

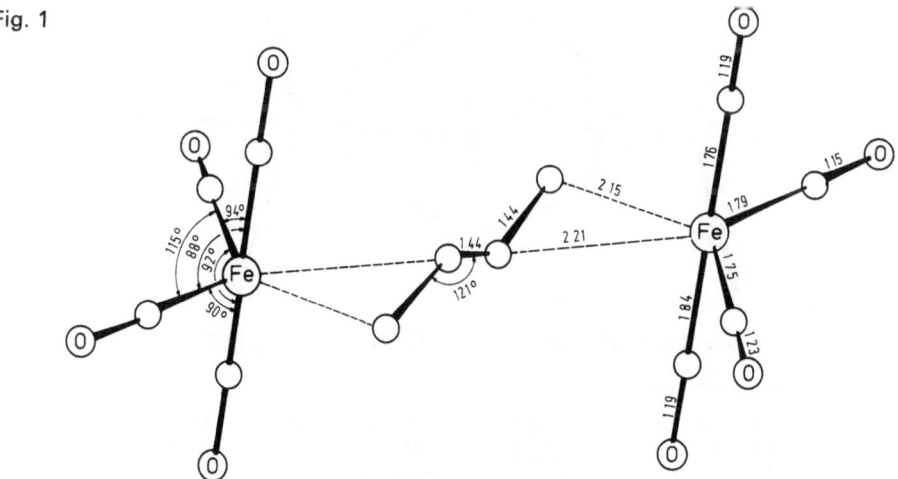

Molecular structure of $C_4H_6Fe_2(CO)_8$ [7].

References on p. 7

The complete IR spectrum in C_2Cl_4 and CS_2 is reported in [2]. The compound is also mentioned in a study of the specific IR intensities of CO groups (integrated intensities per CO group) [3]. – The substance crystallizes in the triclinic system with a = 6.46 (2), b = 6.60 (2), c = 9.80 (3) Å, α = 72.3 (3)°, β = 71.6 (3)°, and γ = 72.3 (3)°; space group P$\bar{1}$ – C$_i^1$. Z = 1 corresponds to D$_c$ = 1.76 g · cm^{-3}. The molecule has a planar trans-butadiene ligand, see **Fig. 1**, p. 3 [7].

Solubilities: low in hexane and benzene, moderate in acetone and methanol. The complex is rather light- and air-stable in the solid state but decomposes readily in solution [2].

$C_6H_8Fe_2(CO)_8$ (Table 1, No. 2). The moderate yield of 20% is only achieved if $Fe_2(CO)_9$ and hexatriene are mixed in a 1:1 mole ratio. Other products are $C_6H_8Fe(CO)_4$, $C_6H_8Fe(CO)_3$, and $C_6H_8Fe_2(CO)_7$ (type $(CO)_4Fe(\mu$-^2L-$^4L)Fe(CO)_3$); at higher mole ratios the last becomes the main product. After removal of the solvent $C_6H_8Fe_2(CO)_8$ crystallizes from the oily mixture on cooling to 0 and −78 °C.

The IR spectrum indicates a coordination of the $Fe(CO)_4$ units to the two terminal double bonds: a $\nu(C=C)$ band occurs at 1622 cm^{-1} but characteristic vinyl vibrations are lacking. A trans configuration for the inner CH=CH group is concluded from the coupling constant J(H,H) = 14.2 Hz, but there is no other relevant data.

The complex dissolves in nonpolar solvents. It is air-stable in the solid state but sensitive to both air and light in solution [5].

Cyclo-$C_8H_{12}Fe_2(CO)_8$ (Table 1, No. 3). By-product of the preparation is cyclo-$C_8H_{12}Fe(CO)_3$. The substance is also formed on storing the unstable liquid cyclo-$C_8H_{12}Fe(CO)_4$ [11].

The complex crystallizes in the triclinic system with a = 10.2628 ± 0.0006, b = 7.082 ± 0.002, c = 6.458 ± 0.001 Å, α = 99.400 ± 0.001 °, β = 96.766 ± 0.001 °, and γ = 96.216 ± 0.001 °; space group P$\bar{1}$ – C$_i^1$. Z = 1 gives D$_c$ = 1.61 g · cm^{-3}. **Fig. 2** shows the chair confor-

Fig. 2

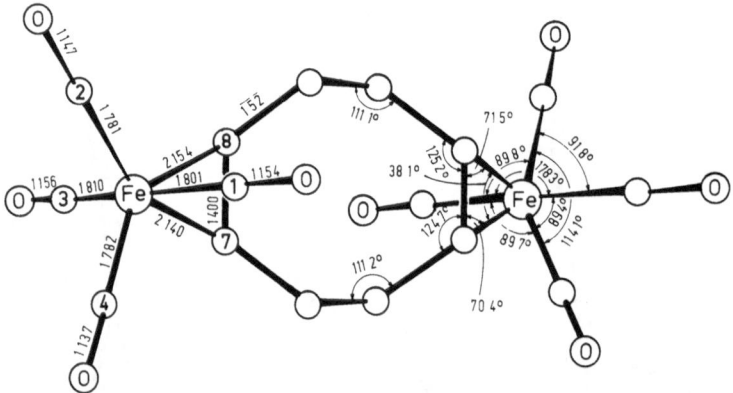

Molecular structure of cyclo-$C_8H_{12}Fe_2(CO)_8$ [13].

Selected angles (°):

C(1)–Fe–C(7)	92.7	C(3)–Fe–C(7)	87.2
C(1)–Fe–C(8)	90.4	C(3)–Fe–C(8)	88.5
C(2)–Fe–C(7)	144.2	C(4)–Fe–C(7)	101.6
C(2)–Fe–C(8)	106.2	C(4)–Fe–C(8)	139.6

References on p. 7

mation of the cycloocta-2,5-diene ligand and the distorted trigonal-bipyramidal environment of the Fe atoms. The axial CO groups are somewhat more distant from the Fe atoms
than the equatorial CO groups (mean values: 1.806 and 1.781 Å, respectively) [13].

The solid compound forms $Fe_3(CO)_{12}$ on slow heating above 77 °C. The reaction
with $P(C_6H_5)_3$ (conditions not reported) yields no substitution product but cycloocta-1,5-
diene (98%), a little cycloocta-1,3-diene and $(CO)_4FeP(C_6H_5)_3$ [11].

Tricyclo-$C_8H_8Fe_2(CO)_8$ (Table 1, No. **5**) is one of the products of the reaction of
syn-tricyclo[4.2.0.0^{2,5}]octa-3,7-diene with $Fe_2(CO)_9$ [18]. The general structure has
been deduced from chemical and spectroscopic evidence [21]. The compound crystallizes
in the orthorhombic system with a =10.021 (2), b =13.986 (2), and c =12.061 (2) Å; space
group $Pca2_1-C_{2v}^5$ or $Pbcm-D_{2h}^{11}$. Z=4 gives D_c=1.73 in agreement with D_m=
1.72 g·cm^{-3}. The rings of the tricyclic ligand are almost planar, see **Fig. 3**. The central

Fig. 3

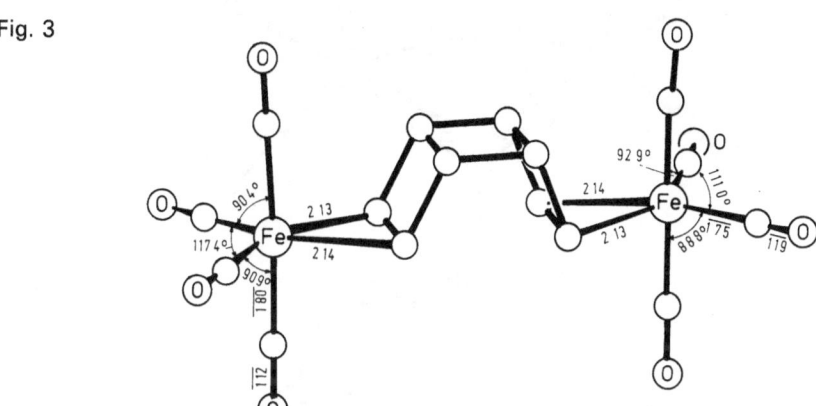

Molecular structure of tricyclo-$C_8H_8Fe_2(CO)_8$ [18].

ring forms angles of 64.1° and 65.1° with the outer rings. The bond lengths within
the rings have the expected values. The coordination around each Fe atom is a distorted
trigonal bipyramid and the olefinic bonds lie almost in the trigonal plane. The maximum
deviation of atoms from a mean plane through the equatorial positions of both the
trigonal bipyramids is 0.09 Å. Differences in Fe-C-O bond lengths can not be demonstrated
due to the high standard deviations and the high thermal motion of the CO groups
at room temperature; average values for each $Fe(CO)_4$ unit are given in Fig. 3 [18].

Spiro-$C_7H_8Fe_2(CO)_8$ (Table 1, No. **6**). This formula had been associated with a deep
red compound [12] which later was found to be $C_7H_8Fe_2(CO)_6$ and having the structure I
[20]. The title compound is formed together with the complexes I, II, and III, yields
were not reported. In solution the compound is converted into II even at room temperature
[20].

$(CO)_2Fe$——$Fe(CO)_4$

I

$(CO)_2Fe$——$Fe(CO)_4$ CH₃

II

$(CO)_2Fe$

III

References on p. 7

Cyclo-$C_5H_4(OC_2H_5)_2Fe_2(CO)_8$ (Table 1, No. **7**). The preparation also yields cyclopentadienone irontricarbonyl, $(\pi-C_5H_4OC_2H_5Fe(CO)_2)_2$, and an unstable unidentified product. These other products also form on refluxing a pentane solution of the title compound; in boiling methylcyclohexane it yields only $(\pi-C_5H_4OC_2H_5Fe(CO)_2)_2$ [14].

Cyclo-$C_4H_4C{=}C(C_6H_5)_2Fe_2(CO)_8$ (Table 1, No. **8**). Other products in the preparation from $Fe_2(CO)_9$ and cyclo-$C_4H_4C{=}C(C_6H_5)_2$ ($=$L) are the complex types $^4LFe(CO)_3$ and $^6LFe_2(CO)_5$. The compound can be recrystallized from benzene/petroleum ether [4]. Cooling of a hexane solution gives small orange rods [19].

Fig. 4

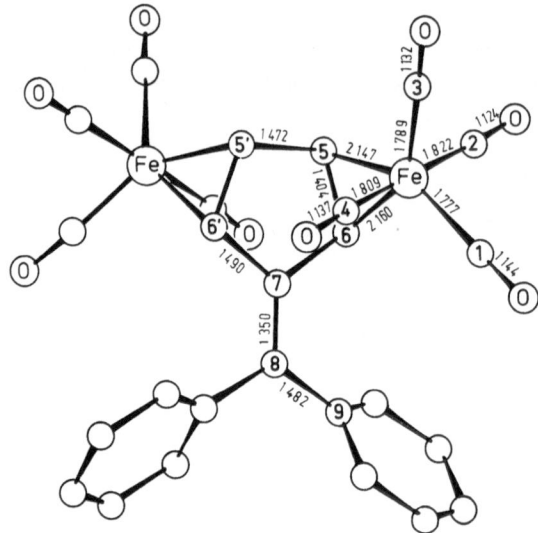

Molecular structure of cyclo-$C_4H_4C{=}C(C_6H_5)_2Fe_2(CO)_8$ [19].

Bond angles (°), "mid" denotes the middle of the olefinic bond:

C(1)–Fe–C(2)	90.5	C(4)–Fe–mid	92.5
C(1)–Fe–C(3)	111.2	Fe–C–O	177.4 (mean)
C(1)–Fe–C(4)	89.9	C(5')–C(5)–C(6)	108.6
C(1)–Fe–mid	125.5	C(5)–C(6)–C(7)	108.8
C(2)–Fe–C(3)	91.4	C(6)–C(7)–C(6')	105.3
C(2)–Fe–C(4)	178.7	C(6)–C(7)–C(8)	127.4
C(2)–Fe–mid	86.3	C(7)–C(8)–C(9)	122.1
C(3)–Fe–mid	123.3		

The crystals belong to the monoclinic system with a$=$22.995(16), b$=$10.096(6), c$=$12.682(8) Å, and $\beta=$124.86(3)°; space group C2/c–C_{2h}^6. Z$=$4 corresponds to $D_c=$1.56 g · cm^{-3}. The two $Fe(CO)_4$ units occupy trans positions relative to the plane of the fulvene ring, see **Fig. 4**. The coordination about the Fe atoms is approximately trigonal–bipyramidal with the olefinic C atoms in the equatorial plane, the angle between this plane and the axis C(5)–C(6) being 5.6°. The plane of the fulvene ring and the equatorial planes form an angle of 112°. The axial Fe–CO distances (mean values) of 1.816 Å are slightly longer than the equatorial distances of 1.783 Å [19], cf. compound No. 3.

The complex is air-stable in the solid state. It is soluble in benzene and acetone, less soluble in ether and petroleum ether. The complex is converted in solution above 60 to 70 °C into the $^4LFe(CO)_3$ type, which is also formed on reacting the complex with the diphenylfulvene [4]. Polarographic reduction occurs irreversibly in two steps at $E_{1/2} = -1.91$ and -2.35 V (in $CH_3OCH_2CH_2OCH_3$ at 22 °C, referred to 10^{-3} M Ag$^+$/Ag) [8].

Tricyclo-$C_8H_8Fe_2(CO)_8$ (Table 1, No. 10) is sensitive to air and heat but can be stored at 0 °C under N_2 without appreciable decomposition [15].

Cyclo-$C_5F_6Fe_2(CO)_8$ (Table 1, No. 11) is formed as a crystalline substance from $Fe(CO)_5$ and the ligand under the influence of UV radiation, 8% yield after chromatographic isolation. The structure is deduced from the ^{19}F NMR spectrum (three pairs of equivalent F atoms) and the IR spectrum (no bridging carbonyls and coordination of both olefinic bonds), but the spectral data are not reported [6].

Bicyclo-$C_6F_6Fe_2(CO)_8$ (Table 1, No. 12). The reaction of $Fe_2(CO)_9$ with "Dewar hexafluorobenzene" yields a complex mixture of products from which only the title compound and bicyclo-$C_6F_6Fe(CO)_4$ could be isolated. There is no band in the IR spectrum assignable to an uncomplexed CF=CF group; other bands are found at 765, 993, 1122, 1235, and 1404 cm^{-1} (in $CHCl_3$) [9].

(HOOC-CH=CH-CO-NH-)$_2$Fe$_2$(CO)$_8$ (Table 1, No. 13) is also obtained by reacting $C_4H_2O_3Fe(CO)_4$ ($C_4H_2O_3$=maleic anhydride) with $N_2H_4 \cdot H_2O$ in CH_3COOH at room temperature for 1 h, 67% yield.

The air-stable complex is soluble in CH_2Cl_2, $CHCl_3$, and CH_3COOH. Dehydration and ring closure of the ligand under the influence of acetic anhydride at 25 to 30 °C (12.5 h) gives compound No. 14, which is isolated by thin layer chromatography on Al_2O_3 with $CHCl_3$ and precipitation from CH_2Cl_2 with hexane, 12.5% yield [17].

References:

[1] E. Weiss, W. Hübel (Angew. Chem. **73** [1961] 298/9). – [2] H.D. Murdoch, E. Weiss (Helv. Chim. Acta **45** [1962] 1156/61). – [3] K. Noack (Helv. Chim. Acta **45** [1962] 1847/59). – [4] E. Weiss, W. Hübel, J. Nielsen, A. Gérondal (Chem. Ber. **95** [1962] 1186/96). – [5] H.D. Murdoch, E. Weiss (Helv. Chim. Acta **46** [1963] 1588/94).

[6] R.E. Banks, T. Harrison, R.N. Hazeldine, A.B.P. Lever, T.F. Smith, J.B. Walton (Chem. Commun. **1965** 30/1). – [7] K.A. Klandermann (Diss. Univ. of Wisconsin 1965; Diss. Abstr. **25** [1964/65] 6253). – [8] R.E. Dessy, R.L. Pohl (J. Am. Chem. Soc. **90** [1968] 1995/2001). – [9] D.J. Cook, M. Green, N. Mayne, F.G.A. Stone (J. Chem. Soc. A **1968** 1771/5). – [10] L. Korec, K. Burger (Acta Chim. [Budapest] **58** [1968] 253/63).

[11] E. Koerner von Gustorf, J.C. Hogan (Tetrahedron Letters **1968** 3191/4). – [12] C.H. DePuy, V.M. Kobal, D.H. Gibson (J. Organometal. Chem. **13** [1968] 266/8). – [13] C. Krüger (J. Organometal. Chem. **22** [1970] 697/706). – [14] A. Eisenstadt, G. Scharf, B. Fuchs (Tetrahedron Letters **1971** 679/82). – [15] D. Ehntholt, A. Rosan, M. Rosenblum (J. Organometal. Chem. **56** [1973] 315/21).

[16] R.S. Case (Diss. Univ. of Texas 1974; Diss. Abstr. Intern. B **35** [1974/75] 3823). − [17] A.N. Nesmeyanov, M.I. Rybinskaya, L.V. Rybin, A.V. Arutyunyan (Zh. Obshch. Khim. **44** [1974] 604/13; J. Gen. Chem. [USSR] **44** [1974] 578/85). − [18] V.N. Narasimhachari (Diss. Univ. of Texas 1975; Diss. Abstr. Intern. B **36** [1975] 745). − [19] U. Behrens (J. Organometal. Chem. **107** [1976] 103/11). − [20] P. Eilbracht, U. Mayser (J. Organometal. Chem. **135** [1977] C26/C28).

[21] W. Sleiger, R. Pettit (personal communication cited in [18]).

2.4.1.1.2   Other Compounds of the $(CO)_3Fe(\mu-^4L-^2D)Fe(CO)_4$ and $(CO)_3Fe(\mu-^2L-^2L)Fe(CO)_3$ Type

The compounds treated in this section are represented by Formulas I to IV. In compounds I to III the heterocyclic rings coordinate to different ironcarbonyl groups through their diene systems and the P or As donor site ($^4L-^2D$ ligand). Compound IV contains a $^2L-^2L$ ligand consisting of two pairs of carbenes bonded to a central tin atom.

I   II   III

IV

$(CO)_3Fe(\mu-C_4H_2(CH_3)_2PC_6H_5)Fe(CO)_4$ (Formula I) is prepared by the reaction of $Fe_3(CO)_{12}$ with the phosphole (2:1 mole) in boiling toluene for 4 h and is isolated by chromatography on SiO_2 with C_6H_{12}/C_6H_6 (7:3), 72% yield.

The yellow−brown crystalline substance melts at 122 °C. The ^1H NMR spectrum ($CDCl_3$) shows resonances at $\delta=2.22$ (CH_3), 2.98 (H-2,5, J(H,P) =24 Hz), 7.09 (C_6H_5, H-ortho), and 7.29 (C_6H_5, H-meta, para) ppm. ^{31}P NMR spectrum ($CDCl_3$): $\delta=20.7$ ppm (referred to external P_4O_6). Seven $v(CO)$ bands in the IR spectrum (decaline) are assigned in the following way: 1994, 2015, 2075 cm^{-1} to $Fe(CO)_3$ and 1939, 1955, 1983, 2053 cm^{-1} to $Fe(CO)_4$. The mass spectrum involves the parent ion $[M]^+$ and the $[M-nCO]^+$ fragments with n=1 to 7.

The spectral data are consistent with structure I, but the position of the ironcarbonyl groups with respect to the ring plane is not yet known [2].

$(CO)_3Fe(\mu-C_4(C_4H_8)_2PC_6H_5)Fe(CO)_4$ (Formula II) has been obtained from $Fe_3(CO)_{12}$ and the phosphole at 80 °C/3 h in about 12% yield besides the $^2DFe(CO)_4$ type and $\pi-C_4(C_4H_8)_2P(O)C_6H_5Fe(CO)_3$. The complex melts at 135 °C with decomposition. The

^{31}P NMR spectrum shows no chemical shift relative to P_4O_6. IR spectrum (decaline): $\nu(CO)$ bands at 1937, 1949, 1978, 1988, 1997, 2020, 2023, 2046, and 2061 cm^{-1}. In the mass spectrum the molecular ion [M]$^+$ occurs with low intensity (0.6%); [M−7CO]$^+$ gives the strongest peak (69%) among the fragments of the successive loss of CO groups [3].

$(CO)_3Fe(\mu\text{-}C_4H_2(CH_3)_2AsC_6H_5)Fe(CO)_4$ (Formula III) is formed in low yield from $Fe_3(CO)_{12}$ and the arsole in boiling toluene; main product (8 to 9% yield) is the $(CO)_4Fe^2D$ type. The yellow compound, crystallized from hexane at 15 °C, melts at about 99 °C. ^1H NMR spectrum (CDCl$_3$): $\delta=1.70$ (s, 2CH$_3$), 5.50 (s, H-3,4), and 7.27 (s, broad, C_6H_5). IR spectrum (decaline): 1944, 1954, 1973, 1980, 1989, 2052, and 2064 cm^{-1}. In the mass spectrum [M]$^+$ and [M−CO]$^+$ are very weak (0.3%); the strongest peaks belong to [M−7CO]$^+$ (23%), [M−7CO−Fe]$^+$ (30.5%), and [M−7CO−2Fe]$^+$ (100%) [4].

$(CO)_3Fe(\mu\text{-}(N(CH_3)_2CO)_2Sn(OCN(CH_3)_2)_2)Fe(CO)_3$ (Formula IV) results from the addition of $Sn(N(CH_3)_2)_4$ to two CO groups on each of two molecules $Fe(CO)_5$. The reaction is carried out in a mole ratio 1:2 in benzene at room temperature for 90 min. After removal of the solvent in a vacuum, a viscous oil remains which crystallizes on stirring with pentane at 0 °C, 90% yield.

The ^1H NMR spectrum (C_6D_6) of the orange–yellow substance shows two singlets at $\delta=2.59$ (cis CH$_3$) and 2.93 (trans CH$_3$) ppm, which coalesce at 38 °C due to a rotation about the C−N axis. The free enthalpy of activation for the rotation has been calculated to be $\Delta G^{\neq}=16.5$ kcal·mol^{-1}. This is lower than in monocarbene complexes and appears to indicate an increased $p_\pi\text{-}p_\pi$ interaction between C and O at the expense of the C−N contribution. Three $\nu(CO)$ bands would be expected for both axial and axial–equatorial coordination of each two carbene ligands, but four bands are observed (in CH$_2$Cl$_2$) at 1936, 1960, 2005, and 2035 cm^{-1}. Other bands are assigned as $\nu(SnO)$ and $\delta(FeCO)$ at 593, 610, 644, 657, and 677 cm^{-1}, $\nu(C-O)$ at 1240 and $\nu(C-N)$ at 1503 cm^{-1}.

The complex decomposes in air only slowly [1].

References:

[1] W. Petz, A. Jonas (J. Organometal. Chem. **120** [1976] 423/32). − [2] F. Mathey, G. Muller (J. Organometal. Chem. **136** [1977] 241/9). − [3] F. Mathey, D. Thavard (Can. J. Chem. **56** [1978] 1952/5). − [4] G. Thiollet, F. Mathey (Inorg. Chim. Acta **35** [1979] L331/L332).

2.4.1.2 Carbonyl Complexes with Fe−Fe Bond

2.4.1.2.1 Compounds Containing Cumulene Ligands

General Literature:

A. Zimniak, Complexes of Butatrienes with Hexacarbonyldiiron and Tetracarbonyliron, Wiad. Chem. **32** [1978] 35/48 (Polish) from C.A. **89** [1978] No. 59909.

The ^4L ligands of the complex type $(CO)_3Fe(\mu\text{-}^4L)Fe(CO)_3$ in 2.4.1.2.1.1 belong to the butatriene and hexapentaene systems. Only four of the six C atoms of the hexapentaenes are coordinated to the Fe atoms. Some phosphine and phosphite monosubstitution products in 2.4.1.2.1.2 have been prepared from butatriene compounds. Intermediates of composition $^4LFe_2(CO)_6{}^2D$ can be isolated if $P(C_4H_9-n)_3$ is used as donor molecule, see 2.4.1.2.1.3.

2.4.1.2.1.1   Cumulene Compounds of the $(CO)_3Fe(\mu-^4L)Fe(CO)_3$ Type

In earlier publications some of the butatriene compounds were described as $^4LFe_2(CO)_5$ complexes [1 to 4]. However, mass spectrometric studies [7, 8, 9, 11, 18] and systematic oxygene analyses [5] revealed the correct formulation with six CO groups.

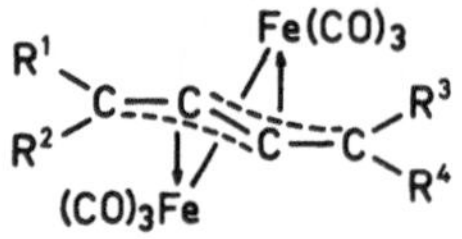

I 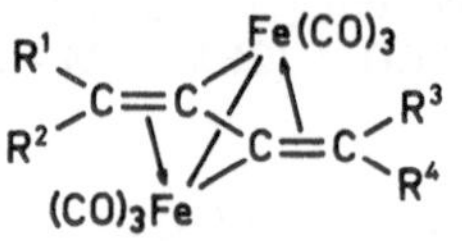   II

The cumulene system, linear in the free state, adopts a nonlinear form with angles of about 130° on the two central C atoms, see **Fig. 6**, p. 20, and **Fig. 7**, p. 23. Two alternative bonding pictures are shown by Formulas I and II. In both cases the ligand acts as a 3–electron donor for each Fe atom. Equivalent Fe atoms are indicated by the ^{57}Fe Mössbauer spectrum of compound No. 1.

Isomeric forms have been observed and partly characterized by ^1H NMR spectroscopy (compounds No. 2, 3, 4, 6, 8, 14, and 15). Other compounds seem to occur only in one form with the bulky substituents occuping the positions R^1 and/or R^4 more remote from the $Fe(CO)_3$ groups [19].

Methods of preparation:

Method I: Dihalogen derivatives of the types III, IV or V with X=Cl or Br are dehaloge-nated in boiling isooctane by Zn dust in the presence of $Fe_3(CO)_{12}$ or without Zn by an excess of $Fe_3(CO)_{12}$. The reaction time is about 3 h. This method permits the preparation of complexes with ligands which are unstable in the free state [3, 4, 5, 16, 19].

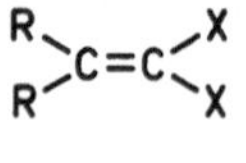

III 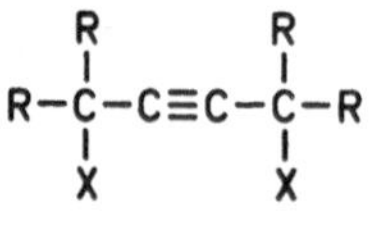   IV 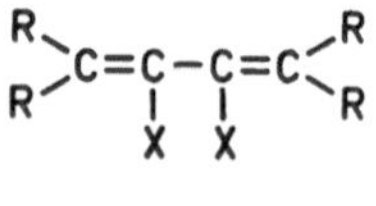   V

Method II: The compounds IV (X=Cl) are reacted with $Na_2[Fe(CO)_4]$ in tetrahydrofuran at reflux for 1 h or with $K_2[Fe_2(CO)_8]$ in methanol/hexane at room temperature for 3 h [5].

Method III: Butynediols of the type VI and $Fe_3(CO)_{12}$ (1:1.4 mole) are heated in tetrahy-drofuran at reflux temperature for 1 h. The yields of the less substituted products (two to four R=H) are low. They can be increased by using the compounds VII as starting materials and treating them with 2.5 moles $Fe_3(CO)_{12}$ [21].

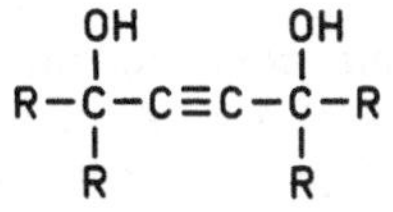

VI

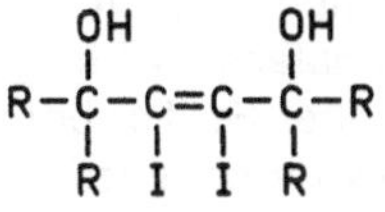

VII

Method IV: $Fe_3(CO)_{12}$ is reacted directly with a sufficiently stable cumulene (2:1 mole) in boiling isooctane or toluene for 2 h [5, 15, 16, 19].

Method V: A fivefold excess of $Fe_2(CO)_9$ is treated with a cumulene in benzene or ether at room temperature for 20 h [16].

The compounds can be isolated and purified by chromatography on SiO_2 or Al_2O_3 and by fractional crystallization. Several products can also be sublimed in vacuum.

^1H NMR and ^{13}C NMR spectra have recently been studied in more detail with the aim of clarifying the orientation of the substituents in the cumulene ligand. The long range coupling values J(CCCH) have confirmed the conclusion based on the chemical shifts that CH_3 and C_6H_5 groups assume opposite orientations at the coordinated double bonds. CH_3 is preferentially trans to carbon while C_6H_5 is preferentially cis, thus compounds VIII A and IX A being the major components [24]. The carbonyl region of the ^{13}C NMR spectra shows only one resonance; thus some process must average the two different types of CO groups. This is not the case for compound No. 21, where the bulky $t-C_4H_9$ groups presumably prevent the necessary motion for the averaging [16].

A

B

VIII

A

B

IX

The mass spectra generally show the usual pattern of the parent ions $[M]^+$ and the fragments from the successive loss of CO groups, $[M - n\ CO]^+$ with n=1 to 6. Frequently shown are also ions of the free cumulene, $[Fe_2]^+$, and $[Fe]^+$ [9, 11].

References on p. 21

The usually red, diamagnetic complexes are air-stable in the solid state; gradual decomposition occurs in solution in contact with the atmosphere [4, 15, 19]. Degradations with Br_2 and I_2 in CH_3OH revealed a decreased reactivity for the bulkier substituents [22] like those in compounds No. 11 to 13. Reactivities of the complexes and the corresponding free cumulenes with I_2/CH_3OH are compared in [23]. CO substitution takes place trans to the Fe–Fe bond, see 2.4.1.2.1.2. Adducts of the composition $^4LFe_2(CO)_6^2D$ can only be obtained with $^2D = P(C_4H_9-n)_3$, see 2.4.1.2.1.3.

The kinetics of CO substitution by $P(C_6H_5)_3$ and $P(OC_6H_5)_3$ have been investigated for compounds No. 1, 13, and 19. The substitution in compound No. 1 follows a second-order rate law, $v = k_2[complex][^2D]$. However, the reactions of Nos. 13 and 19 are more complex as the measurements imply a rate law which also contains a first-order term, $v = k_1[complex] + k_2[complex][^2D]$. The constant k_1 probably characterizes a rate determining dissociative step $(\rightarrow {}^4LFe_2(CO)_5 + CO)$ while k_2 arises from a pathway which forms an unstable intermediate adduct, presumably by formation of a Fe–P bond and simultaneous insertion of a CO ligand into an iron–butatriene carbon bond. The following selected data are for reactions in decaline at 100 to 110 °C:

Complex	^2D	$k_1 \times 10^5$ s^{-1}	$k_2 \times 10^3$ $l \cdot mol^{-1} \cdot s^{-1}$	ΔH^* $kcal \cdot mol^{-1}$	ΔS^* $cal \cdot mol^{-1} \cdot K^{-1}$
No. 1	$P(C_6H_5)_3$	–	0.42(2)	16 (1)	−32 (2)
	$P(OC_6H_5)_3$	–	0.118(7)	20 (1)	−24 (2)
No. 13	$P(C_6H_5)_3$	1.2(3)	0.10(1)	24 (1)	−15 (4)
	$P(OC_6H_5)_3$	1(1)	0.34(2)	17 (1)	−29 (3)
No. 19	$P(C_6H_5)_3$	4(1)	2.8(1)	14 (1)	−32 (1)
	$P(OC_6H_5)_3$	4(1)	2.0(2)	16 (1)	−27 (3)

The activation parameters for dissociation of CO in compound No. 13 are $\Delta H^* = 32(2)$ kcal · mol^{-1} and $\Delta S^* = 40(5)$ cal · mol^{-1} · K^{-1} [20].

Reactions of the complexes No. 1, 13, and 19 with $P(C_4H_9-n)_3$ are very rapid and yield the adducts $^4LFe_2(CO)_6P(C_4H_9-t)_3$. The rate determining second-order reaction could only be measured for compound No. 13 at 40 to 60 °C: at 50 °C $k_2 = 4.0(1) \times 10^{-3}$ l · mol^{-1} · s^{-1}, $\Delta H^* = 13(1)$ kcal · mol^{-1}, and $\Delta S^* = -30(3)$ cal · mol^{-1} · K^{-1}. The substantial rate difference between this reaction and the reactions with $P(C_6H_5)_3$ and $P(OC_6H_5)_3$ are reflected only in the enthalpies of activation [20].

Explanations to Table 2: Isomer A designates the major component of an isomeric mixture. Other isomers have not been obtained in a pure form, but their spectra have been extracted from the spectra of the isomeric mixtures. — The numbering of the C atoms and substituents of the cumulene ligand is given in Formula X.

X

References on p. 21

Table 2
Cumulene compounds of the $(CO)_3Fe(\mu-^4L)Fe(CO)_3$ Type.
Further information on compounds preceded by an asterisk is given at the end of the table.
For abbreviations and dimensions see p. 170.

No. Cumulene ligand Method of preparation (yield in %)		Properties and further remarks Explanations on p. 12	Ref.
*1	$\overset{H}{\underset{H}{>}}\overset{1}{C}=\overset{2}{C}=\overset{3}{C}=\overset{4}{C}\overset{H}{<}_{H}$ I (17), II (11), III (14)	red orange; m.p. 65 to 66°, 69 to 70° ^{1}H NMR (CDCl$_3$): 4.05, 4.69 (d's, J=1.7) ^{13}C NMR (CDCl$_3$): 70.6 (C-1,4), 123.3 (C-2,3), 209.6 (CO) ^{57}Fe-γ (77 K): δ=0.275, Δ=0.98 IR (C$_6$H$_{14}$): 1983, 2001, 2012, 2040, 2079 UV (C$_2$H$_5$OH): λ_{max}(lg ε) =346(3.734), 448(3.356)	[4, 5, 13, 16, 21, 24]
2	$\overset{CH_3}{\underset{H}{>}}C=C=C=C\overset{H}{<}_{H}$ IV (−) 2 isomers	m.p. 64 to 66° ^{1}H NMR (CDCl$_3$), isomer A: 1.83 (CH$_3$-t), 4.06 (H-t), 4.70 (H-c'), 5.86 (H-c); isomer B: 2.01 (CH$_3$-c), 4.06 (H-t'), 4.63 (H-c'), 5.10 (H-t) ^{13}C NMR (CDCl$_3$), isomer A: 19.4 (CH$_3$-t), 92.3, 123.1, 124.6, 71.1 (C-1 to 4), 210.4 (CO); isomer B: 26.9 (CH$_3$-c), 94.4, 121.1, 123.3, 71.1 (C-1 to 4), 210.6 (CO)	[24]
3	$\overset{C_6H_5}{\underset{H}{>}}C=C=C=C\overset{H}{<}_{H}$ III (41) 2 isomers	m.p. 77 to 78° ^{1}H NMR (CDCl$_3$), isomer A: 4.22 (H-t'), 4.91 (H-c', d's, J=1.9), 6.13 (H-t), 7.31 (C$_6$H$_5$-c); isomer B: 4.18 (H-t'), 4.91 (H-c', d's, J=1.9), 6.55 (H-c), 7.25 (C$_6$H$_5$-t) ^{13}C NMR (CDCl$_3$), isomer A: 97.1, 116.6, 124.2, 70.7 (C-1 to 4), 140.1 (C$_6$H$_5$-c), 209.8 (CO); isomer B: 96.6, 120.1, 125.6, 71.8 (C-1 to 4), 136.9 (C$_6$H$_5$-t), 209.6 (CO) IR (C$_6$H$_{14}$): 1982, 2000, 2004, 2035, 2075	[21, 24]
*4	$\overset{CH_3}{\underset{H}{>}}C=C=C=C\overset{CH_3}{<}_{H}$ I (4.4), III (38) 3 isomers	red orange; m.p. 104° ^{1}H NMR (CDCl$_3$), isomer A: 1.87 (CH$_3$-t', CH$_3$-t, d's, J=6.3), 5.86 (H-c', H-c); isomer B: 1.85 (CH$_3$-t'), 2.03 (CH$_3$-c), 5.14 (H-t), 5.86 (H-c'); isomer C: 2.01 (CH$_3$-c, CH$_3$-c'), 5.16 (H-t, H-t')	[3, 21, 24]

[continued on p. 14]

Table 2 [continued]

No.	Cumulene ligand Method of preparation (yield in %)	Properties and further remarks Explanations on p. 12	Ref.

*4 [continued]

^{13}C NMR (CDCl$_3$), isomer A: 19.3 (CH$_3$-t, CH$_3$-t'), 92.1, 123.3, 123.2, 92.1 (C-1 to 4), 210.5 (CO); spectra not resolved for minor isomers B and C
IR (C$_6$H$_{14}$): 1977, 1993, 2000, 2032, 2071
UV (C$_2$H$_5$OH): λ_{max}(lg ε) = 298(3.83), 342(3.728), 452(3.303)

5   CH$_3$\
     >C=C=C=C<$\overset{H}{_{H}}$
    CH$_3$/

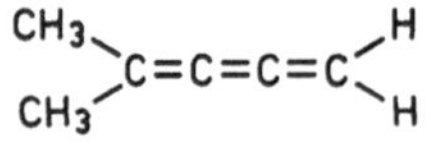

III (14)

m.p. 46 to 47°   [21,
^{1}H NMR (CDCl$_3$): 1.96, 2.21 (s's,   24]
CH$_3$), 4.10 (H-t'), 4.67 (H-c'), d's, J=1.1
^{13}C NMR (CDCl$_3$): 26.8 (CH$_3$-t), 35.4 (CH$_3$-c), 71.9, 118.5, 121.1, 124.1 (C-1 to 4), 210.6 (CO)
IR (C$_6$H$_{14}$): 1980, 1997, 2003, 2034, 2076

6   C$_6$H$_5$\
     >C=C=C=C<$\overset{C_6H_5}{_{H}}$
    H/

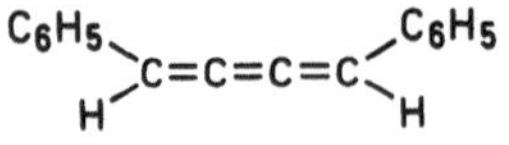

IV (−)   2 isomers

m.p. 119 to 120°   [24]
^{1}H NMR (CDCl$_3$), isomer A: 6.24 (H-t, H-t'), ≈7.4 (C$_6$H$_5$-c, C$_6$H$_5$-c'); isomer B: 6.24 (H-t'), no assignment made for the other protons
^{13}C NMR (CDCl$_3$), isomer A: 97.3, 117.3, 117.3, 97.3 (C-1 to 4), 140.0 (C$_6$H$_5$-c, C$_6$H$_5$-c'), 209.7 (CO); isomer B: 96.8, 120.5, 118.4, 98.7 (C-1 to 4), 136.9 (C$_6$H$_5$-t), 140.0 (C$_6$H$_5$-c'), 209.7 (CO)

*7   C$_6$H$_5$\
      >C=C=C=C<$\overset{H}{_{H}}$
    C$_6$H$_5$/

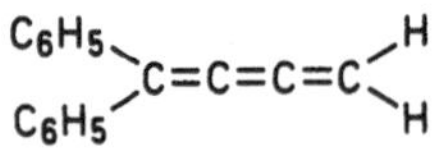

I (1), III (38)

orange; m.p. 81 to 82°   [19,
^{1}H NMR (CDCl$_3$): 4.05 (H-t'), 4.89   21,
(H-c', d's, J=1.4), 7.14 to 7.66   24]
(C$_6$H$_5$)
^{13}C NMR (CDCl$_3$): 69.9, − (?), 121.8, 124.2 (C-1 to 4), 140.7 (C$_6$H$_5$-t), 141.3 (C$_6$H$_5$-c), 209.6 (CO)
IR (C$_6$H$_{14}$): 1983, 1995, 2006, 2034, 2071

References on p. 21

Table 2 [continued]

No.	Cumulene ligand Method of preparation (yield in %)	Properties and further remarks Explanations on p. 12	Ref.
8	CH_3, H–C=C=C=C–CH_3, CH_3 III (29) 2 isomers	m.p. 74 to 76° ^{1}H NMR (CDCl$_3$), isomer A: 1.86 (CH_3-t, d, J=6.3), 1.96 (CH_3-t′), 2.17 (CH_3-c′), 5.84 (H–c, q, J=6.3); isomer B: 1.96 (CH_3-t′), 1.99 (CH_3-c, d, J=6.3), 2.13 (CH_3-c′), 5.19 (H–t, q, J=6.3) ^{13}C NMR (CDCl$_3$), isomer A: 19.7 (CH_3-t), 93.3, 122.8, 120.8, 119.2 (C–1 to 4), 27.1 (CH_3-t′), 35.2 (CH_3-c′) 211.4 (CO); isomer B: 27.1 (CH_3-c, CH_3-t′), 96.0, 121.4, 118.9, 118.9 (C–1 to 4), 35.2 (CH_3-c′), 211.4 (CO) IR (C_6H_{14}): 1977, 1995, 2002, 2033, 2071	[21, 24]
*9	CH_3, CH_3–C=C=C=C–CH_3, CH_3 I (4 to 11), II (36) III (66), IV (65)	orange red; m.p. 115 to 117°, 118° ^{1}H NMR (CDCl$_3$): 1.95 (CH_3-t, CH_3-t′), 2.14 (CH_3-c, CH_3-c′) ^{13}C NMR (CDCl$_3$): 27.2 (CH_3-t, CH_3-t′), 35.0 (CH_3-c, CH_3-c′), 120.3, 120.1 (C–1 to 4), 211.6 (CO) IR (C_6H_{14}): 1965, 1983, 1997, 2004, 2062 UV (C_2H_5OH): λ_{max}(lg ε) =299(3.86), 340 (sh), 470(3.299)	[3, 5, 21, 24]
*10	t-C_4H_9, CH_3–C=C=C=C–CH_3, C_4H_9-t IV (61)	blood red; dec. at 143 to 146° ^{1}H NMR (CCl$_4$): 1.26 (C_4H_9-t, s), 1.97 (CH_3, s) IR (CCl$_4$): 1972, 1987, 2000, 2028, 2070	[19]
*11	t-C_4H_9, C_6H_5–C=C=C=C–C_6H_5, C_4H_9-t IV (68)	orange red; dec. at 190 to 195° ^{1}H NMR (CCl$_4$): 1.29 (C_4H_9-t, s), 7.15 to 7.50 (C_6H_5, m) IR (CCl$_4$): 1984, 1994, 2020, 2040, 2083	[15, 19]
*12	C_6H_5, C_6H_5–C=C=C=C–C_6H_5, C_4H_9-t IV (72)	dark red; dec. at 182 to 186° ^{1}H NMR (CCl$_4$): 1.45 (C_4H_9-t, s), 7.30 (C_6H_5, m) IR (CCl$_4$): 1970, 1985, 2002, 2025	[19]

References on p. 21

Table 2 [continued]

No.	Cumulene ligand Method of preparation (yield in %)	Properties and further remarks Explanations on p. 12	Ref.
*13	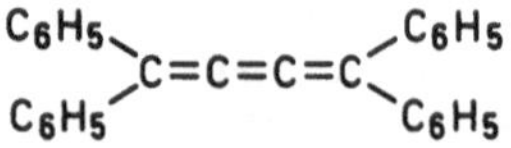 C_6H_5 C=C=C=C $C_6H_5$$$$C_6H_5$,$C_6H_5$ IV (79), V (94)	red; dec. at 206° IR (CCl_4): 1984, 2000, 2033, 2065 UV (C_2H_5OH): no distinctive maxima	[4, 5, 6]
14	4-$CH_3C_6H_4$ C=C=C=C $C_6H_5$$$$C_6H_5$, $C_6H_4CH_3$-4 IV (−)$$2 isomers	dark red oil; dec. at 90 to 100° ^{1}H NMR (CCl_4): 2.20, 2.39 (CH_3, s's), 7.30 (C_6H_5, C_6H_4, m) IR (CCl_4): 1980, 1995, 2005, 2030, 2070 mixture of isomers	[15, 19]
15	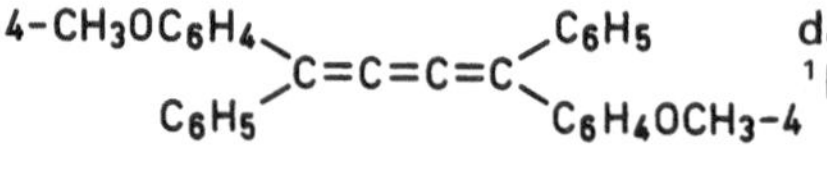 4-$CH_3OC_6H_4$ C=C=C=C $C_6H_5$$$$C_6H_5$, $C_6H_4OCH_3$-4 IV (−)$$2 isomers	dark red oil; dec. at 70 to 80° ^{1}H NMR (CCl_4): 3.75, 3.79 (OCH_3, s's), 6.70, 7.40 (C_6H_5, C_6H_4, m) IR (CCl_4): 1980, 1995, 2005, 2030, 2070 mixture of isomers	[15, 19]
*16	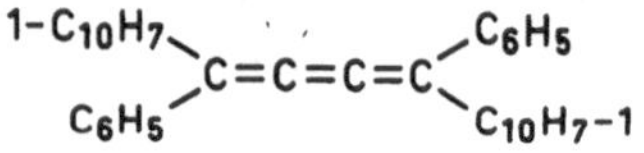 1-$C_{10}H_7$ C=C=C=C $C_6H_5$$$$C_6H_5$, $C_{10}H_7$-1 IV (26)	orange red; dec. at 201° IR (CCl_4): 2000, 2022, 2051, 2087	[15, 19]
*17	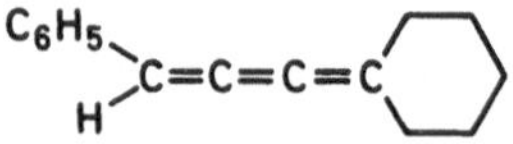 C_6H_5 C=C=C=C⟨cyclohexyl⟩ H III (41)	m.p. 124 to 126° ^{1}H NMR ($CDCl_3$): 1.82 to 2.47 (C_5H_{10}, m), 6.13 (H–t, s), 7.25 (C_6H_5–c, m) ^{13}C NMR ($CDCl_3$): 26.4 to 47.3 (C_5H_{10}), 98.4, 119.9, 117.0, (C–1 to 4), 140.5 (C_6H_5–c), 210.6 (CO) IR (C_6H_{14}): 1976, 1995, 2000, 2033, 2065	[21, 24]
*18	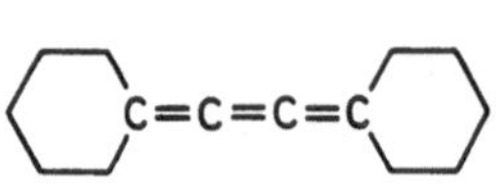 ⟨cyclohexyl⟩C=C=C=C⟨cyclohexyl⟩ III (76)	m.p. 112 to 113° ^{1}H NMR ($CDCl_3$): 1.72 to 2.35 (C_5H_{10}, m) ^{13}C NMR ($CDCl_3$): 26.5 to 47.2 (C_5H_{10}), 129.8, 117.8, 117.8, 129.8 (C–1 to 4), 211.6 (CO) IR (C_6H_{14}): 1966, 1981, 1992, 2025, 2061	[21, 24]

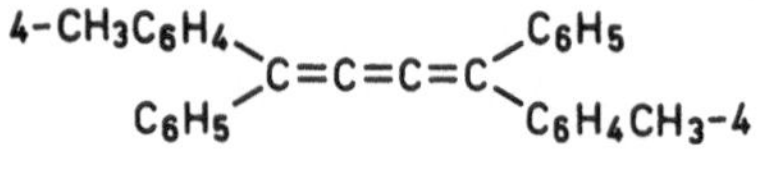

References on p. 21

Table 2 [continued]

No.	Cumulene ligand Method of preparation (yield in %)	Properties and further remarks Explanations on p. 12	Ref.
*19	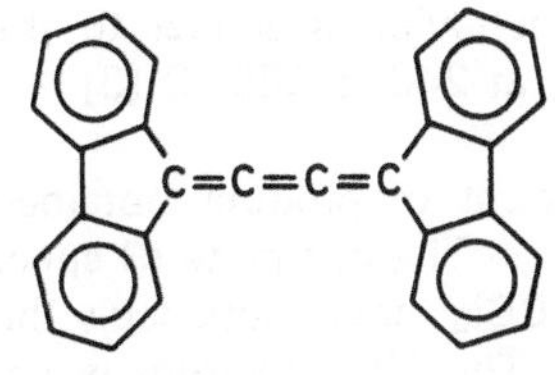 (fluorenylidene cumulene structure) C=C=C=C IV (72)	dark red; dec. at 176° IR (CCl$_4$): 1976, 2008, 2033, 2066	[5]
20	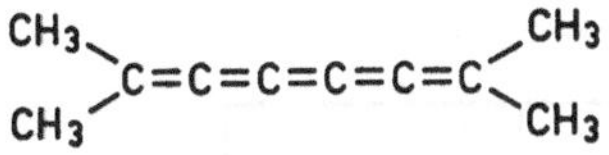 CH$_3$, CH$_3$ \ C=C=C=C=C=C / CH$_3$, CH$_3$ I (2)	red; m.p. 117 to 119° ^{1}H NMR (CDCl$_3$): 2.22 (s) IR (C$_6$H$_{14}$): 1984, 1992, 1998, 2005, 2009, 2031, 2059, 2078; the complex spectrum suggests a mixture of isomers	[16]
*21	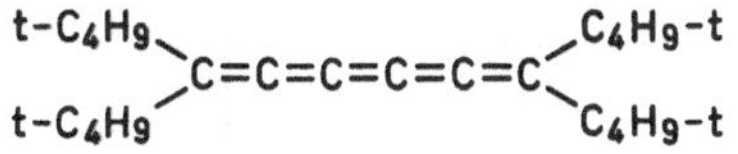 t-C$_4$H$_9$, C$_4$H$_9$-t \ C=C=C=C=C=C / t-C$_4$H$_9$, C$_4$H$_9$-t IV (41), V (7)	red; m.p. 114 to 116° ^{1}H NMR (C$_6$D$_6$): 1.34. (s) ^{13}C NMR (CDCl$_3$): 33.0 (CH$_3$), 37.8 (tertiary C), 83.1 (C–2,5), 139.7 (C–3,4), 161.7 (C–1,6), 207.0 (2 CO trans to Fe–Fe), 211.8 (4 CO) IR (C$_6$H$_{14}$): 1980, 1992, 2006, 2038, 2077	[16]

*Further information:

H$_2$C$_4$H$_2$Fe$_2$(CO)$_6$ (Table 2, No. 1) sublimes at 40 to 50 °C/3 Torr [3]. ^{1}H NMR shifts measured in solutions of CCl$_4$ [4] and CS$_2$ [5] differ only slightly from the values given in the table. IR spectra recorded in CCl$_4$ [5] or CHCl$_3$ [20] show only four CO bands close to 1988, 2002, 2035, and 2080 cm^{-1}; the full spectrum (KBr and CCl$_4$) outside the carbonyl region is given in [5].

The compound crystallizes in the orthorhombic system with a = 12.03(5), b = 8.12(3), and c = 12.25(5) Å; space group Pna2$_1$ – C$_{2v}^9$ or Pnma – D$_{2h}^{16}$. Z = 4 corresponds to D$_c$ = 1.84 g·cm^{-3}, D$_m$ = 1.79 g·cm^{-3}. The molecular structure has not been determined as the substance is not stable in X-rays [14].

The complete mass spectrum is reported by [11], see also [7, 8, 9, 18]. The molecular ion [M]$^+$ undergoes the stepwise loss of six CO groups to give [C$_4$H$_4$Fe$_2$]$^+$. This further fragments to [Fe$_2$]$^+$ and [Fe]$^+$. A complete series of metastable ions is observed for all these processes [11]. – The solid is surprisingly stable in air. Thermal decomposition in air does not occur until 200 to 230 °C [3, 4]. Even a solution in an organic solvent is stable in air for a few days [4]. The kinetics of the CO substitution by donors has been discussed in the introduction on p. 12.

$H(CH_3)C_4(CH_3)HFe_2(CO)_6$ (Table **2**, No. **4**). The preparation by method III leads to three isomers [21]. The melting point and the ^1H NMR chemical shifts indicate that the major isomer A is identical with the product obtained from method I [3]. The complex sublimes at 80 °C/3 Torr [3]. In the ^1H NMR spectra of the three isomers the CH_3 protons appear as doublets and the cumulene protons as quartets with coupling constants between 6.3 and 6.6 Hz [21]. Only four $v(CO)$ bands of the IR spectrum (solvent not specified) are given in [3]. The mass spectroscopic behavior is similar to that of compound No. 1 [9]. The substance decomposes under N_2 at 240 to 250 °C [3].

$(C_6H_5)_2C_4H_2Fe_2(CO)_6$ (Table **2**, No. **7**). The small amount of product obtained by method I was purified by sublimation at 70 to 100 °C/1 Torr. — The complete IR spectrum (in KBr) has been reported; $v(CO)$ bands measured in CCl_4 have noticeably higher wave numbers: 2000, 2011, 2022, 2049, and 2087 cm^{-1}. The UV spectrum is shown in **Fig. 5**. — The compound decomposes at 165 °C [19].

Fig. 5

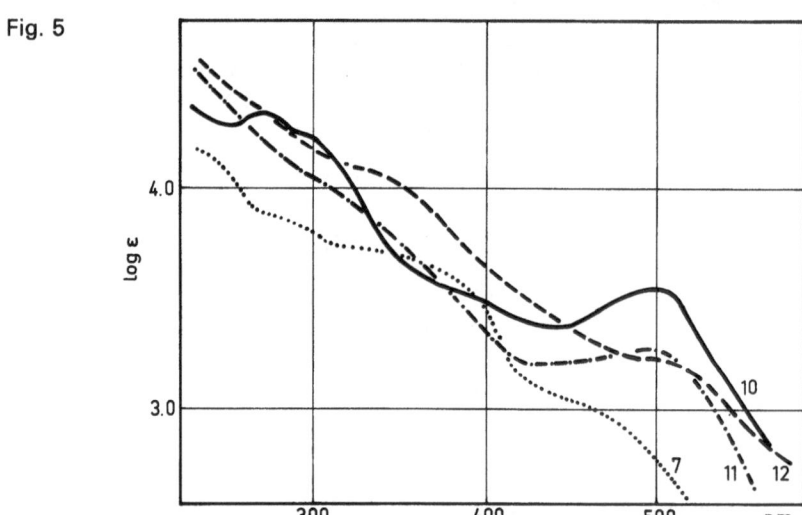

UV spectra of compounds No. 7, 10, 11, and 12 (in cyclohexane) [19].

$(CH_3)_2C_4(CH_3)_2Fe_2(CO)_6$ (Table **2**, No. **9**) forms deep orange needles which start to decompose at 108 °C and then melt at 118 °C [5]. It sublimes at 90 to 100 °C/3 Torr [3].

The complete IR spectrum (KBr and CCl_4) is given in [5]; only four $v(CO)$ bands are found in CCl_4 solution: 1976, 2008, 2024, and 2066 cm^{-1} [5]. The complex crystallizes in the monoclinic system with a=15.17(6), b=14.87(6), c=15.56(6) Å, and $\beta=$ 114.63(2)°; space group $P2_1n - C_{2h}^5$. Z=4 corresponds with $D_c = D_m = 1.62$ g·cm^{-3} [14].

The complex decomposes in a N_2 atmosphere at 240 to 250 °C [3]. The pyrolysis at 190 °C in vacuum does not give free butatriene, and attempts to displace the ligand by o-phenylene diamine or $P(C_4H_9-n)_3$ also failed [5].

References on p. 21

$CH_3(t-C_4H_9)C_4(C_4H_9-t)CH_3Fe_2(CO)_6$ (Table 2, No. **10**) has been prepared by method IV with a mixture of cis and trans butatriene; nevertheless, a uniform reaction product is indicated by the spectra. For steric reasons the $t-C_4H_9$ groups likely occupy the positions more distant from the Fe atoms, i.e., R^1 and R^4 in Formulas I and II on page 10. The UV spectrum is shown in Fig. 5 [19].

$C_6H_5(t-C_4H_9)C_4(C_4H_9-t)C_6H_5Fe_2(CO)_6$ (Table 2, No. **11**). The high yield in method IV has been achieved by using the trans-butatriene. If the reaction is run with the cis isomer, the yield amounts to only 23%. As the reaction mixture contains about 20% of the trans-butatriene, isomerization must occur during complexation. In the complex the $t-C_4H_9$ groups are equivalent (cf. ^1H NMR spectrum) and probably occupy the positions more distant from the Fe atoms [19]. The UV spectrum is shown in Fig. 5.

The degradation of the compound with excess Br_2 in CH_3OH at room temperature gives $CH_3O(C_6H_5)(t-C_4H_9)C-C\equiv C-C(C_4H_9-t)(C_6H_5)OCH_3$ in 62% yield. No reaction takes place with I_2, even in boiling CH_3OH, which is interpreted as a more effective screening by the $t-C_4H_9$ groups towards the larger halogen [22].

$(C_6H_5)_2C_4(C_4H_9-t)C_6H_5Fe_2(CO)_6$ (Table 2, No. **12**). Isomeric forms have not been observed [19]. The UV spectrum is shown in Fig. 5. The reaction with excess Br_2 in CH_3OH at room temperature yields $CH_3O(C_6H_5)_2C-C\equiv C-C(C_4H_9-t)(C_6H_5)OCH_3$ (62%). Treatment with 1.4 mole of I_2 in boiling CH_3OH gives only a 35% yield of the same product [22], however, the complex is completely degraded. Degradation with I_2 in boiling CCl_4 yields 20% of the fulvenes XI [23].

XI

$(C_6H_5)_2C_4(C_6H_5)_2Fe_2(CO)_6$ (Table 2, No. **13**) has also been prepared from $Fe(CO)_5$ (in excess) and tetraphenylbutatriene in ethylcyclohexane at reflux for 14 h, 26% yield [1, 4]. Sublimation at 3 Torr requires 180 °C where some decomposition occurs [3].

XII   XIII   XIV

The complete IR spectrum (KBr) is reported in [5]. The monoclinic crystals have the parameters a =17.27(6), b =10.18(5), c =17.29(6) Å, and β=110.67(5)°; space group $P2_1/n-C_{2h}^5$, Z=4 gives D_c=1.49; D_m=1.48 g·cm^{-3}. Because of poorly formed crystals the molecular structure could not completely be resolved [14]; some bond lengths reported by [10] are given in formula XII.

The crystalline solid is stable in air [1, 4, 5]. It is soluble in benzene and slightly soluble in hexane; it can be recrystallized from a mixture of these two solvents [1, 4, 5]. Pyrolysis (conditions not given) yields the butatriene, CO, and compound XIII [1, 4]. Compound XIV is the product of treatment with excess Br_2 in CH_3OH at room

References on p. 21

temperature (21% yield) [22]. Complete degradation with 1 mole I_2 in CH_3OH at room temperature or at the boiling point gives a high yield of tetraphenylbutatriene (up to 90%) and small amounts of $CH_3O(C_6H_5)_2C-C\equiv C-C(C_6H_5)_2OCH_3$ (9% yield) [22, 23]. $P(C_6H_5)_3$ does not attack the complex at 80 °C in benzene [4]. Substitution of one CO group by $P(C_6H_5)_3$ or $P(OC_6H_5)_3$ takes place in decaline above 100 °C whereas $P(C_4H_9-n)_3$ reacts already at about 50 °C [20]. The kinetics of these reactions have been investigated and are discussed in the introduction on page 12.

$C_6H_5(1-C_{10}H_7)C_4(C_{10}H_7-1)C_6H_5Fe_2(CO)_6$ (Table 2, No. 16, $C_{10}H_7$=naphthyl) is obtained in only one isomeric form from a cis, trans mixture of the respective butatrienes, cf. compound No. 11. – Unlike the other butatriene complexes, the title compound is unstable at ambient temperature and decomposes within a few days even in the solid state, possibly due to the steric crowding of the naphthyl groups [19].

$H(C_6H_5)C_4(CH_2)_5Fe_2(CO)_6$ and $(CH_2)_5C_4(CH_2)_5Fe_2(CO)_6$ (Table 2, No. 17 and 18). The cyclopentamethylene substituents have been incorrectly named in [21] as "cyclohexadienyl".

Fig. 6

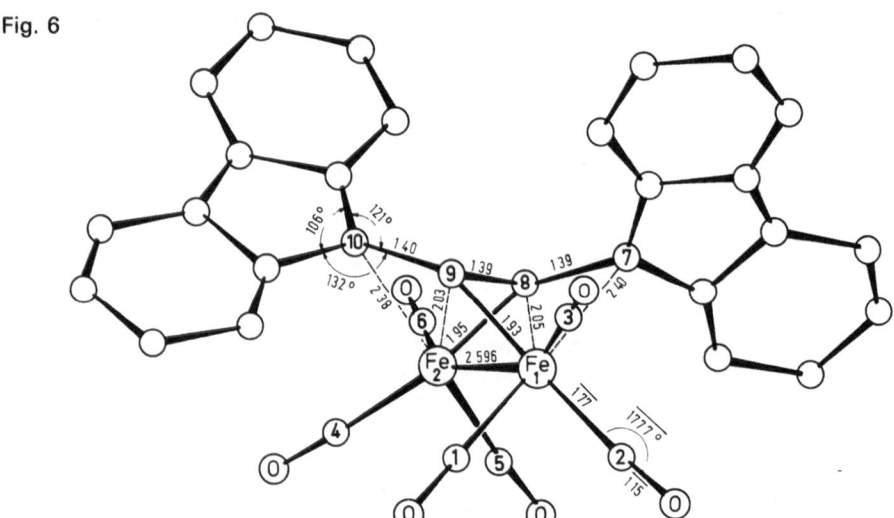

Molecular structure of $C_{12}H_8C_4C_{12}H_8Fe_2(CO)_6$ [14].

Selected bond angles (°), mean values of corresponding angles:

C(7)-C(8)-C(9)	128.5		C(1)-Fe(1)-C(9)	103.6
C(7)-Fe(1)-C(8)	35.6		C(2)-Fe(1)-C(9)	151.2
C(7)-Fe(1)-C(9)	70.0		C(3)-Fe(1)-C(9)	100.8
C(8)-Fe(1)-C(9)	40.8			
C(1)-Fe(1)-C(7)	167.5		C(1)-Fe(1)-C(2)	93.5
C(2)-Fe(1)-C(7)	88.7		C(1)-Fe(1)-C(3)	95.1
C(3)-Fe(1)-C(7)	96.5		C(2)-Fe(1)-C(3)	102.0
C(1)-Fe(1)-C(8)	133.4		Fe-C-O	177.7
C(2)-Fe(1)-C(8)	111.5			
C(3)-Fe(1)-C(8)	116.8		Fe(1)-C(9)-Fe(2)	80.9

$C_{12}H_8C_4C_{12}H_8Fe_2(CO)_6$ (Table 2, No. 19). The complete IR spectrum (KBr) is reported in [5]. The compound crystallizes in the monoclinic system with a = 16.36(6), b = 10.83(5), c = 16.47(6) Å, and β = 110.13(5)°; space group $P2_1/n - C_{2h}^5$. Z = 4 corresponds to D_c = 1.49 g·cm^{-3}; D_m = 1.51 g·cm^{-3}. The central atoms of the butatriene chain, C(8) and C(9), occupy slightly asymmetric bridging positions with respect to the Fe–Fe vector, see **Fig. 6**; the mean Fe–C bridging distances are 1.94 and 2.04 Å. The C_4 chain takes up a zig-zag configuration with bond angles of 128° and 129° at the central atoms. The Fe–C distances to the terminal atoms of the chain, C(7) and C(10), are much longer (mean 2.39 Å), while the three C–C bonds of the butatriene are equal in length (mean 1.39). The bonding can be regarded as a very asymmetric double π-allyl-metal arrangement. The twist about the C(8)–C(9) bond (torsion angle 34°) increases through the molecule until the angle between the $C_{12}H_8$ groups becomes 80° [14].

The reactions with donor molecules are discussed in the introduction to this chapter on p. 12 and the two following sections.

$(t\text{-}C_4H_9)_2C_6(C_4H_9\text{-}t)_2Fe_2(CO)_6$ (Table 2, No. 21) can be sublimed at about 100 °C/ 0.01 Torr. – The ^{13}C NMR spectrum indicates that the two outer double bonds of the hexapentaene ligand are not involved in the bonding to the $Fe_2(CO)_6$ unit [16].

References:

[1] A. Nakamura, P.J. Kim, N. Hagihara (Bull. Chem. Soc. Japan **37** [1964] 292). – [2] N. Hagihara (Ann. N.Y. Acad. Sci. **125** [1965] 98/101). – [3] A. Nakamura (Bull. Chem. Soc. Japan **38** [1965] 1868/73). – [4] A. Nakamura, P.J. Kim, N. Hagihara (J. Organometal. Chem. **3** [1965] 7/15). – [5] K.K. Joshi (J. Chem. Soc. A **1966** 594/7).

[6] K.K. Joshi (J. Chem. Soc. A **1966** 598/9). – [7] R.B. King (J. Am. Chem. Soc. **88** [1966] 2075/7). – [8] A. Nakamura, P.J. Kim, N. Hagihara (J. Organometal. Chem. **6** [1966] 420). – [9] S. Otsuka, A. Nakamura, T. Yoshida (Bull. Chem. Soc. Japan **40** [1967] 1266/7). – [10] O.S. Mills (unpublished in: G.E. Coates, M.L.H. Green, K. Wade, Organometallic Compounds, Vol. 2, London 1968, p. 60).

[11] R.B. King, A. Efraty (Org. Mass Spectrom. **2** [1969] 657/79). – [12] H.A. Brune, W. Schwab, H.P. Wolf (Z. Naturforsch. **25b** [1970] 892/3). – [13] R.B. King, L.M. Epstein, E.W. Gowling (J. Inorg. Nucl. Chem. **32** [1970] 441/5). – [14] D. Bright, O.S. Mills (J. Chem. Soc. Dalton Trans. **1972** 2465/9). – [15] A. Zimniak, W. Jasiobedzki (Roczniki Chem. **48** [1974] 365/7).

[16] R.B. King, C.A. Harmon (J. Organometal. Chem. **88** [1975] 93/100). – [17] N. Hagihara (Mem. Inst. Sci. Ind. Res. Osaka Univ. **32** [1975] 45/58). – [18] E. Sappa (Atti Accad. Sci. Torino Classe Sci. Fis. Mat. Nat. **109** [1975] 623/31). – [19] A. Zimniak, W. Jasiobedzki (Roczniki Chem. **49** [1975] 759/70). – [20] J.N. Gerlach, R.M. Wing, P.C. Ellgen (Inorg. Chem. **15** [1976] 2959/64).

[21] R. Victor (J. Organometal. Chem. **127** [1977] C25/C28). – [22] A. Zimniak, W. Jasiobedzki, H. Koseda (Bull. Acad. Polon. Sci. Ser. Sci. Chim. **25** [1977] 87/92). – [23] A. Zimniak, W. Jasiobedzki, H. Koseda (Bull. Acad. Polon. Sci. Ser. Sci. Chim. **25** [1977] 921/5). – [24] R. Victor, I. Ringel (Org. Magn. Resonance **11** [1978] 31/33).

2.4.1.2.1.2   Cumulene Complexes of the $^4LFe_2(CO)_5{}^2D$ Type

The compounds in Table 3 are prepared from the parent $^4LFe_2(CO)_6$ complexes ($^4L=$ butatriene ligand) and the donor molecules ^2D (1:1 mole ratio) in decaline at temperatures between 80 and 130 °C. One CO group trans to the Fe–Fe bond is replaced by the donor. The products are purified and isolated by chromatography on SiO_2 with $C_6H_6/CHCl_3$ as eluent; yields were not reported [2]. According to [1] it was not possible to prepare compound No. 5 in benzene at reflux temperature. The kinetics of formation have been investigated, see introduction to Chapter 2.4.1.2.1.1 on p. 12.

The $P(C_4H_9-n)_3$ derivatives are oily substances; all other compounds can be obtained as dark red to black crystals from $CH_2Cl_2/hexane$ by slow cooling [2]. Except for No. 2, the compounds have been characterized only by the $v(CO)$ bands in the IR spectra.

Table 3

Cumulene compounds of the $^4LFe_2(CO)_5{}^2D$ Type from [2].
Further information on compound No. 2 is given at the end of the table.

No.	^4L ligand	^2D ligand	IR spectrum (in CH_2Cl_2) $v(CO)$ in cm^{-1}
1		$P(C_4H_9-n)_3$	1925, 1954, 2034
*2	H_2CCCCH_2	$P(C_6H_5)_3$	1933, 1960, 1984, 2047
3		$P(OC_6H_5)_3$	1930, 1975, 1992, 2051
4		$P(C_4H_9-n)_3$	1929, 1958, 1990, 2043
5	$(C_6H_5)_2CCCC(C_6H_5)_2$	$P(C_6H_5)_3$	1935, 1962, 1986, 2048
6		$P(OC_6H_5)_3$	1933, 1970, 1995, 2050
7		$P(C_4H_9-n)_3$	1938, 1965, 1995, 2050
8	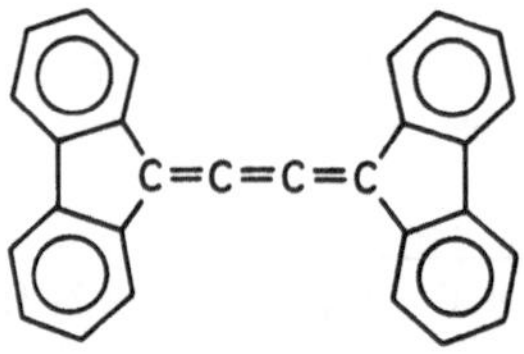 [structure: C=C=C=C with fluorenyl-type groups]	$P(C_6H_5)_3$	1935, 1957, 1992, 2049
9		$P(OC_6H_5)_3$	1940, 1961, 2000, 2055

*Further information:

$H_2C_4H_2Fe_2(CO)_5P(C_6H_5)_3$ (Table 3, No. 2). The red crystals belong to the monoclinic system with $a=11.277(4)$, $b=13.349(6)$, $c=17.604(7)$ Å, and $\beta=107.64(1)°$; space group $P2_1/c-C_{2h}^5$, $Z=4$ gives $D_c=1.488$ g · cm^{-3}; $D_m=1.50$ g · cm^{-3}. The molecule contains a nonlinear butatriene ligand with practically identical C–C bond lengths, see **Fig. 7**, and also similar angles of $130.1(9)°$ at C(1) and $128.9(9)°$ at C(2). The twofold π–allyl bond is very asymmetric as revealed by the different distances Fe(1)–C(2) and Fe(1)–C(3). The almost identical bond angles at the two Fe atoms indicate that their environment is only slightly distorted by the phosphine substitution [2].

Fig. 7

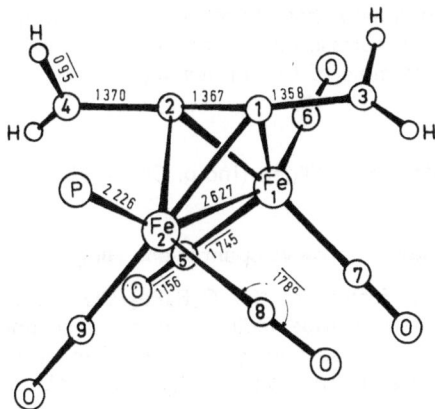

Molecular structure of $H_2C_4H_2Fe_2(CO)_5P(C_6H_5)_3$ [2].

Other selected bond lengths (in Å) and angles (°):

Fe(1)–C(1)	2.033(10)	Fe(2)–C(2)	2.025(10)
Fe(1)–C(2)	1.948(11)	Fe(2)–C(1)	1.940(11)
Fe(1)–C(3)	2.208(13)	Fe(2)–C(4)	2.205(12)
C(3)–C(1)–C(2)	130.1(9)	Fe(2)–Fe(1)–C(5)	87.0(6)
C(4)–C(2)–C(1)	128.9(9)	Fe(2)–Fe(1)–C(7)	107.2(6)
C(1)–Fe(1)–C(2)	40.2 mean	Fe(1)–Fe(2)–C(8)	89.1(5)
C(1)–Fe(1)–C(3)	37.5 mean	Fe(1)–Fe(2)–C(9)	106.3(5)
C(2)–Fe(1)–C(3)	72.8 mean	P–Fe(2)–C(1)	102.9(4)
Fe(1)–C(1)–Fe(2)	82.8 mean	P–Fe(2)–C(2)	111.8(4)
Fe(1)–Fe(2)–C(1)	50.0 mean	P–Fe(2)–C(8)	96.3(5)
Fe(1)–Fe(2)–C(2)	47.3 mean	P–Fe(2)–C(9)	99.5(5)
Fe(1)–Fe(2)–P	152.8(2)	C(8)–Fe(2)–C(9)	96.6(6)
Fe(2)–Fe(1)–C(6)	152.9(7)		

References:

[1] A. Nakamura, P.J. Kim, N. Hagihara (J. Organometal. Chem. **3** [1965] 7/15). –
[2] J.N. Gerlach, R.M. Wing, P.C. Ellgen (Inorg. Chem. **15** [1976] 2959/64).

2.4.1.2.1.3 Cumulene Complexes of the $^4LFe_2(CO)_6P(C_4H_9-n)_3$ Type

The title adducts are rapidly formed below 60 °C from $P(C_4H_9-n)_3$ and the parent compounds, cf. 2.4.1.2.1.1; details are not reported. Reactions with other less basic donor molecules require higher temperatures and thus result immediately in a substitution of CO.

The adducts have a strong IR absorption in the 1520 to 1570 cm^{-1} region. Therefore, a structure has been suggested in which a CO ligand inserts between one Fe atom and one of the butatriene C atoms. The following complexes are mentioned and character-ized by the IR spectra (measured in CH_2Cl_2):

$H_2C_4H_2Fe_2(CO)_6P(C_4H_9-n)_3$ 1568, 1925, 1954, 2034 cm^{-1};

$(C_6H_5)_2C_4(C_6H_5)_2Fe_2(CO)_6P(C_4H_9-n)_3$ 1529, 1958, 1990, 2043 cm^{-1};

$C_{12}H_8C_4C_{12}H_8Fe_2(CO)_6P(C_4H_9-n)_3$ 1520, 1938, 1965, 1995, 2050 cm^{-1}; for the cumulene ligand see Table 3, compounds No. 7 to 9.

Mass spectra have confirmed these formulations. A strong parent peak is observed, and the other strong peaks correspond to the stepwise loss of up to six CO groups. The compounds evolve one mole of CO on heating.

Reference:

J.N. Gerlach, R.M. Wing, P.C. Ellgen (Inorg. Chem. **15** [1976] 2959/64).

2.4.1.2.2   Compounds with a Ferracyclopentadiene Ring

The common structural feature of the $C_4R_4Fe_2(CO)_6$ complex type in 2.4.1.2.2.1 is an 1,1,1-tricarbonyl-1-ferracyclopentadiene ring and a second $Fe(CO)_3$ group bonded to this ring. In the literature these compounds are often called "Ferroles". Some few donor substitution products of the type $C_4R_4Fe_2(CO)_5^2D$ are described in Chapter 2.4.1.2.2.2.

2.4.1.2.2.1   Ferroles of the $C_4R_4Fe_2(CO)_6$ Type

Since the first preparation of a ferrole [1], the structure of these compounds has been the subject of numerous investigations. Yet incorrect ideas on their structure are found in earlier publications [2 to 4, 6, 8, 16, 21]. Several X-ray studies have revealed the structure shown by Formulas I and II [20, 44, 61, 65, 67, 73, 78, 83, 86, 94].

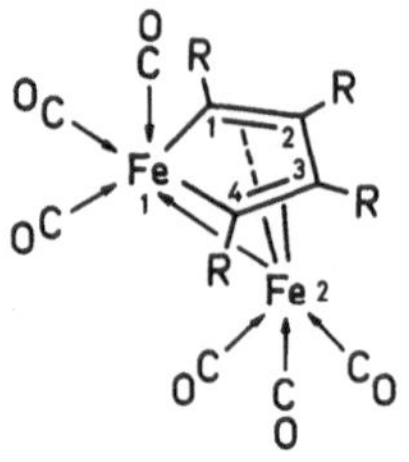

I 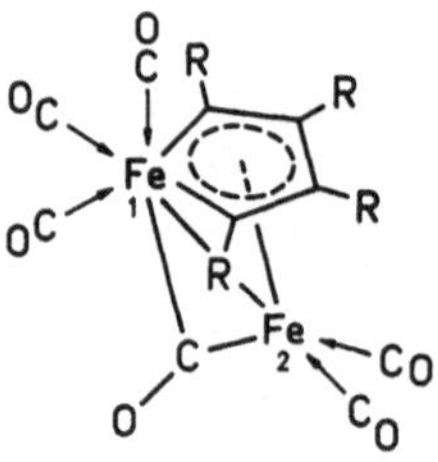   II

The ferracyclopentadiene unit is approximately planar. Two CO groups lie in this plane, the third is perpendicular to the ring. The orientation of the other $Fe(CO)_3$ group is such that one CO ligand points towards Fe-1 as shown in II. Compound No. 49 in Table 3 is an exception.

Formulas I and II are used to describe the bonding in the ferroles. Formula I emphasizes the σ bonds between Fe-1 and the ring C atoms and the π bonding between Fe-2 and the diene system. In order to explain the diamagnetism of the ferroles [11, 24] and to give the Fe atoms the rare gas configuration a dative Fe-Fe bond is assumed. Formula II shows the close approach of one CO ligand to Fe-1. Some interaction has to be expected. Unlike the other CO ligands, this CO group is significantly twisted. Therefore, it is considered to be a "semi-bridging carbonyl". Formula II also indicates fractional multiple bonding between Fe-1 and the ring C atoms, for the measured Fe-C distances are clearly shorter than would be expected for Fe-C single bonds [83]. The electronic structure of the ferroles has also been included in a systematic molecular orbital study of complexes containing $M_2(CO)_6$ transition-metal fragments [93, 95]. For earlier MO discussions see [8, 9].

References on p. 57

Asymmetrically substituted ferroles can occur as optically active isomers, in particular compounds No. 47, 54, and 55.

Due to the different bonding of the $Fe(CO)_3$ groups, the ferroles show four to six $\nu(CO)$ bands in the IR spectra. The nonequivalent Fe atoms, however, cannot be resolved by normal ^{57}Fe Mössbauer spectra of the neat solids. Resolution has only been achieved by taking the spectra in glassy matrices of $C_6H_5C_4H_9$-n at 85 ± 3 K. Two doublets are observed: The doublet with the more positive isomer shift and Δ in the range 1.22 to 1.26 mm·s^{-1} can be assigned to the π-bonded Fe(2) whereas the other doublet with Δ in the range 0.94 to 1.02 mm·s^{-1} can be assigned to Fe-1 in the ring [75].

From a comparative ^1H NMR study, it has been concluded that there is some π-electron delocalization in the ferrole ring. This is consistent with the usually small chemical shift difference of the 1,4- and 2,3-protons [45], the resonance of the 1,4-protons being at higher field.

The ^{13}C chemical shifts of the ring atoms can be widely different (δ up to 100 ppm), depending on whether the atoms are σ- and π-bonded to the two Fe atoms (C-1,4 in Formula I) or π-bonded to only one Fe atom (C-2,3). The shifts at lower field have been assigned to the atoms C-1 and C-4. All shifts are strongly influenced by the substituents. Symmetrically substituted ferroles show at room temperature three CO resonances in the intensity ratio 2:1:3 corresponding to two CO ligands on Fe(1) coplanar with the ring system, one CO ligand on Fe-1 perpendicular to the ring plane, and three CO ligands on Fe-2. The equivalence of the CO groups on Fe-2 may be due to a rotation of the $Fe(CO)_3$ moiety about the Fe-Fe bond. Bulky substituents on C-1 and C-4, for example compound No. 8, appear to hinder this rotation as the signal splits at low temperature into two peaks with the intensity ratio 1:2 [81]. The signal of the unique CO group on Fe-1 disappears on substitution by a phosphine [87].

The mass spectra show generally the parent ion [M]$^+$ and the ions [M −n CO]$^+$ with n=1 to 6. The further fragmentation depends strongly on the substituents on the ferracyclopentadiene ring. In the following, the mass spectra are not described if the usual fragmentation scheme is obeyed. For mass spectroscopic studies of several ferroles see [58, 60, 80].

The crystalline ferroles are stable in air. In solution, they decompose on admission of air. Their solubility depends on the kind of substituents in the ferracyclopentadiene ring. Substituents with functional groups can be modified by various chemical reactions without destroying the ferrole skeleton. Examples have been cited particularly for the type $C_4R_2(OH)_2Fe_2(CO)_6$ (compounds No. 33 to 41) and are given in Scheme 1 on p. 26. Degradations of the ferrole skeleton have been carried out with $C_4(C_6H_5)_4Fe_2(CO)_6$ (compound No. 20), but can also be applied to other ferroles. They are summarized in Scheme 2 on p. 27.

One CO ligand can be replaced by phosphines [24, 25, 87] or SO_2 [25]. Degradation reactions appeared to indicate that substitution occurs on the π-bonded $Fe(CO)_3$ unit [24, 25]. However, an X-ray study of $C_4(C_6H_5)_4Fe_2(CO)_5P(C_6H_5)_3$ demonstrated that the phosphine ligand was attached to the ferrole iron [87], cf. 2.4.1.2.2.2, p. 60.

Ferroles are formed in numerous ways. As a rule, they always occur together with other types of organoiron carbonyls and must be separated by chromatography on SiO_2 or Al_2O_3. The following methods are of a general nature. Other unique procedures of preparation are described in the further information on p. 39 and are indicated in Table 3 by the term "special".

Scheme 1

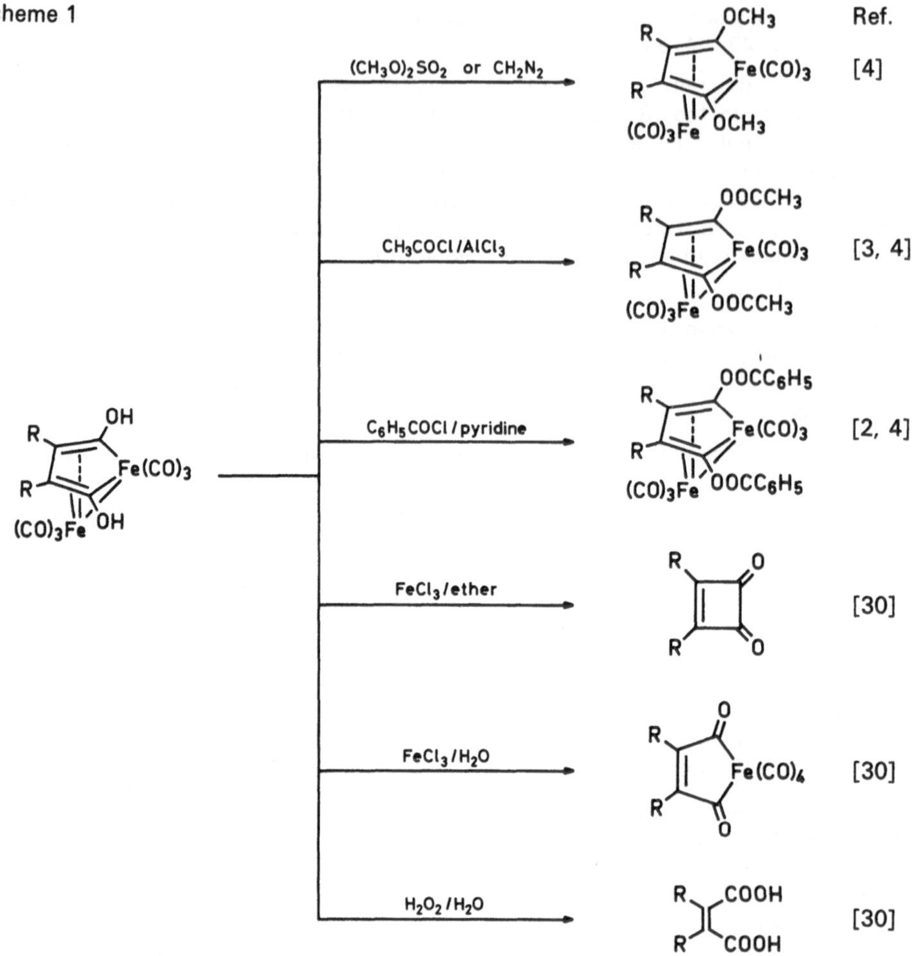

Method I: Ferroles are formed with other products in the reaction of $Fe_3(CO)_{12}$ with alkynes. The yields depend on the type of the alkyne and the conditions. The alkynes $HOOCC\equiv CCOOH$, $(CH_3)_3SiC\equiv CSi(CH_3)_3$, and $4\text{-}NO_2\text{-}C_6H_4C\equiv CC_6H_4NO_2\text{-}4$ do not yield ferroles at all [11]. Volatile alkynes like $HC\equiv CH$ are reacted in an autoclave under pressure [13, 24]. Unsymmetrically substituted alkynes do not react in a stereospecific manner, as assumed earlier, but give isomeric ferroles which differ in the position of the substituents in the ferracyclopentadiene ring [27, 34, 80, 81, 88].

Method II: Reactions of alkynes with benzalacetonetricarbonyliron, $C_6H_5CH=CH-CO-CH_3Fe(CO)_3$, shown to be a convenient source of $Fe(CO)_3$ groups, proceed readily in boiling C_6H_6 to give the corresponding ferroles. The product is essentially free from other organoiron carbonyl derivatives [69].

Method III: $Fe_3(CO)_{12}$ is treated with thiophene [18, 83] or monosubstituted thiophenes [90]. This results in removal of the S atom from the thiophene ring and its substitution by the Fe atom of an irontricarbonyl group [18]. $Fe_3(CO)_{12}$

References on p. 57

Scheme 2

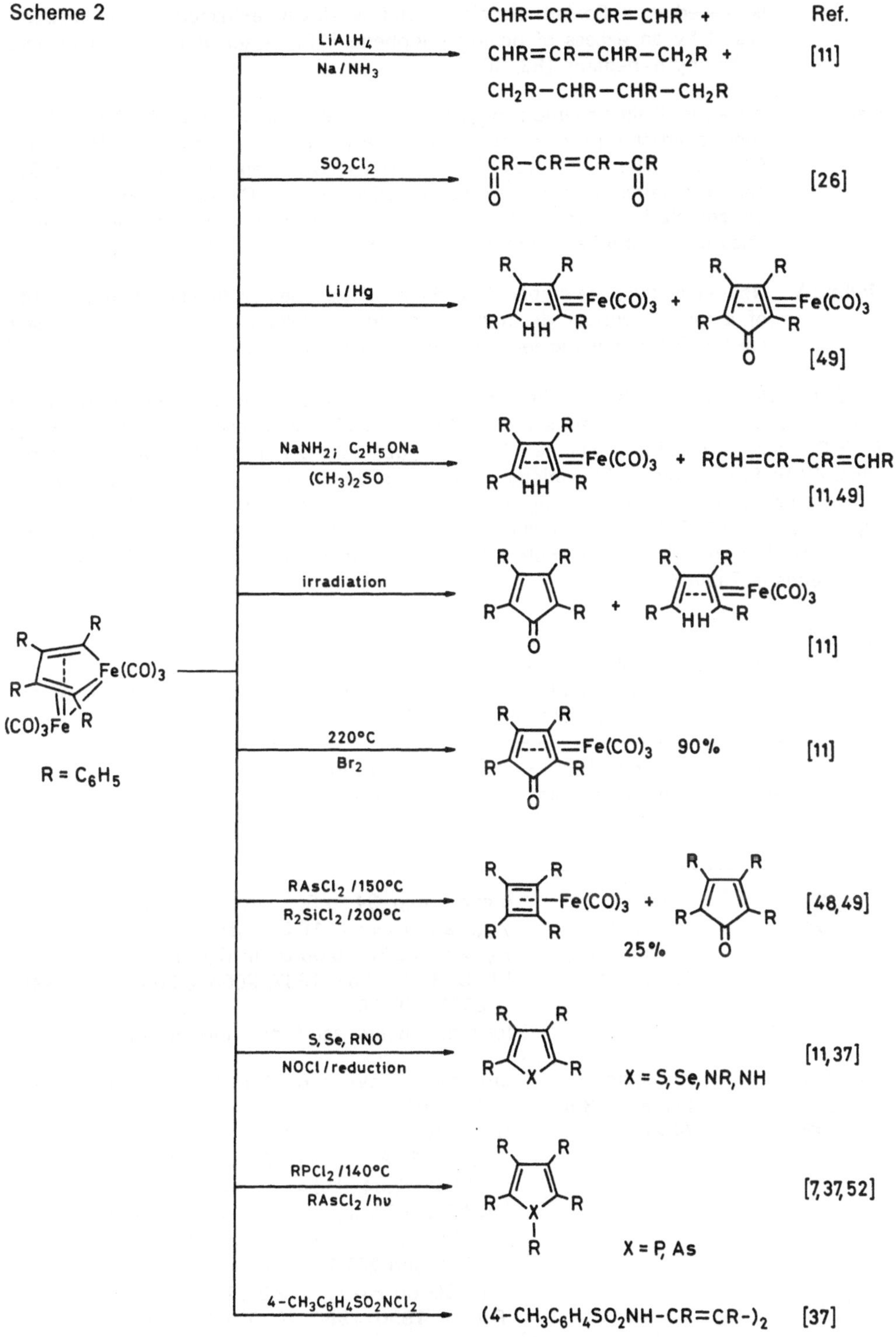

Reagents / conditions (left to right) and products:

LiAlH$_4$ / Na/NH$_3$ →
CHR=CR—CR=CHR +
CHR=CR—CHR—CH$_2$R +
CH$_2$R—CHR—CHR—CH$_2$R   [11]

SO$_2$Cl$_2$ →
CR—CR=CR—CR (with =O groups)   [26]

Li/Hg → (two Fe(CO)$_3$ complexes)   [49]

NaNH$_2$; C$_2$H$_5$ONa / (CH$_3$)$_2$SO → (Fe(CO)$_3$ complex) + RCH=CR—CR=CHR   [11,49]

irradiation → + (Fe(CO)$_3$ complex)   [11]

220°C / Br$_2$ → (Fe(CO)$_3$ complex) 90%   [11]

RAsCl$_2$ /150°C / R$_2$SiCl$_2$ /200°C → (Fe(CO)$_3$ complex) + 25%   [48,49]

S, Se, RNO / NOCl / reduction → X = S, Se, NR, NH   [11,37]

RPCl$_2$ /140°C / RAsCl$_2$ /hν → X = P, As   [7,37,52]

4-CH$_3$C$_6$H$_4$SO$_2$NCl$_2$ → (4-CH$_3$C$_6$H$_4$SO$_2$NH—CR=CR-)$_2$ [37]

R = C$_6$H$_5$

is placed in a Soxleth apparatus and is slowly extracted in the reaction vessel by an excess of boiling thiophene [18] or the thiophene derivative in boiling n-heptane [83, 90].

Method IV: Aqueous alkaline solutions of $[HFe(CO)_4]^-$ are reacted with $HC\equiv CH$ at atmospheric pressure [2, 4] or higher pressure [1] to give the ferrole $C_4H_2(OH)_2$-$Fe_2(CO)_6$. The complexes $C_4HR(OH)_2Fe_2(CO)_6$ and $C_4R_2(OH)_2Fe_2(CO)_6$ can be obtained from the alkynes $RC\equiv CH$ and $RC\equiv CR$ [3, 4]. These ferroles are soluble in aqueous alkali and are isolated by precipitation with an acid. They can be purified by repeating this precipitation [1, 4].

Method V: New derivatives are prepared by chemical reactions on functional substituents of a ferrole without destroying the ferrole skeleton. Most frequently used are the OH containing ferroles mentioned above in method IV.

Explanations to Table 3: The compounds in Table 3 are arranged in the following manner: ferroles with carbon substituents (Nos. 1 to 32), complexes of the $C_4R_2(OH)_2$-$Fe_2(CO)_6$ type (Nos. 33 to 41), ferroles with substituents bonded through an O atom (Nos. 42 to 50), ferroles with various substituents but in unknown positions (Nos. 51 to 61), and ferroles containing ring systems condensed onto the ferracyclopentadiene ring (Nos. 62 to 74). Table 3 closes with three compounds recently reported. The numbering of the ferrole ring indicated in Table 3 for compound No. 1 is also valid for all other compounds in Table 3.

Table 3
Ferracyclopentadiene complexes of the $C_4R_4Fe_2(CO)_6$ type.
Further information on compounds preceded by an asterisk is given at the end of the table. For abbreviations and dimensions see p. 170.

No. Ferracyclo-pentadiene ring	Method of prep-aration, reaction conditions (yield in %)	Properties and further remarks Explanations see above	Ref.
*1	I and III, see further information (up to 17)	orange; m.p. 54 to 55°, subl. at 40°/10^{-2} χ_{mol} (293 K) = $(-116.3 \pm 1.8) \times 10^{-6}$ μ_D (25°) = 2.86 ± 0.08 D (in C_6H_6) IR (C_6H_{14}): 1962, 1997, 2005.5, 2040, 2061, 2077.5 for other spectra see further information	[13, 24, 77, 83]
*2 CH$_3$	III, boiling heptane/13 h (4.2)	brownish yellow; m.p. 59 to 60°, subl. at 30°/0.01 ^{1}H NMR (CDCl$_3$): 2.41 (s, CH$_3$), 5.94 (dd, H-2), 6.18 (t, H-3), 6.66 (dd, H-4); J(2,3) = 2.34, J(2,4) = 2.34, J(3,4) = 5.26 ^{13}C NMR (CDCl$_3$): 32.1 (CH$_3$), 110.3 (C-3), 110.6 (C-2), 152.9 (C-4), 180.8 (C-1); 206.9 and 207.2 (2 CO on Fe-1), 209.8 (1 CO on Fe-1), 212.3 (3 CO on Fe-2) IR (−): 1950, 1990, 2000, 2035, 2075	[90, 91]

Table 3 [continued]

No.	Ferracyclo-pentadiene ring	Method of preparation, reaction conditions (yield in %)	Properties and further remarks Explanations on p. 28	Ref.
*3	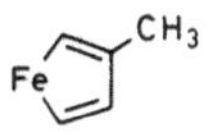	III, boiling heptane/7 h (10.7)	orange red; m.p. 40°, subl. at 40°/0.01 ^{1}H NMR (CDCl$_3$): 2.25 (s, CH$_3$), 6.14 (dd, H–3), 6.37 (d, H–1), 6.84 (d, H–4); J(1,3) =2.73, J(3,4) =5.18 ^{13}C NMR (CDCl$_3$): 19.4 (CH$_3$), 112.3 (C–3), 131.1 (C–2), 151.8 (C–4), 154.7 (C–1), 207.1 (2 CO on Fe–1), 209.8 (1 CO on Fe–1), 211.7 (3 CO on Fe–2) IR (–): 1930, 1975, 2000, 2035, 2045, 2080	[90, 91]
*4	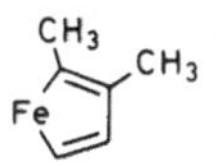	special	orange red; m.p. 71°, subl. at 70°/5 ^{1}H NMR (CCl$_4$): 2.21 and 2.36 (s's, CH$_3$), 6.28 and 6.70 (d's, 1 H each, J =7.5) IR (CCl$_4$): 1951, 1990, 2033, 2115	[64]
*5	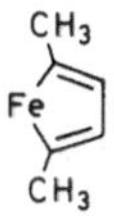	I, no details given (–) special	orange; m.p. 45° ^{1}H NMR (CDCl$_3$): 2.30 (s, CH$_3$), 5.82 (s, H–2,3) ^{13}C NMR (CDCl$_3$): 32.2 (CH$_3$), 110.4 (C–2,3), 178.3 (C–1,4), 206.8 (2 CO on Fe–1), 210.4 (1 CO on Fe–1), 213.7 (3 CO on Fe–2)	[81, 91]
6	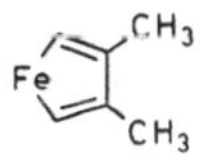	I, no details given (–)	^{13}C NMR (CDCl$_3$): 20.2 (CH$_3$), 130.3 (C–2,3), 153.6 (C–1,4, J(C,H) =160); CO shifts not assigned not obtained as pure sample, see No. 5	[81]
7	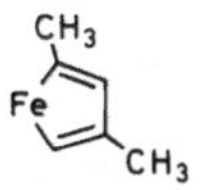	I, no details reported (–)	^{13}C NMR (CDCl$_3$): 19.6 (CH$_3$–3), 32.2 (CH$_3$–1), 113.2 (C–2, J(C,H) =166), 131.6 (C–3), 150.1 (C–4, J(C,H) =161), 181.1 (C–1); CO resonances not assigned not obtained as pure sample, see No. 5	[81]
8	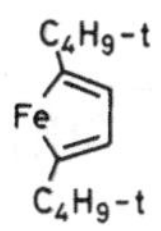	I, boiling heptane/25 min (–)	^{1}H NMR (CCl$_4$): 1.28 (s, C$_4$H$_9$-t), 6.24 (s, H–2,3) ^{13}C NMR (CDCl$_3$): 33.7 (CH$_3$), 43.1 (C–5,6), 109.7 (C–2,3, J(C,H) =166), 197.0 (C–1,4), 208.9 (2 CO on Fe–1), 211.6 (1 CO on Fe–1), 216.7 (3 CO on Fe–2) ^{13}C NMR (CH$_2$Cl$_2$, at –100°): 209.1 (2 CO on Fe–1), 211.8 (1 CO on Fe–1), 214.1 (2 CO on Fe–2), 222.2 (1 CO on Fe–2)	[80, 91]

References on p. 57

Table 3 [continued]

No.	Ferracyclo-pentadiene ring	Method of prep-aration, reaction conditions (yield in %)	Properties and further remarks Explanations on p. 28	Ref.
9	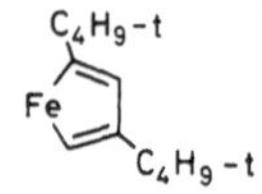 C_4H_9-t, Fe, C_4H_9-t	I, boiling heptane/25 min (−)	^{1}H NMR (CCl$_4$): 1.15 and 1.30 (s's, C_4H_9-t), 5.30 and 8.45 (s's, 2 H)	[80]
10	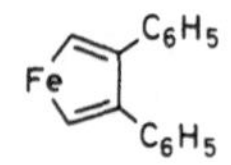 C_6H_5, Fe, C_6H_5	I, from Fe$_3$(CO)$_{12}$ and $C_6H_5C≡CH$ (1:6 mole) in petroleum ether/ benzene at 50 to 65° (0.4)	orange needles (from petroleum ether); m.p. 125 to 127° ^{1}H NMR (CCl$_4$): 6.80 (s, H−1,4), 7.15 (m, C_6H_5) IR (KBr): 1934, 1972, 2024, 2062 readily soluble in common organic sol-vents, formation of 3,4−diphenylthiophene with excess S$_x$ at 200°/15 min	[34, 80]
*11	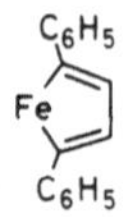 C_6H_5, Fe, C_6H_5	I, like No. 10 (−)	orange yellow (from C_2H_5OH); m.p. ≈180° (dec.), subl. above 100° in vacuum ^{1}H NMR (CCl$_4$): 6.28 (s, H−2,3), 7.15 (m, C_6H_5) ^{13}C NMR (CDCl$_3$): 112.6 (C−2,3, J(C,H) =165), 127.7 to 128.9, 149.9 (C_6H_5), 176.5 (C−1,4), 206.4 (2 CO on Fe−1), 210.0 (1 CO on Fe−1), 214.1 (3 CO on Fe−2) IR (KBr): 1923, 1949, 2000, 2041, 2075	[6, 11, 16, 21, 34, 37, 80, 81, 87]
12	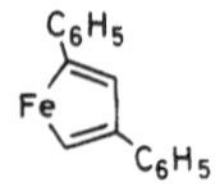 C_6H_5, Fe, C_6H_5	I, no details reported (−)	^{1}H NMR (CCl$_4$): 6.75 and 7.05 (s's, H−1,3), 7.40 to 7.80 (m, C_6H_5)	[80]
*13	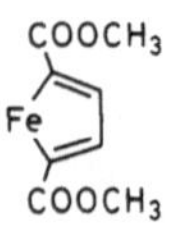 COOCH$_3$, Fe, COOCH$_3$	special	yellow red; m.p. 56 to 58°, 61.5 to 62°, subl. above 50°/0.1 ^{1}H NMR (CDCl$_3$): 3.74 (CH$_3$), 6.54 (H−2,3) IR (−): 1972, 2009, 2022, 2052, 2084; v(C=O) at 1722	[22, 45, 55, 72]
*14	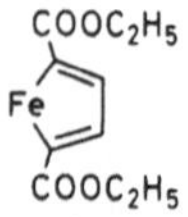 COOC$_2$H$_5$, Fe, COOC$_2$H$_5$	special	red; m.p. 60.5 to 61°, subl. at 80°/0.2 ^{1}H NMR (CDCl$_3$): 1.28 (CH$_3$), 4.20 (CH$_2$), 6.55 (H−2,3)	[45]
*15	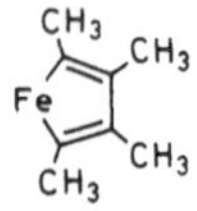 CH$_3$, CH$_3$, Fe, CH$_3$, CH$_3$	I, no information (−) II, boiling C_6H_6 (25)	yellow; m.p. 110 to 125°, 149 to 151°, 156 to 157°, 160° (dec.) ^{1}H NMR (CDCl$_3$): 2.06 and 2.26 (s's, CH$_3$) ^{57}Fe−γ (85.5 K): $\delta = -0.010$ (Fe), $\Delta=1.157$ IR (CCl$_4$): 1986, 2024, 2063	[49, 53, 69, 75, 96]

References on p. 57

Table 3 [continued]

No.	Ferracyclo-pentadiene ring	Method of preparation, reaction conditions (yield in %)	Properties and further remarks Explanations on p. 28	Ref.
*16	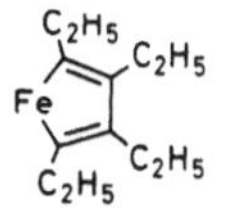 C$_2$H$_5$ / C$_2$H$_5$ / Fe / C$_2$H$_5$ / C$_2$H$_5$	I, no information (−) II, boiling C$_6$H$_6$ (50)	yellow; m.p. 100 to 175°, above 150° (dec.) ^{57}Fe-γ (87 K): $\delta = -0.006$ (Fe), $\Delta = 1.163$	[37, 49, 69, 75]
17	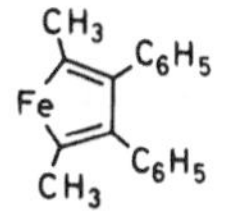 CH$_3$ / C$_6$H$_5$ / Fe / CH$_3$ / C$_6$H$_5$	no information	yellow; m.p. 122° formation of 2,5-dimethyl-3,4-diphenyl-thiophene (11% yield) with excess S$_x$ at 180 to 190°/7 min	[37, 49]
18	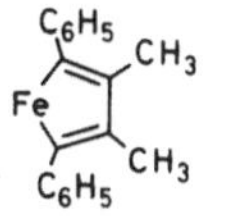 C$_6$H$_5$ / CH$_3$ / Fe / C$_6$H$_5$ / CH$_3$	no information	yellow; m.p. 157 to 158° (dec.)	[49]
*19	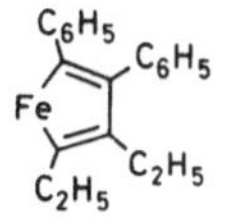 C$_6$H$_5$ / C$_6$H$_5$ / Fe / C$_2$H$_5$ / C$_2$H$_5$	special	orange; m.p. 141 to 142°	[48, 49]
*20	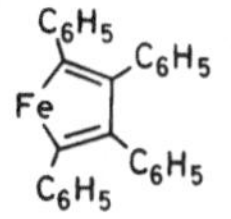 C$_6$H$_5$ / C$_6$H$_5$ / Fe / C$_6$H$_5$ / C$_6$H$_5$	I, from Fe$_3$(CO)$_{12}$ and C$_6$H$_5$C≡CC$_6$H$_5$ (1:3 mole) in petroleum ether at 80 to 90°/2 h (19) II, boiling C$_6$H$_6$ (50)	orange; m.p. above 150° (dec.), 174 to 178° (dec.), above about 200°, subl. above 130°/vacuum $\chi_{mol} = (260 \pm 15) \times 10^{-6}$ $\mu_D = 3.3 \pm 0.2$ D (in C$_6$H$_6$) ^{57}Fe-γ (86 K): $\delta = -0.001$ (Fe), $\Delta = 1.152$ IR (Nujol): 1925, 1970, 2000, 2010, 2070 for other spectra see further information	[11, 14, 69, 75]
*21	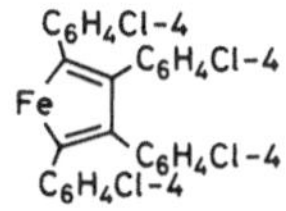 C$_6$H$_4$Cl-4 / C$_6$H$_4$Cl-4 / Fe / C$_6$H$_4$Cl-4 / C$_6$H$_4$Cl-4	I, from Fe$_3$(CO)$_{12}$ and 4-ClC$_6$H$_4$C≡C-C$_6$H$_4$Cl-4 (1:2.4 mole) in petroleum ether at 80°/3 h (20 to 25)	orange; two crystal modifications with m.p. 133 to 135° and 185 to 188° (dec.) $\mu_D = 1.44 \pm 0.15$ D IR (KBr): 1923, 1988, 2037, 2070	[6, 12, 21, 35]
*22	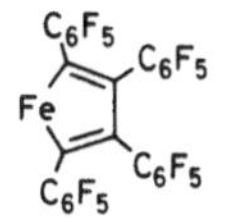 C$_6$F$_5$ / C$_6$F$_5$ / Fe / C$_6$F$_5$ / C$_6$F$_5$	I, from Fe$_3$(CO)$_{12}$ and C$_6$F$_5$C≡CC$_6$F$_5$ (3:7 mole) in C$_6$H$_6$ (sealed tube) at 140°/20 h (45)	yellow; dec. at 200° IR (Nujol): 1969, 2012, 2024, 2045, 2066, 2100 UV (C$_6$H$_{14}$): λ_{max}(lg ε) = 425 (3.55)	[41]

References on p. 57

Table 3 [continued]

No.	Ferracyclo-pentadiene ring	Method of preparation, reaction conditions (yield in %)	Properties and further remarks Explanations on p. 28	Ref.
23	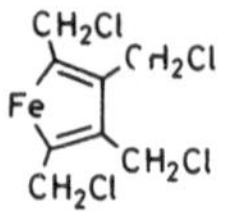 CH₂Cl / CH₂Cl / Fe / CH₂Cl / CH₂Cl	I, no information (−)	yellow	[80]
*24	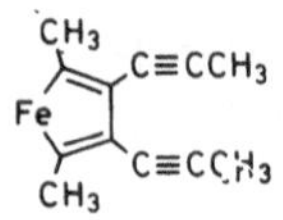 CH₃ / C≡CCH₃ / Fe / C≡CCH₃ / CH₃	special	yellow ^{1}H NMR (−): 2.18, 2.46 (s's, equal intensities) IR (CCl₄): 1950, 1992, 2009, 2036, 2072; ν(C≡C) at 2243	[88]
25	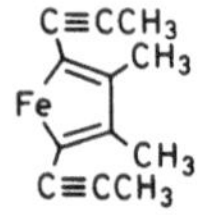 C≡CCH₃ / CH₃ / Fe / CH₃ / C≡CCH₃	special, see No. 24	yellow; m.p. 123 to 123.5° ^{1}H NMR (−): 2.11, 2.25 (s's, equal intensities) IR (CCl₄): 1957, 2000, 2006, 2037, 2073; ν(C≡C) at 2207	[88]
*26	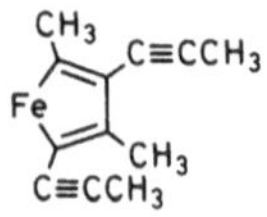 CH₃ / C≡CCH₃ / Fe / CH₃ / C≡CCH₃	special, see No. 24 (<0.5)	red orange; m.p. 113.5 to 114.5° ^{1}H NMR (−): 2.08, 2.13, 2.41, 2.48 (s's, equal intensities) IR (CCl₄): 1951, 1997, 2004, 2037, 2072; ν(C≡C) at 2210 and 2236	[88]
*27	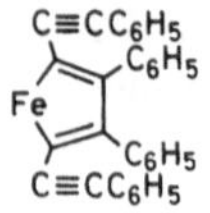 C≡CC₆H₅ / C₆H₅ / Fe / C₆H₅ / C≡CC₆H₅	I, from Fe₃(CO)₁₂ and C₆H₅C≡C-C≡CC₆H₅ (1:3 mole) in petroleum ether at 80 to 100°/1 h (total yield of No. 27 to 29 up to 15)	orange yellow; m.p. 195 to 200° (dec.) IR (KBr): 1927, 2004, 2041, 2075; ν(C≡C) at 2183	[27]
28	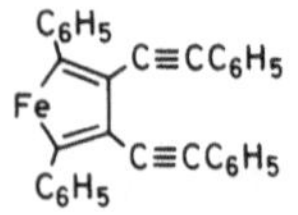 C₆H₅ / C≡CC₆H₅ / Fe / C≡CC₆H₅ / C₆H₅	I, see No. 27	yellow orange; m.p. 225 to 226° (dec.) IR (KBr): 1946, 2004, 2049, 2075; ν(C≡C) at 2217	[27]
29	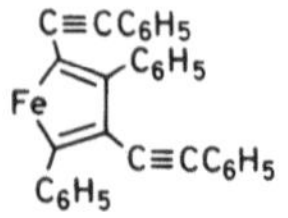 C≡CC₆H₅ / C₆H₅ / Fe / C≡CC₆H₅ / C₆H₅	I, see No. 27 (main product of the isomers No. 27 to 29)	yellow needles; m.p. 180 to 182° (dec.) IR (KBr): 1946, 2004, 2049, 2075; ν(C≡C) at 2188 and 2222	[27]

References on p. 57

Table 3 [continued]

No.	Ferracyclo-pentadiene ring	Method of preparation, reaction conditions (yield in %)	Properties and further remarks Explanations on p. 28	Ref.
30	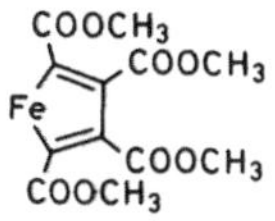 COOCH₃ COOCH₃ Fe COOCH₃ COOCH₃	I, from Fe₃(CO)₁₂ and CH₃OOCC≡C-COOCH₃ (1:4 mole) in petroleum ether at 80 to 100°/30 min (20)	yellow needles; m.p. 110 to 112° (dec.), subl. above 70°/vacuum IR (CCl₄): 1990, 2021, 2041, 2067, 2090, 2094; ν(C=O) at 1736, 1750, 1776 UV irradiation with SCl₂ in CCl₄ for 1 h gives tetracarbomethoxythiophene	[12, 21, 37, 53, 60]
31	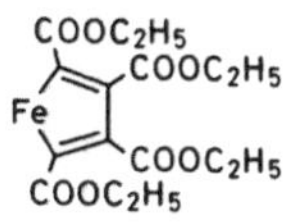 COOC₂H₅ COOC₂H₅ Fe COOC₂H₅ COOC₂H₅	I, from Fe₃(CO)₁₂ and C₂H₅OOC-C≡CCOOC₂H₅ (1:3 mole) in petroleum ether at 40 to 70°/8 h (−)	yellow; from heptane at −10° IR (CCl₄): 1978, 2019, 2037, 2065, 2091; ν(C=O) at 1713, 1738	[60]
*32	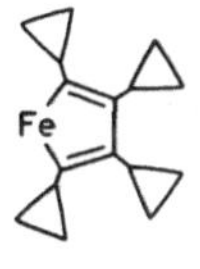 Fe	I, from Fe₃(CO)₁₂ and dicyclo-propylacetylene in C₆H₆ at 80°/5 h (6)	yellow–orange plates (from pentane); m.p. 106 to 107° ¹H NMR (CDCl₃): 0.54 to 1.22 (4 H), 1.34 to 2.12 (1 H) IR (C₆H₁₄): 1979, 1992, 1997, 2020, 2061	[92]
*33	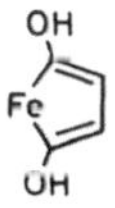 OH Fe OH	IV, normal pressure at room temperature/9 h (71) IV, 15 atm HC≡CH at 45 to 50°/16 h (45)	yellow; m.p. 80 to 85° (dec.), subl. at 70 to 80°/0.1 ¹H NMR (CDCl₃): 4.90 (s, H-2,3), 5.83 (OH) IR (CS₂): 1998, 2033, 2073; ν(OH) at 3463, 3565; ν(C-O) at 1093, 1174, 1271	[1, 2, 4, 72]
34	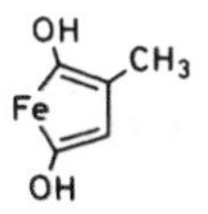 OH CH₃ Fe OH	IV, normal pressure at room temperature/ 3 weeks (18)	yellow; m.p. 100 to 105° (dec.), subl. at 70°/10⁻⁵ for chemical reactions see compounds No. 46, 51, and 52	[3,4]

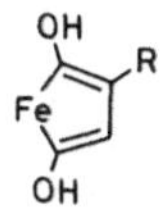

OH R Fe OH

*35	R=C₂H₅	IV, no details reported	compounds have been used as starting materials for oxidations, see Scheme 1 on p. 26	[3, 10, 30]
36	R=n-C₄H₉			[10, 30]
*37	R=t-C₄H₉			[23, 30]
38	R=C₆H₅			[10, 30]

Table 3 [continued]

No.	Ferracyclo-pentadiene ring	Method of preparation, reaction conditions (yield in %)	Properties and further remarks Explanations on p. 28	Ref.
	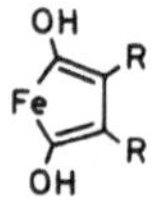 ![structure with OH, R, Fe, R, OH]			
*39	R=CH$_3$			[3, 10, 30]
		IV, no details reported	compounds have been used as starting materials for oxidations, see Scheme 1 on p. 26	
*40	R=C$_2$H$_5$			[10, 30]
41	R=C$_6$H$_5$			[10, 30]
*42	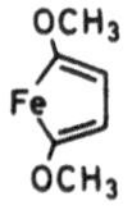 OCH$_3$ / Fe / OCH$_3$	V, from No. 33 and (CH$_3$O)$_2$SO$_2$ in aqueous alkali (66)	bright orange; m.p. 155 to 156° ^{1}H NMR (CDCl$_3$): 3.57 (s, CH$_3$), 4.84 (s, H-2,3) IR (−): 1979, 1988, 2009, 2031, 2074	[3, 4, 72]
43	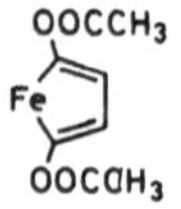 OOCCH$_3$ / Fe / OOCCH$_3$	V, from No. 33 with CH$_3$COCl/AlCl$_3$ (−)	^{1}H NMR (CD$_3$COCD$_3$): 2.81 (s, CH$_3$), 6.04 (s, H-2,3) IR (−): 1956, 1973, 2008, 2042, 2077	[72]
44	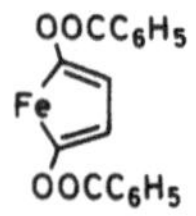 OOCC$_6$H$_5$ / Fe / OOCC$_6$H$_5$	V, from No. 33 with C$_6$H$_5$COCl in pyridine (100)	yellow, microcrystals from C$_2$H$_5$OH/H$_2$O; m.p. 155 to 160° (dec.)	[2, 4]
45	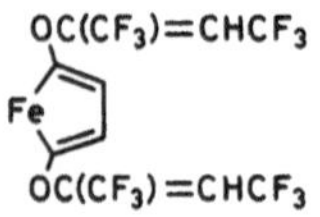 OC(CF$_3$)=CHCF$_3$ / Fe / OC(CF$_3$)=CHCF$_3$	V, from No. 33 and CF$_3$C≡CCF$_3$ (1:2.3 mole) in THF in a sealed tube at 90°/15 h (32)	orange-yellow needles; m.p. 101 to 103° ^{1}H NMR (CCl$_4$): 4.82 (s, H-2,3), 5.99 (q, CF$_3$CH, J(F,H)=7) ^{19}F NMR (−): 59.77 (d, J(F,H)=7), 69.37 (s) IR (−): 2003, 2016, 2026, 2051, 2088	[72]
46	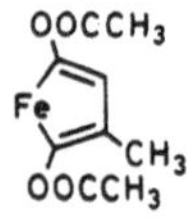 OOCCH$_3$ / Fe / CH$_3$ / OOCCH$_3$	V, by acetylation of No. 34, no details reported (−)	hot CH$_3$OH converts the compound into a monoacetate, see No. 51	[3]

References on p. 57

Table 3 [continued]

No.	Ferracyclo-pentadiene ring	Method of preparation, reaction conditions (yield in %)	Properties and further remarks Explanations on p. 28	Ref.
*47	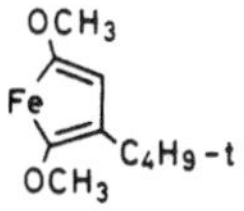 OCH₃ / Fe / C₄H₉-t / OCH₃	V, see further information (−)	no data reported	[23]
*48	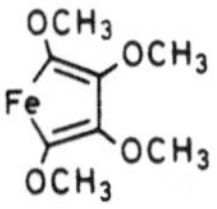 OCH₃ / OCH₃ / Fe / OCH₃ / OCH₃	special	yellow oil ^{1}H NMR (CCl₄): 3.68, 3.90 (s's of equal intensity) IR (CCl₄): 1985, 2000, 2038, 2080	[71]
*49	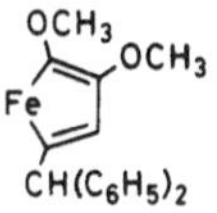 OCH₃ / OCH₃ / Fe / CH(C₆H₅)₂	special	red prisms from petroleum ether; m.p. 148° ^{1}H NMR (CCl₄): 3.64, 3.83 (s's, CH₃), other resonances not reported	[54, 61]
*50	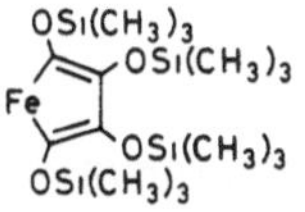 OSi(CH₃)₃ / OSi(CH₃)₃ / Fe / OSi(CH₃)₃ / OSi(CH₃)₃	special	yellow microcrystals by sublimation at 80 to 90°/vacuum; m.p. 118 to 120° ^{1}H NMR (C₆H₁₂): 0.27, 0.34 (s's of equal intensity) IR (C₆H₁₂): 1934, 1962, 1976, 2000, 2022, 2066	[56, 59]
*51	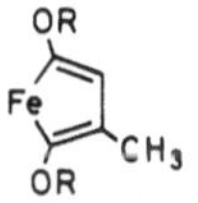 OR / Fe / CH₃ / OR / R = H and COCH₃	V, see further information (−)	−	[3]
*52	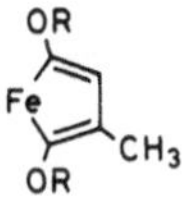 OR / Fe / CH₃ / OR / R = COCH₃ and COCH₂Cl	V, see further information (−)	−	[3]
*53	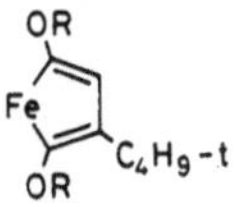 OR / Fe / C₄H₉-t / OR / R = H and CH₃	V, see further information (−)	−	[23]
*54	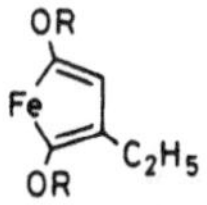 OR / Fe / C₂H₅ / OR / R = H and C₁₂H₂₁O₂	V, see further information (−)	$C_{12}H_{21}O_2 =$ 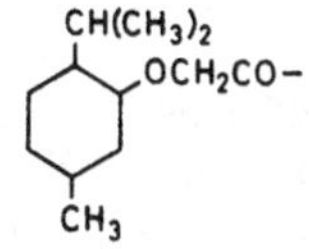 (structure: cyclohexane ring with CH(CH₃)₂, OCH₂CO−, CH₃ substituents)	[23]

References on p. 57

Table 3 [continued]

No.	Ferracyclo-pentadiene ring	Method of prep-aration, reaction conditions (yield in %)	Properties and further remarks Explanations on p. 28	Ref.
*55	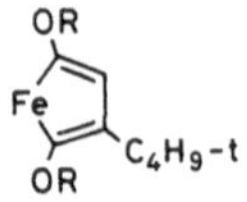 OR Fe C₄H₉-t OR R=CH₃ and C₁₀H₁₅O₃S	V, see further information (−)	CH_3 CH $C_{10}H_{15}O_3S =$ CH_2 O $-SO_2$	[23]
56	R R Fe R R 2 R = C₆H₅ 2 R = COOCH₃	I, from Fe₃(CO)₁₂ and C₆H₅C≡C-COOCH₃ in petroleum ether at 75°/2 h (−)	yellow; m.p. 114 to 122° (dec.) IR (CH₂Cl₂): 1961, 1992, 2020, 2045, 2083; ν(C=O) at 1712	[36]
*57	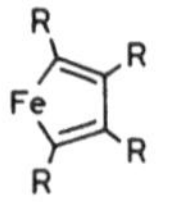 R R Fe R R 2 R = H 2 R = 4-BrC₆H₄	I, from Fe₃(CO)₁₂ and 4–BrC₆H₄C≡CH (2:1 mole) in petroleum ether at 65°/3 h (2)	yellow; m.p. 208 to 215° (dec.), subl. in vacuum at 120 to 150° IR (KBr): 1927, 1992, 2016, 2041, 2075	[12, 34]
58	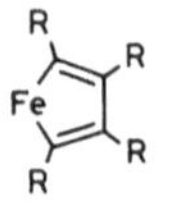 R R Fe R R 2 R = C₆H₅ 2 R = Si(CH₃)₃	I, no further information (−)	orange; two crystal modifications with m.p. 128° (dec.) and 146 to 148° (dec.)	[12, 49]
*59	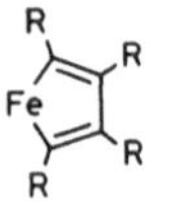 R R Fe R R 2 R = C₆H₅ 2 R = t-C₄H₉	special	brick red; 155° (dec.)	[50]
60	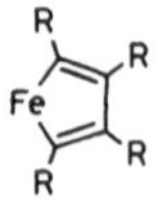 R R Fe R R 2 R = H 2 R = N(CH₃)₂	I, from Fe₃(CO)₁₂ and (CH₃)₂NC≡CH (2:3 mole) in hexane at reflux for 15 h (8)	orange; m.p. 61 to 64° ¹H NMR (CDCl₃): 2.77 (s, CH₃), 5.89 (s, 2 H) ¹³C NMR (CDCl₃): 40.8 (CH₃), 123.0 (CH–ring), 147.7 (CN–ring), 212.0 (CO) IR (C₆H₁₂): 1939, 1972, 1978, 2027, 2072	[85]

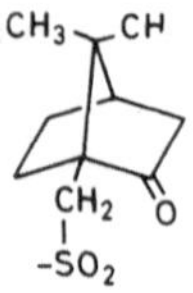

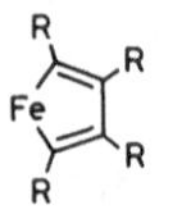

References on p. 57

Table 3 [continued]

No.	Ferracyclo-pentadiene ring	Method of preparation, reaction conditions (yield in %)	Properties and further remarks Explanations on p. 28	Ref.
*61	R R Fe R R 2R = OH 2R = OSi(CH$_3$)$_3$	V, reaction of No. 50 with dry HCl in hexane (quantitative)	bright yellow; dec. at $\approx$200° without melting ^{1}H NMR (C$_6$H$_{12}$): 0.30 (CH$_3$); (CDCl$_3$): 4.97 (broad signal, OH) IR (Nujol): 1932, 1952, 1972, 1990, 2020, 2064; ν(OH free) at 3578, 3596, ν(OH, H–bonded) at 3382, 3488	[56, 59]
*62	CH$_3$ Fe CH$_3$	special	yellow; m.p. 140° (dec.), subl. at 70 to 75°/0.2 ^{1}H NMR (CDCl$_3$): 2.29 (s, CH$_3$), 2.81 (m, CH$_2$) IR (C$_6$H$_{12}$): 1938, 1945, 1975, 2020, 2058	[89]
*63	CH$_3$ Fe CH$_3$	special	yellow; m.p. 82° ^{1}H NMR (CDCl$_3$): 1.73 (s, CH$_3$), 1.86 (m, 2 CH$_2$), 2.48 (br m, 2 CH$_2$) IR (C$_6$H$_{12}$): 1980, 2000, 2065	[89]
*64	Fe	special	orange; m.p. 97 to 98° ^{1}H NMR (CS$_2$): $\approx$1.7 (m, 8 H), $\approx$2.5 (m, 4 H), $\approx$2.8 (m, 4 H) ^{57}Fe-γ (82 K): δ=0.088 (Fe), Δ=1.226 for Fe-2; δ= −0.053 (Fe), Δ=0.951 for Fe-1 IR (C$_6$H$_{14}$): 1917, 1979, 1985, 2023, 2063	[62, 70, 75]
65	(CH$_2$)$_4$ (CH$_2$)$_4$ Fe (CH$_2$)$_4$ (CH$_2$)$_4$	II, boiling C$_6$H$_6$ (21)	m.p. 174 to 177° ^{57}Fe-γ (84 K): δ=0.005 (Fe), Δ=1.126; (in glassy C$_6$H$_5$C$_4$H$_9$-n at 84 k): δ= −0.078 (Fe), Δ=1.007, δ= +0.039 (Fe), Δ=1.241	[69, 75]
*66	(CH$_2$)$_m$ (CH$_2$)$_n$ Fe (CH$_2$)$_n$ (CH$_2$)$_m$ m, n =4, 5 or 5, 4	II, boiling C$_6$H$_6$ (30)	yellow; m.p. 128 to 129°, 133° ^{1}H NMR (CS$_2$): $\approx$1.6 (m, 16 H), $\approx$2.2 (m, 20 H) ^{57}Fe-γ (85.5 K): δ= −0.003 (Fe), Δ=1.139 IR (C$_6$H$_{12}$): 1935, 1982, 2027, 2063	[69, 70, 75]
67	(CH$_2$)$_5$ (CH$_2$)$_5$ Fe (CH$_2$)$_5$ (CH$_2$)$_5$	II, boiling C$_6$H$_6$ (4)	m.p. 173 to 175° ^{57}Fe-γ (86.5 K): δ= −0.009 (Fe), Δ=1.122	[69, 75]

References on p. 57

T a b l e 3 [continued]

No.	Ferracyclo-pentadiene ring	Method of prep-aration, reaction conditions (yield in %)	Properties and further remarks Explanations on p. 28	Ref.
68	(image) m, n = 4, 6 or 6, 4	II, boiling C_6H_6 (23)	^{57}Fe-γ (86.5 K): $\delta = -0.009$ (Fe), $\Delta = 1.140$ there are ambiguities in the positions of the unequal CH_2 chains	[69, 75]
69	(image) m, n = 5, 6 or 6, 5	II, boiling C_6H_6 (8)	m.p. 168 to 170° ^{57}Fe-γ (84 K): $\delta = -0.012$ (Fe), $\Delta = 1.108$ see remarks to No. 68	[69, 75]
*70	(image)	special	deep red; 120 to 121° ^{1}H NMR (CDCl$_3$): 6.64 (d, H-1), 6.97 (m, H-6), 7.10 (d, H-2), 7.29 (m, H-7), 7.66 (d, H-5), 7.80 (d, H-8); J(1,2) = 5.8, J(5,6) ≈ J(7,8) ≈ 8.5 IR (C_6H_{14}): 1993, 2000, 2036, 2070; (KBr): ν(C=C) at 755 and 1600	[57, 76, 84]
*71	(image)	special	orange; m.p. 79.5 to 80.5°, 86 to 87° ^{1}H NMR (CS$_2$): 7.2 to 7.45 (m), 7.3 (s) IR (Skelly B): 1955, 2000, 2005, 2043, 2078	[73, 76]
*72	(image) C_6H_5	I, Fe$_3$(CO)$_{12}$ and $C_6H_5C \equiv CC_6H_5$ (1:3 mole) in petroleum ether at 80 to 90°/1.5 to 2 h (7.9)	red; m.p. 146°, 150°; subl. above 100° in vacuum ^{1}H NMR (−): all protons between 7.0 and 7.7 IR (KBr): 1923, 1980, 2024, 2062	[11, 16, 21, 35, 38, 80]
*73	(image) C_6H_4Cl-4 Cl	I, like No. 72 with 4-ClC$_6$H$_4$C≡C-C$_6$H$_4$Cl-4 (0.5 to 2)	red; m.p. 175° IR (KBr): 1923, 1988, 2037, 2075	[12, 35]

Table 3 [continued]

No.	Ferracyclo-pentadiene ring	Method of preparation, reaction conditions (yield in %)	Properties and further remarks Explanations on p. 28	Ref.
*74	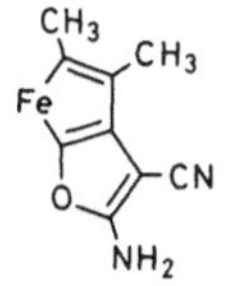 CH₃, CH₃, Fe, O, —CN, NH₂	special	yellow ¹H NMR (CDCl₃): 2.31 and 2.40 (s's, CH₃), 5.9 (br, NH₂) IR (CCl₄): 1993, 2002, 2034, 2073; ν(C≡N) at 2215	[94]

Supplements:

No.	Ferracyclo-pentadiene ring	Method of preparation, reaction conditions (yield in %)	Properties and further remarks	Ref.
*75	C₂H₅, Fe	III, from Fe₃(CO)₁₂ and 2-ethylthio-phene in boiling heptane/16 h (−)	red-orange oil ¹H NMR (C₆D₆): 0.90 (t, CH₃, J=7.5), 2.14 (m, CH₂), 5.11 (dt, H-2), 5.34 (dd, H-4), 6.27 (dd, H-3), J(2,3)=J(2,4)=2.4, J(3,4)=5.4 IR (C₆H₁₄): 1950, 1995, 2000, 2035, 2075	[99]
*76	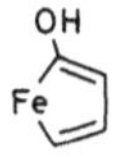 OH, Fe	special	yellow-brown oil ¹H NMR (C₆D₆): 4.40 (t, H-2), 4.90 (dd, H-4), 5.33 (s, OH), 6.12 (dd, H-3), J(2,3)=J(2,4)=2.7, J(3,4)=5.4 IR (C₆H₁₄): 1970, 2005, 2035, 2075; ν(OH) at 3580	[99]
77	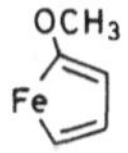 OCH₃, Fe	V, from No. 76 with (CH₃O)₂SO₂ in aqueous alkaline solution (70)	yellow oil ¹H NMR (C₆D₆): 2.91 (s, CH₃), 4.40 (t, H-2), 5.04 (dd, H-4), 6.23 (dd, H-3), J(2,3)=J(2,4)=2.8, J(3,4)=5.6 IR (C₆H₁₄): 1950, 1997, 2011, 2038, 2081	[99]

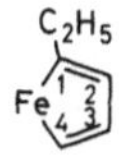

*Further information:

$C_4H_4Fe_2(CO)_6$ (Table 3, No. 1) is obtained in yields up to 12 wt.% together with several other complexes by method I in petroleum ether at 65 °C and under pressures of 10 atm of C_2H_2 and 20 atm of CO. Repeated chromatography on Al_2O_3 is necessary to separate $C_4H_4Fe_2(CO)_6$ from a complex of the composition $(C_2H_2)_3Fe_2(CO)_6$ [24]. An easier preparation is method III carried out with $Fe_3(CO)_{12}$ and thiophene (1:2.5 mole) in boiling n-heptane [83]. The complex can also be obtained along with $(CO)_9Fe_3Te_2$ and $C_4H_4TeFe_2(CO)_6$ by the reaction of $Fe_3(CO)_{12}$ with tellurophene in benzene at reflux temperature. It is formed from $C_4H_4TeFe_2(CO)_6$ on heating above the melting point or heating in solution [63]. Small amounts of the complex have been observed in the products of a cocondensation of Fe atoms with thiophene at −196 °C and subsequent warming of the condensate in a CO atmosphere [82, 91]. The complex is formed in the rapid thermal degradation of compound III (p. 40) in a vacuum at 130 °C (3% yield) [77]. $Fe(CO)_5$ reacts with HC≡CH (saturated solution in tetrahydrofuran at 20 °C) at 110 °C and 9000 to 10000 atm to give $C_4H_4Fe(CO)_3$, $C_4H_4Fe_2(CO)_6$ and quinhydrone as the main product [66].

$$\text{III} \qquad\qquad \text{IV}$$

The ^{1}H NMR spectrum (compare Formula IV) can be interpreted as a degenerated AA'XX' system with $J(1,4) = 0$ Hz [77, 83], see also [15]:

Solvent	δ in ppm		Coupling constants $J(H,H)$ in Hz			Ref.
	H-1,4	H-2,3	H-1,2	H-1,3	H-2,3	
CCl_4	6.8	6.2	5.3	2.3	2.4	[83, 91]
CCl_4	6.85	6.25	−	−	−	[80]
C_6D_6	6.27	5.18	5.4	2.3	2.5	[77]
CS_2	6.77	6.19	−	−	−	[15]

^{13}C NMR spectrum (compare Formula IV):

Solvent	δ in ppm					Ref.
	C-1,4	C-2,3	CO^a	CO^b	CO^c	
CD_2Cl_2	154.9	110.6	206.9	209.0	211.1	[87]
−	153.0	108.7	205.2	207.3	209.5	[90]
$C_6D_5CD_3$	156.5	111.4	one signal at 211.5			[77]

The C,H coupling on C-1,4 and C-2,3 amounts to 155 and 170 Hz, respectively [77].

The ^{57}Fe Mössbauer spectrum indicates two nonequivalent Fe atoms. Three peaks are observed, see **Fig. 8**, and are interpreted as two quadrupole doublets of similar but unequal splitting [31].

Fig. 8

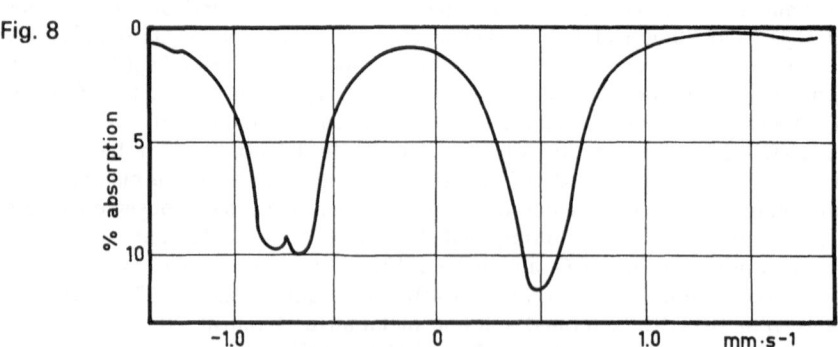

^{57}Fe Mössbauer spectrum of $C_4H_4Fe_2(CO)_6$ [31].

For older reports on the IR spectrum, measured in CS_2, CCl_4, and $CCl_2=CCl_2$ see [15, 18]. Outside the $\nu(CO)$ region the spectrum shows bands at 844, 874, 954, 1031, 1079, 1096, 1224, 1247, 1388, 1399, 1543, 2990, and 3060 cm^{-1} [15].

The complex crystallizes from hexane in monoclinic prisms with a=11.61(1), b= 6.40(1), c=16.35(1) Å, and β=107.6(2)°; space group $P2_1/c-C_{2h}^5$, Z=4. The four C atoms of the ring lie exactly in a plane while the atom Fe(1) is placed 0.23 Å outside the plane. The rather short Fe–C distances in the ring, see **Fig. 9**, indicate fractional multiple bonding, which explains the Fe(1)–C(5,7) bonds of the coplanar carbonyl groups being significantly longer then the bond Fe(1)–C(6). The different formal charges at the Fe atoms are compensated by the highly unsymmetrical "semi bridging carbonyl" (Fig. 9b) which is present in a bent form [83].

Fig. 9

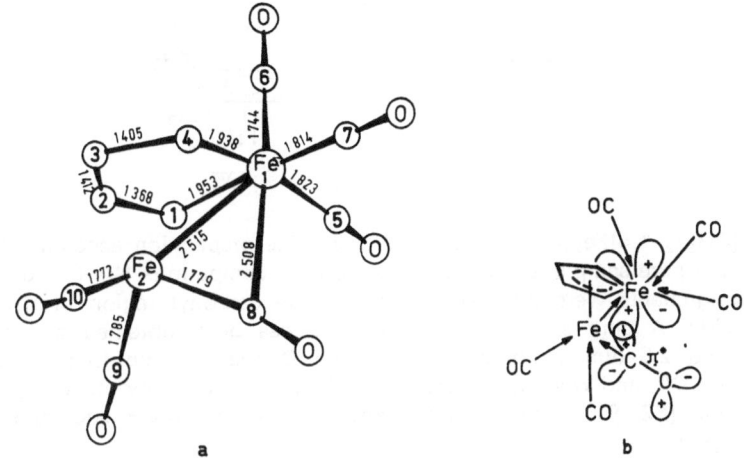

Molecular structure (a) and bonding model (b) of $C_4H_4Fe_2(CO)_6$ [83].

Other selected bond lengths (in Å) and angles (°):

Fe(2)–C(1)	2.080(4)	Fe(2)–C(3)	2.112(7)
Fe(2)–C(2)	2.111(7)	Fe(2)–C(4)	2.082(7)
C–O	1.136 mean		

C(1)–Fe(1)–C(4)	79.3(3)	C(1)–Fe(1)–C(5)	92.8(3) (?)
Fe(1)–C(1)–C(2)	117.5(5)	C(1)–Fe(1)–C(6)	92.0(3) (?)
C(1)–C(2)–C(3)	113.0(7)	C(4)–Fe(1)–C(7)	91.6(3) (?)
C(2)–C(3)–C(4)	112.5(7)	C(5)–Fe(1)–C(7)	94.4(3)
C(3)–C(4)–Fe(1)	116.6(6)	C(6)–Fe(1)–C(7)	96.7(3)
Fe(1)–C(8)–Fe(2)	169.1(6)	C(8)–Fe(2)–C(9)	97.0(4)
Fe(2)–C(8)–O	167.3(7)	C(8)–Fe(2)–C(10)	95.7(4)
Fe–C–O (others)	178.1 mean	C(9)–Fe(2)–C(10)	93.7(4)

The mass spectrum of the compound has been described several times [51, 80, 83]. The fragmentation of $[C_4H_4Fe_2]^+$, formed by successive loss of CO groups, has been reported in a scheme by [83].

The complex is very soluble in all common organic solvents. On heating at 220 °C/5 h under 160 atm CO pressure, it forms cyclopentadienone iron tricarbonyl (65% yield) and $Fe(CO)_5$ [24]. For the reaction with $P(C_6H_5)_3$ see 2.4.1.2.2.2, p. 60.

$(C_4H_3CH_3\text{-}1)Fe_2(CO)_6$ and $(C_4H_3CH_3\text{-}2)Fe_2(CO)_6$ (Table **3**, No. **2** and **3**) can also be obtained in small amounts by cocondensation of 2- or 3-methylthiophene and Fe atoms at $-196\,°C$ [82, 91], see compound No. 5. The products of this method are described as orange oils [91]. Data on the ^1H and ^{13}C NMR spectra which are somewhat different from the values in Table 3 are reported in [90]. The ^{13}C NMR spectrum remains unchanged on cooling the complexes to $-95\,°C$ [91]. Compound No. 2 dissolves in hexane, benzene, and chloroform. It is hygroscopic and air-stable only for a short period of time. Compound No. 3 is quite soluble in hexane, benzene, ether, acetone, and chloroform. It is air-stable for a few days [90].

$(C_4H_2(CH_3)_2\text{-}1,2)Fe_2(CO)_6$ (Table **3**, No. **4**) has been prepared from an excess $Fe_2(CO)_9$ and compound V in tetrahydrofuran at 40 to $60\,°C/7$ h. It can be separated from the by-product VI (2% yield) by chromatography on Al_2O_3, 10% yield [64].

$(C_4H_2(CH_3)_2\text{-}1,4)Fe_2(CO)_6$ (Table **3**, No. **5**). The preparation according to method I did not give a pure sample, but rather mixtures with compounds No. 6 and 7. However, spectral assignments were possible, except for the carbonyl region of the ^{13}C NMR spectrum [81]. A pure sample of the complex has been obtained by cocondensing Fe atoms and 2,5-dimethylthiophene at $-196\,°C$, warming up to room temperature under 0.5 atm CO, and working the product up by chromatography on Al_2O_3 with n-hexane, 5% yield [82, 91]. There is no change in the ^{13}C NMR spectrum on cooling to $-95\,°C$ [91].

$C_4H_2(C_6H_5)_2Fe_2(CO)_6$ (Table **3**, No. **11**). The preparation by method I is carried out with an excess phenylacetylene as the reaction medium at $75\,°C$ for 30 min. Five other, mononuclear iron carbonyl complexes are formed. They are separated by repeated chromatography on Al_2O_3. $C_4H_2(C_6H_5)_2Fe_2(CO)_6$ is eluted with petroleum ether. It is accompanied by 1,2,4-triphenylbenzene, from which it can be separated by recrystallization [11]. Slightly different chemical shifts in the ^{13}C NMR spectrum are reported in [87]. — The compound is very soluble in all common solvents [11]. The reaction with an excess S_x at $200\,°C/5$ min yields 2,5-diphenylthiophene in 65% yield [34].

$C_4H_2(COOR)_2Fe_2(CO)_6$ (Table **3**, No. **13** and **14**, $R=CH_3$ and C_2H_5). The two derivatives are formed in low yield in the reaction of $Na_2[Fe(CO)_4]$ with the esters $ROOC\text{-}CCl=CH\text{-}CH=CCl\text{-}COOR$ in boiling ether for 24 h [45]. The earlier formulation of compound No. 13 as a mononuclear complex [22] was corrected in [55].

The ^1H NMR spectrum of compound No. 13, measured in C_6F_6, shows chemical shifts at $\delta=3.82$ (CH_3) and 6.64 (H-2,3) ppm.

Compound No. 14, which is separated by chromatography on SiO_2, with CH_2Cl_2, must be purified by repeated sublimation and chromatography. The two complexes are soluble in all common solvents [45].

$C_4(CH_3)_4Fe_2(CO)_6$ (Table **3**, No. **15**) has also been obtained in 9% yield from the reaction of $Fe(CO)_5$ with $(cyclo\text{-}C_4(CH_3)_4NiCl_2)_2$ in boiling C_6H_6 saturated with H_2O. It is separated from two other mononuclear complexes by chromatography on Al_2O_3.

References on p. 57

Crystallization from light petroleum yields a product having the melting point 156 to 157 °C. The IR spectrum of the solid (KBr) has $v(CO)$ bands at 1915, 1975, 2010, 2035, and 2070 cm^{-1} [97]. All fragments of the mass spectrum are reported in [58].

$C_4(C_2H_5)_4Fe_2(CO)_6$ (Table 3, No. 16). The ^{57}Fe Mössbauer spectrum measured in a glassy n-butylbenzene matrix at 83.5 K could be fit to two quadrupole doublets with $\delta = -0.005$ and $+0.020$ (referred to Fe) and $\Delta = 1.019$ and 1.257 mm·s^{-1}. The doublet with the smaller quadrupole splitting has been assigned to the Fe atom of the ferrole ring [75]. The mass spectrum is completely given in [58].

The complex yields tetraethylthiophene on reaction with Na_2S_x in boiling C_2H_5-OC_2H_4OH for 3 h. Similarly, tetraethylselenophene is formed with K_2Se_x in CCl_4 under UV irradiation (1 h) [37, 49].

$C_4(C_2H_5)_2(C_6H_5)_2Fe_2(CO)_6$ (Table 3, No. 19) is formed in the reaction of the unstable $C_6H_5C_2C_6H_5Fe_2(CO)_7$ with $C_2H_5C\equiv CC_2H_5$ at 20 °C, 25% yield [48, 49].

$C_4(C_6H_5)_4Fe_2(CO)_6$ (Table 3, No. 20). For the formation by method I see also [6, 16, 21, 43]. Other preparations from iron carbonyls are the reactions of $C_6H_5C\equiv CC_6H_5$ with excess $Fe(CO)_5$ in petroleum ether in a sealed tube at 160 °C for 4 h (28% yield) [6, 16, 21] or in boiling C_6H_6 under UV irradiation for 20 h (42% yield) [14] and the reaction of $C_6H_5C\equiv CC_6H_5$ with $Fe_2(CO)_9$ (2.5:1 mole) in C_6H_6 at 70 °C for 5 min (15% yield) [6, 16, 21]. $C_4(C_6H_5)_4Fe_2(CO)_6$ also forms frequently on treatment of other organoiron carbonyl complexes with $C_6H_5C\equiv CC_6H_5$: From $C_4(C_6H_5)_4Fe_3(CO)_8$ on heating in a sealed tube in $C_6H_5CH_3$ at 170 to 185 °C/84 h (70% yield) [14] or on degradation of the same complex with NaOH in CH_3OH/C_6H_6 at reflux (80% yield) [14], from cyclo-$C_8H_8Fe(CO)_3$ in boiling xylene [29], from cyclo-$C_7H_8Fe(CO)_3$ in tetrahydrofuran (no more details) [74, 78], and from $(C_6H_5)_2C_2SFe_2(CO)_6$ in a sealed tube at 150 °C/1 h (43% yield) [98]. The formation together with $C_4(C_6H_5)_4Fe(CO)_3$ from $(CO)_4FeBr_2$ and $Li(C_6H_5)C=C(C_6H_5)-C(C_6H_5)=C(C_6H_5)Li$ has been reported by [48, p. 292/4]. Reactions of diphenylcyclopropenone with $Fe_2(CO)_9$ in C_6H_6 at room temperature or with $Fe_3(CO)_{12}$ in light petroleum at 90 °C give a mixture of complexes which contains $C_4(C_6H_5)_4Fe_2$-$(CO)_6$ [40].

The highest melting point so far reported, 203 °C with decomposition, has been measured for a sample crystallized from methanol [40]. Two sets of data are given for the ^{13}C NMR spectrum. For the assignment see formula IV on p. 40:

Solvent	δ in ppm					Ref.
	C-1,4	C-2,3	CO^a	CO^b	CO^c	
CDCl$_3$	173.8	129.2	204.5	212.3	216.2	[81]
CD$_2$Cl$_2$	172.3	147.6	203.2	210.6	214.7	[87]

The difference in the C-2,3 values in [81] and [87] may be due to [87] incorrectly assigning a resonance of the C_6H_5 groups to C-2,3. According to [81], the phenyl carbon atoms bonded to C-1,4 and C-2,3 lie at $\delta = 148.6$ and 136.9 ppm, respectively, while the other phenyl carbons occur at $\delta = 127.1$ to 132.0 ppm. The CO region of the spectrum remains unchanged between -125 and $+95$ °C. Unlike for compound No. 16, the nonequivalent Fe atoms are not resolved by ^{57}Fe Mössbauer spectroscopy in a glassy matrix of n-butylbenzene at 82 K, and only one quadrupole doublet with $\delta = 0.023$ (referred to Fe) and $\Delta = 1.098$ mm·s^{-1} is observed [75]. The X-ray K absorption edge is shown in a diagram in [96]. The principal maximum is at 17 eV. It has a weak shoulder at 4.5 eV [11, 96].

References on p. 57

The complex crystallizes from ether/methanol in monoclinic plates with a=16.519(8), b=7.895(5), c=11.386(5) Å, and β=98.38(2)°; space group $P2_1 - C_2^2$. Z=2 gives D_c= 1.44 g·cm^{-3}. The densities D_m=1.44 [11], 1.448 [78], and 1.42 [86] g·cm^{-3} have been measured. For somewhat different cell parameters see [6, 16, 21, 78]. The molecular structure shown in **Fig. 10** comes from two independent X-ray studies. Bond lengths and angles found in these references differ only slightly [78, 86].

Fig. 10

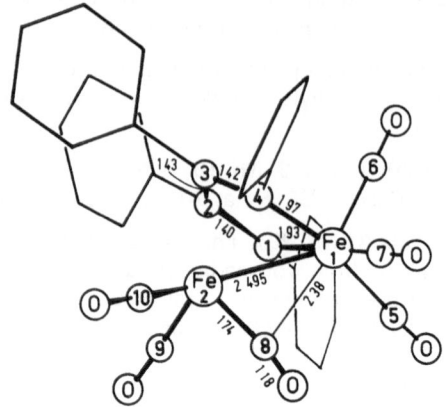

Molecular structure of $C_4(C_6H_5)_4Fe_2(CO)_6$ [86].

The positions of the phenyl rings are only schematically indicated.

Other selected bond lengths (in Å) and angles (°):

Fe(2)–C(1)	2.07(2)	Fe–CO (the rest)	1.78 mean
Fe(2)–C(2)	2.09(2)	C–O (the rest)	1.13 mean
Fe(2)–C(3)	2.15(2)	C–C(phenyl)	1.52 mean
Fe(2)–C(4)	2.07(1)		

C(1)–Fe(1)–C(4)	80.1(6)	C(5)–Fe(1)–C(4)	170.5(8)
Fe(1)–C(1)–C(2)	117.3(11)	C(5)–Fe(1)–C(6)	95.6(11)
C(1)–C(2)–C(3)	113.9(13)	C(5)–Fe(1)–C(7)	92.2(8)
C(2)–C(3)–C(4)	111.0(12)	C(6)–Fe(1)–C(7)	97.4(11)
C(3)–C(4)–Fe(1)	116.5(11)	C(7)–Fe(1)–C(1)	166.5(9)
Fe(1)....C(8)–O	121.7(15)	Fe(1)–Fe(2)–C(8)	65.7(6)
Fe(2)–C(8)–O	165.5(12)	C(8)–Fe(2)–C(9)	97.1(10)
Fe–C–O (the rest)	175.2 mean	C(8)–Fe(2)–C(10)	96.3(9)
		C(9)–Fe(2)–C(10)	87.7(10)

The mass spectrum is completely reported in [58]. The title compound has been used for studying the chemical behaviour of the $C_4R_4Fe_2(CO)_6$ complex type. Several reactions are compiled in [19, 46, 49], see Scheme 2 on p. 27.

The compound is soluble in all common organic solvents [6, 16, 21]. The thermal decomposition in toluene in a closed system at 180 to 195 °C for 68 h gives the cyclopenta-dienone VII (42% yield) and its iron tricarbonyl complex VIII (50% yield). The same compounds (VIII in 90% yield) and $Fe(CO)_5$ are formed in decaline under 200 atm of CO at 250 °C for 4 h. The complexes VII and IX have been obtained in a photolytic degradation with UV light in C_6H_6 for 3 h [11].

References on p. 57

The stoichiometric bromination in warm CH_3COOH leads only to tetracyclone VII, but decomposition products of VII are obtained, if an excess of Br_2 is employed [14]. However, according to [11] the degradation with Br_2 in about tenfold excess in CH_3COOH at room temparature gives predominantly complex VIII and little of VII. Compound X has been isolated from a reaction of the complex in C_6H_6 with an excess of Cl_2 gas [32].

Hydrogenation on Pt or Raney Ni failed. Other reductive degradations had the following results: Na in liquid NH_3 in the presence of C_2H_5OH at −60 °C gives 1,2,3,4-tetraphenyl-butane, $NaNH_2$ in liquid NH_3 at −60 °C/15 min gives 1,2,3,4-tetraphenylbutadiene (14%), VII (10%), and IX (31%), $LiAlH_4$ in ether at room temperature for 1 h and at reflux for 1 h gives 1,2,3,4-tetraphenylbut-1-ene (66%), and $LiAlH_4$ in tetrahydrofuran at 0 °C for 20 h gives 1,2,3,4-tetraphenylbut-1-ene (20%), 1,2,3,4-tetraphenylbutane (31%), some IX and traces of VII [11].

The degradation with NaOH in CH_3OH/C_6H_6 at 50 °C for 30 min and a work up in aqueous medium give the complex IX (75%) and small amounts of tetracyclone VII [14].

	No.	XI	XII	XIII	XIV
	E =	S	Se	NH	NC_6H_5
	No.	XV	XVI	XVII	XVIII
	E =	PC_6H_5	$PCH_2C_6H_5$	$P(O)CH_2C_6H_5$	AsC_6H_5

$C_4(C_6H_5)_4Fe_2(CO)_6$ has been suggested as a starting material for the synthesis of the heterocyclic compounds XI to XVIII [11, 37, 52]. The reaction partners and conditions are briefly summarized as follows: UV photolysis in the presence of an excess S_x in C_6H_6 for 16 h gives XI (50%), heating of a finely powdered mixture of the complex and S_x at 200 °C for 10 min gives XI (80%) [11, 37], or with Se_x at 200 °C for 20 min gives XII (60%), UV photolysis in the presence of amorphous Se_x in piperidine gives XII (25%), NOCl between −70 °C and room temperature for 3 h and reduction with Zn dust in CH_3COOH/HCl overnight gives XIII (34%), UV photolysis with C_6H_5NO in C_6H_6 for 6 h gives XIV (51%) [37], with an excess of $C_6H_5PCl_2$ at 140 °C for 30 min gives XV (66%) [7, 37, 52], with $C_6H_5CH_2PBr_2$ in C_6H_6 in a closed system at 170 °C for 15 h gives XVI and XVII, probably by subsequent air oxidation [37, 52], and UV photolysis with $C_6H_5AsCl_2$ in CH_2Cl_2 for 22 h gives XVIII (3%) [37, 52]. The thermal reaction with $C_6H_5AsCl_2$ at 150 °C or with $(C_6H_5)_2SiCl_2$ at 200 °C yields only tetracyclone VII and $C_4(C_6H_5)_4Fe(CO)_3$ [48].

Oxidative degradation to $O=C(C_6H_5)-C(C_6H_5)=C(C_6H_5)-C(C_6H_5)=O$ has been carried out with boiling SO_2Cl_2 (25% yield of ketone) and has also been observed on treatment with Cl_2O [26]. The photolytic degradation in the presence of $4-CH_3C_6H_4SO_2NCl_2$

in C_6H_6 at 30 °C for 2 h yields $4-CH_3C_6H_4SO_2NH-C(C_6H_5)=C(C_6H_5)-C(C_6H_5)=$ $C(C_6H_5)-NHSO_2-C_6H_4CH_3-4$ [37]. The C_6H_5 groups of the ring can be mono- and diacetylated by $CH_3COCl/AlCl_3$, but the reaction conditions were not reported [49].

$C_4(C_6H_5)_4Fe_2(CO)_6$ has been used to prove that ferroles are not intermediates in the cyclic trimerization of alkynes to substituted benzenes [17] as had been suggested earlier [3]. The trimerization appears to be catalyzed by iron carbonyl fragments originating in the decomposition of the ferroles at elevated temperature [17].

For phosphine substitution products see 2.4.1.2.2.2, p. 60.

$C_4(C_6H_4Cl-4)_4Fe_2(CO)_6$ (Table 3, No. 21) can also be prepared from $Fe_2(CO)_9$ and $4-ClC_6H_4C\equiv CC_6H_4Cl-4$ (1:1 mole) in C_6H_6 at 70 °C for 5 to 10 min [6, 21]. The two crystal modifications have different Debye–Scherrer diagrams but identical IR spectra [16]. The compound is very soluble in all common solvents [6, 12, 16, 21]. Polarographic reduction in $CH_3OCH_2CH_2OCH_3$ occurs irreversibly at $E_{1/2}=-1.87$ V (referred to 10^{-3} M Ag^+/Ag) [47].

$C_4(C_6F_5)_4Fe_2(CO)_6$ (Table 3, No. 22) is formed in 21% yield from $Fe(CO)_5$ and $C_6F_5C\equiv CC_6F_5$ in petroleum ether under UV irradiation at 40 to 60 °C for 18 h. The unsufficient solubility does not permit the recording of an useful ^{19}F NMR spectrum. At least six lines have been observed indicating two structurally different C_6F_5 groups.

The complex can be recrystallized from petroleum ether. The reaction with excess Br_2 in boiling CCl_4 for 3 h gives tetrakis(pentafluorphenyl)cyclopentadienone. UV photolysis in the presence of S_x in boiling C_6H_6 for 20 h yields 21% tetrakis(pentafluorphenyl)-thiophene [41].

$(C_4(CH_3)_2-1,4(C\equiv CCH_3)-2,3)Fe_2(CO)_6$ (Table 3, No. 24) is formed together with the isomers No. 25 and 26 in the reaction of $Fe(CO)_5$ with $CH_3C\equiv CC\equiv CCH_3$ in xylene at 120 to 130 °C for 45 min. The main isomer No. 26 was separated by cooling a hot CH_3CN solution of the mixture to room temperature and was recrystallized from a saturated CH_3CN solution at -10 °C. Repeated chromatography on Al_2O_3 with hexane was required to separate the two other isomers. Compound No. 24 can not be obtained in a pure form. It has been mentioned that method II (in C_6H_6 at 90 °C for 70 min) also gives a mixture of the three isomers in apparently better yield [88].

$(C_4(CH_3)_2-1,3(C\equiv CCH_3)_2-2,4)Fe_2(CO)_6$ (Table 3, No. 26). An unambiguous assignment of the ^{13}C NMR resonances to each individual C atom has not been possible. The following signals (in $CDCl_3$) have been measured, cf. Formula XIX:

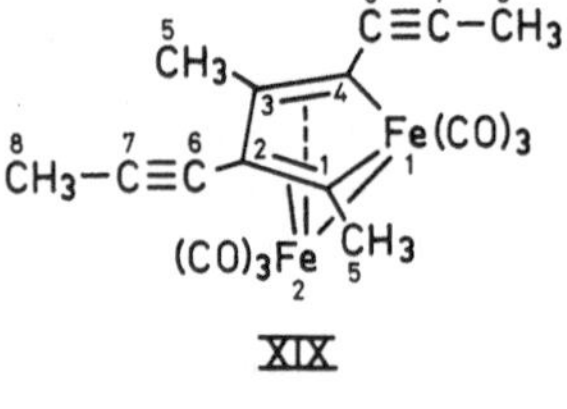

δ (ppm) =4.3, 5.0 (C-8, J(C,H) =132 Hz), 17.8, 30.6 (C-5, J(C,H) =130 and 128 Hz), 92.6, 100.6 (C-7, J(C,H) =10 and 12 Hz), 73.8, 113.6 (C-6), 84.9, 176.4 (C-1,3, J(C,H) =4 and 7 Hz), 131.5, 140.0 (C-2,4) ppm. CO groups on Fe-1: δ=205.4, 206.6, 210.1 ppm; the CO groups on Fe-2 give only one signal at δ=212.4 ppm which does not split at -78 °C [88].

XIX

$(C_4(C_6H_5)_2-2,3(C\equiv CC_6H_5)_2-1,4)Fe_2(CO)_6$ (Table 3, No. 27). The preparation by method I affords all three isomers No. 27 to 29. They are separated by repeated fractional crystallization from benzene/petroleum ether. The mixture of isomers is also obtained from $Fe(CO)_5$ and $C_6H_5C\equiv CC\equiv CC_6H_5$ in petroleum ether at 150 °C/20 h in an autoclave or from $Fe_2(CO)_9$ and the alkyne (5:3 mole) in benzene/petroleum ether at 40 to 45 °C/ 1 h. The total yields are 14% and 5.7%, respectively [27].

$C_4(C_3H_5-cyclo)_4Fe_2(CO)_6$ (Table 3, No. **32**) is formed together with several other organoiron carbonyls on UV irradiation of $Fe(CO)_5$ and dicyclopropylacetylene in hexane at room temperature for 45 min. The kind of products and their relative amounts depend on the concentration of the starting materials. Solutions 0.1 M in $Fe(CO)_5$ and 0.05 M in alkyne give the complex in 6% yield. A similar distribution of the major products was obtained from $Fe_2(CO)_9$ (0.14 M) and the alkyne (0.1 M) in refluxing pentane for 1 h, giving an 8% yield for the ferrole. A quite different product distribution results from the reaction of $Fe_3(CO)_{12}$ (0.12 M) and the alkyne (0.1 M) in refluxing benzene for 5 h, giving an 6% yield for the ferrole. Chromatographic separation on SiO_2 with pentane elutes the ferrole first. The ferrole has also been detected among the products which are formed in photochemical reactions of the alkyne with the complexes $C_8H_{10}CO$-$Fe_2(CO)_6$ and $(C_8H_{10})_2COFe_2(CO)_6$ [92].

No reaction occurs when the complex is irradiated with $Fe(CO)_5$ in pentane for 1 h or is heated with $Fe_3(CO)_{12}$ in refluxing C_6H_6 for 2 h [92].

$C_4H_2(OH)_2Fe_2(CO)_6$ (Table 3, No. **33**). The first structure proposals for this compound [2, 3, 4] proved to be false. The compound is hygroscopic and decomposes in air under the influence of light, in an inert atmosphere it remains unchanged in daylight. It is very soluble in CH_3OH, C_2H_5OH, acetone, acetic acid, acetic ester, soluble in H_2O, dilute acids, C_6H_6, and insoluble in petroleum ether [1]. The following solubilities (in g per 100 cm³ of solvent) have been reported: 0.41 in hexane, 1.63 in CS_2, and 5.82 in C_6H_6 [4]. According to [4] the complex is only sparingly soluble in H_2O.

The compound crystallizes from aqueous solutions as a monohydrate which forms light yellow to orange needles, m.p. 104 to 110 °C [4] or 110 to 111 °C (dec.) [1]. The monohydrate is considerably less soluble than the anhydrous complex, 0.014 in hexane, 0.025 in CS_2, and 0.188 in C_6H_6 [4]. It can be dehydrated in vacuum [1] or over P_2O_5 in vacuum for 4 to 8 h [4]. The monohydrate is much more stable than the anhydrous compound. Thus it can be stored under He at room temperature for six months without decomposition whereas the anhydrous complex shows signs of decomposition after 12 h, even when stored in the dark in the absence of air [4].

The two OH groups of the ferrole ring are acidic, $pK_a^1=6.30$ and $pK_a^2=9.14$ [2, 4]. Only one acidic H atom has been observed by titration in [1] which may have been due to the presence of oxygen [4]. Because of the acidity the complex dissolves very readily in aqueous alkali. These solutions are stable in daylight under N_2 [1] but are rather sensitive toward oxygen [1, 4]. The precipitation from alkaline solutions with dilute acids has been used to purify the complex as monohydrate [1, 4]. In the absence of air, the alkaline solutions show little decomposition on heating to 85 to 90 °C [1].

Oxidation in acidic media, preferably with $FeCl_3$, leads to compound XX [10, 30]. The reaction of an alcoholic solution of the complex with $HC{\equiv}CH$ under pressure at 120 to 150 °C gives the acrylic acid ester and small amounts of hydroquinone [1]. For transformations of the OH groups into other substituents see compounds No. 42 to 45.

XX

$C_4H(C_2H_5)(OH)_2Fe_2(CO)_6$ (Table **3**, No. **35**) can be converted into its mono-(−)-menthoxyacetate (compound No. 54). The alkaline hydrolysis of this optically active product yields the title compound, but in the racemic form. Extrapolation from other measurements suggests that the energy barrier for racemization of the dianion [C_4H-$(C_2H_5)(O^-)_2Fe_2(CO)_6$] can not be higher than about 13 kcal·mol^{-1} [23].

$C_4H(C_4H_9\text{-}t)(OH)_2Fe_2(CO)_6$ (Table **3**, No. **37**). For chemical reactions on the OH groups see compounds No. 47 and 55.

$C_4(CH_3)_2(OH)_2Fe_2(CO)_6$ (Table **3**, No. **39**) was the first ferrole whose structure could be elucidated by X-ray diffraction. It crystallizes in the monoclinic system with the cell parameters $a=12.26$, $b=7.47$, $c=15.70$ Å, and $\beta=97.5°$; space group $P2_1/c-C_{2h}^5$. $Z=4$ gives $D_c=1.83$ g·cm^{-3}, $D_m=1.79$ g·cm^{-3}. The molecule, shown in **Fig. 11**, has a mirror plane through the Fe atoms perpendicular to the C_4Fe ring. Small deviations from the mirror plane can be explained by packing effects in the crystal. The metal–carbon separations can be classified into three groups of average length 2.13 for Fe(2)–C(ring), 1.95 for Fe(1)–C(1,4), and 1.78 Å for Fe–CO. Carbon–oxygen separations correspond to the normal carbonyl bonding (1.14 Å) and to the phenolic C–OH length (1.37 Å). Each Fe atom can acquire a rare gas shell if the Fe–Fe single bond is regarded as a dative bond Fe(2)→Fe(1). The bent ligand C(8)–O lies in the plane of symmetry but with the O atom forced away from the other Fe(CO)$_3$ group. C(8) occupies the sixth octahedral site about Fe(1) but with the long distance of 2.48 Å, which excludes any simple interaction. The molecules pack in infinite chains parallel to the b direction, being linked by hydrogen bonds between the C–OH groups of successive molecules (O(5)....O(6) separation 2.85 Å). The presence of hydrogen bonds is also clearly seen in the IR spectrum of the solid [5, 20].

Fig. 11

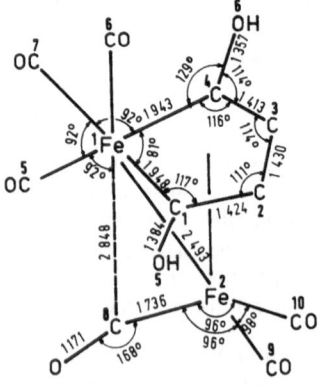

Bond lengths (in Å) and angles (°) of $C_4(CH_3)_2(OH)_2Fe_2(CO)_6$ [20].

Other angles (°):

C(5)–Fe(1)–C(6)	96		C(1)–Fe(1)–C(6)	92
C(7)–Fe(1)–C(6)	99		C(4)–Fe(1)–C(6)	93

The complex can not be acetylated under mild conditions, and more vigorous conditions yield a diacetate. Neither the reaction conditions nor the product have been described [3].

References on p. 57

$C_4(C_2H_5)_2(OH)_2Fe_2(CO)_6$ (Table **3**, No. **40**) also forms in 5% yield along with several other complexes from $Fe_3(CO)_{12}$ and $C_2H_5C\equiv CC_2H_5$ in refluxing n–heptane. The structure has recently been reported. It shows the two $Fe(CO)_3$ units in an eclipsed configuration. There is no semibridging carbonyl [100, 101].

$C_4H_2(OCH_3)_2Fe_2(CO)_6$ (Table **3**, No. **42**) reacts in boiling SO_2 under UV irradiation by substitution of one CO ligand [25], see 2.4.1.2.2.2. Attempts to synthesize a pyrrole derivative by degradation with $4-CH_3C_6H_4SO_2NCl_2$ failed. Instead, dimethyl fumarate, $4-CH_3C_6H_4SO_2NH_2$, and compound XXI were obtained in CH_2Cl_2 at room temperature. At -10 °C, this reaction leads to complex XXII (14.3% yield), dimethyl fumarate, and $4-CH_3C_6H_4SO_2NH_2$ [42], see also [26, 37]. UV irradiation in C_6H_6 in the presence of $(C_6H_5)_2CN_2$ results in the formation of complex XXIII (by insertion of diphenylcarbene, 4% yield) and complex No. 49 (6% yield). The reaction mechanism is difficult to account for as it involves a skeletal rearrangement [54].

$$4-CH_3C_6H_4SO_2N=C-CH_2CH_2-C=NSO_2C_6H_4CH_3-4$$
$$\underset{\displaystyle OCH_3}{|} \qquad\qquad \underset{\displaystyle OCH_3}{|}$$

XXI

XXII   XXIII

$C_4H(C_4H_9-t)(OCH_3)_2Fe_2(CO)_6$ (Table **3**, No. **47**) has been obtained from the optically active complex No. 55 by hydrolysis with CH_3ONa/CH_3OH at -40 °C and subsequent methylation with CH_2N_2. The products of different experiments showed rotations between $[\alpha]_D = +10$ and $+25°$. This racemized at 76.0 °C with the rate $k = (1.1 \pm 0.1) \times 10^{-3}$ s^{-1} and at 52.4 °C with $k = (5.1 \pm 0.5) \times 10^{-5}$ s^{-1}. Thus, $E_a = 29 \pm 2$ kcal·mol^{-1} and lg A = 14.6 ± 0.6 [23].

$C_4(OCH_3)_4Fe_2(CO)_6$ (Table **3**, No. **48**) is formed from $Fe_2(CO)_9$ and $CH_3OC\equiv COCH_3$ between -60 °C and room temperature. The yields are low because decomposition of the rather unstable alkyne competes with the iron carbonyl reaction at higher temperature [71].

$C_4H(CH(C_6H_5)_2)(OCH_3)_2Fe_2(CO)_6$ (Table **3**, No. **49**) occurs together with the isomeric compound XXIII in the reaction of complex No. 42 with an excess $(C_6H_5)_2CN_2$ in C_6H_6 under irradiation for 5 h. The two compounds are readily isolated by chromatography and crystallization from petroleum ether [54].

The substance crystallizes in the monoclinic system with a = 14.75, b = 8.95, c = 19.58 Å, and $\beta = 108.2°$; space group $P2_1/c - C_{2h}^5$. Z = 4, $D_c = 1.509$ and $D_m = 1.495$ g·cm^{-3}. The ferracyclopentadiene ring of the molecule is nearly planar, see **Fig. 12**, p. 50. The distances and angles correspond largely to the values found in other ferrole structures. However,

References on p. 57

the shortest nonbonded distance between Fe(1) and a carbonyl carbon atom on Fe(2) is 2.86 Å [Fe(1)...C(9)], the bond angle at this CO group being 173°. These values suggest that there is no interaction of the type "semi bridging carbonyl" [61].

Fig. 12

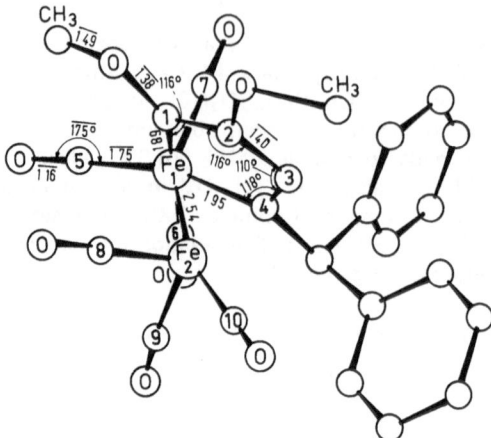

Molecular structure of $C_4H(CH(C_6H_5)_2)(OCH_3)_2Fe_2(CO)_6$ [61].

Other selected bond lengths (in Å) and angles (°):

Fe(2)–C(1)	2.21	Fe(2)–C(3)	2.16
Fe(2)–C(2)	2.13	Fe(2)–C(4)	2.08

C(1)–Fe(1)–C(4)	80	C(6)–Fe(1)–C(7)	99
C(1)–Fe(1)–C(5)	98	Fe(2)–Fe(1)–C(1)	58
C(1)–Fe(1)–C(6)	168	Fe(2)–Fe(1)–C(4)	53
C(1)–Fe(1)–C(7)	91	Fe(1)–Fe(2)–C(8)	106
C(4)–Fe(1)–C(5)	156	Fe(1)–Fe(2)–C(9)	81
C(4)–Fe(1)–C(6)	91	Fe(1)–Fe(2)–C(10)	157
C(4)–Fe(1)–C(7)	100	C(8)–Fe(2)–C(9)	95
C(5)–Fe(1)–C(7)	104	C(8)–Fe(2)–C(10)	97
C(5)–Fe(1)–C(6)	87	C(9)–Fe(2)–C(10)	97

$C_4(OSi(CH_3)_3)_4Fe_2(CO)_6$ (Table 3, No. 50) has been prepared from $Na_2[Fe(CO)_4]$ and $(CH_3)_3SiI$ [56, 59] or $(CH_3)_3SiBr$ [65] (1:2 mole) in tetrahydrofuran at room temperature for 30 min, 21.7% yield [59]. For this substance, incorrect empirical formulas and structures were initially reported, e.g. $[(Si(CH_3)_3)_2Fe(CO)_4]_2$ [56] and $[Si(CH_3)_3Fe(CO-Si(CH_3)_3)(CO)_3]_2$ [59]. The correct ferrole structure [65] is shown in Fig. 13.

On sublimation in an evacuated sealed tube, the complex forms triclinic crystals with a=11.535(6), b=12.288(6), c=12.543(6) Å, α=72.01(3)°, β=87.31(4)°, and γ= 88.08(4)°; space group P$\bar{1}$–C$_i^1$. Z=2 corresponding to D$_c$=1.34 and D$_m$=1.32 g·cm^{-3}. The geometry of the Fe(CO)$_3$C$_4$Fe(CO)$_3$ portion of the molecule is almost the same as that of compound No. 33. The siloxy substituents are arranged in a manner that minimizes intramolecular repulsion, with each Si–O–C(ring) plane approximately perpendicular to the FeC$_4$ plane, and with Si atoms alternately up and down with respect to this plane. Intermolecular forces may be responsible for the large range of Si–O–C(ring) angles (124 to 146°) [65].

References on p. 57

Fig. 13

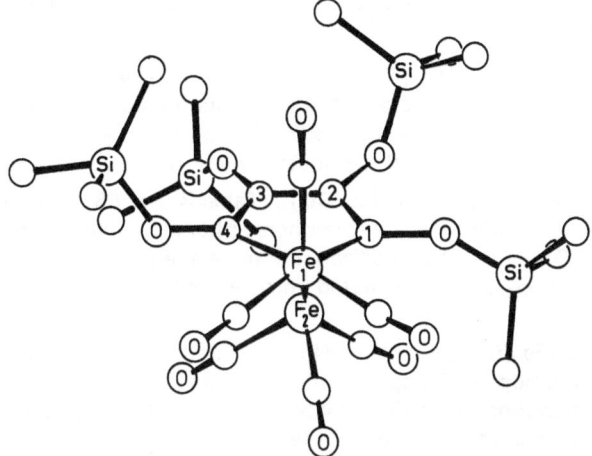

Molecular structure of $C_4(OSi(CH_3)_3)_4Fe_2(CO)_6$ [65].

Selected bond lengths (in Å) and angles (°):

Fe(1)–Fe(2)	2.500(3)	Fe(2)–CO	1.72(2) mean
Fe(1)–C(1,4)	1.92(2) mean	C–OSi	1.39(3) mean
Fe(1)–CO	1.83(2) mean	O–Si	1.66(2) mean
Fe(2)–C(1,4)	2.13(2) mean	Si–CH₃	1.86(2) mean
Fe(2)–C(2,3)	2.19(2) mean		

C(1)–O–Si	146(1)	C(3)–O–Si	138(1)
C(2)–O–Si	124(1)	C(4)–O–Si	127(1)

Other bands of the IR spectrum between 572 and 1466 cm^{-1} are listed and assigned. The mass spectrum is also completely reported [59].

The complex has good thermal stability in the vacuum, it could be recovered essentially unchanged after 1 h at 225 °C. Extended heating at this temperature (18 h) gave some $(CH_3)_3SiOSi(CH_3)_3$ and a black crystalline solid. The decomposition in O_2 (638 Torr) at room temperature proceeds very slowly giving after 24 h small amounts of $Fe(CO)_5$ and $(CH_3)_3SiOSi(CH_3)_3$. No reaction was detectable between the solid complex and H_2O, HCl gas, $N(CH_3)_3$, $P(CH_3)_3$, and $[N(C_4H_9-n)_4]Cl$ at room temperature. 78% of the $Si(CH_3)_3$ groups were found as $(CH_3)_3SiOSi(CH_3)_3$ when the compound was hydrolyzed in ether solution with H_2O at room temperature for 1 h. The attack by CH_3OH in cyclohexane is much slower. The composition $[Si(CH_3)_3Fe(COH)(CO)_3]_2$ had been attributed to a yellow precipitate formed from the title complex in hexane and an excess of HCl gas [59]. This is not likely to be correct [65], see compound No. 61.

$C_4H(CH_3)(OH)(OOCCH_3)Fe_2(CO)_6$ and $C_4H(CH_3)(OOCCH_3)(OOCCH_2Cl)Fe_2(CO)_6$ (Table 3, No. **51** and **52**). Compound No. 51 exists in two isomeric forms. One isomer has been obtained by acetylation of No. 34 under mild conditions, the other isomer results from treatment of the diacetate No. 46 with hot CH_3OH. The two isomeric monoacetates give with $(CH_2Cl-CO)_2O$ two isomers of No. 52 [3]. No experimental details have been reported for these reactions and products.

$C_4H(C_4H_9-t)(OH)(OCH_3)Fe_2(CO)_6$ (Table 3, No. **53**). A very rough estimate has been made of the rate of racemization of the $[C_4H(C_4H_9-t)(O^-)(OCH_3)Fe_2(CO)_6]$, giving

References on p. 57

$k = (1.5 \pm 0.5) \times 10^{-4}$ s^{-1} which would correspond to an energy barrier of about 21 kcal·mol^{-1} [23].

$C_4H(C_2H_5)(OH)(C_{12}H_{21}O_3)Fe_2(CO)_6$ (Table 3, No. 54) has been prepared from compound No. 35. The initial reaction product changed its rotation on repeated crystallization in alcohol at 18 °C from $[\alpha]_D = -44°$ to $-397°$. The alkaline hydrolysis yielded the parent complex No. 35 but without detectable optical activity [23].

$C_4H(C_4H_9\text{-}t)(OCH_3)(C_{10}H_{15}O_4S)Fe_2(CO)_6$ (Table 3, No. 55) has been prepared from compound No. 37. The primary $(+)$-camphorsulphonate methyl ether had a rotation of $[\alpha]_D = +14° \pm 1°$. On repeated crystallization a more sharply melting product with $[\alpha]_D = -153° \pm 3°$ formed. It could be converted into an optically active dimethyl ether, see No. 47 [23].

$C_4H_2(C_6H_4Br\text{-}4)_2Fe_2(CO)_6$ (Table 3, No. 57). Method I gives four other organoiron complexes, tris(4-bromophenyl)troponeiron tricarbonyl being the main product (20% yield). The title compound is readily soluble in C_6H_6, CH_2Cl_2, and $CHCl_3$, and less soluble in ether, alcohols, and petroleum ether [34].

$C_4(C_4H_9\text{-}t)_2(C_6H_5)_2Fe_2(CO)_6$ (Table 3, No. 59) is prepared from $Fe_2(CO)_9$ and $t\text{-}C_4H_9C\equiv CC_6H_5$ (1:3 mole) in cyclohexane at 50 °C for 10 h and is isolated by chromatography on Al_2O_3 with petroleum ether [50].

$C_4(OH)_2(OSi(CH_3)_3)_2Fe_2(CO)_6$ (Table 3, No. 61). Other empirical formulas attributed to this compound [56, 59] are improbable. The by-product of the preparation, $(CH_3)_3SiCl$, and the spectroscopic data strongly suggest that two $(CH_3)_3SiO$ groups of the parent compound No. 50 had been replaced by OH groups. — For other IR bands between 568 and 1526 cm^{-1} and the complete mass spectrum see [59].

The compound is insufficiently soluble in C_6H_6, CH_2Br_2, CH_3CN, C_2H_5OH, $CH_3COC_2H_5$, and H_2O to determine the molecular weight. The treatment with excess $(CH_3)_3SiCl$ and $N(CH_3)_3$ at room temperature for 2 h regenerates the parent compound No. 50 [59].

$C_4(CH_3)_2(C_2H_4)Fe_2(CO)_6$ (Table 3, No. 62) has been obtained in very low yield (1%) together with a dimer of the diyne ($<1\%$) from $Fe(CO)_5$ and $CH_3C\equiv CCH_2CH_2C\equiv CCH_3$ in toluene at 110 °C for 24 h. A large number of minor products is also formed. They are separated by repeated chromatography. On sublimation at 70 °C/0.2 Torr, the compound condenses as a fine yellow powder. In the ^1H NMR spectrum the CH_2 multiplet appears to be a quartet of doublets ($J = 15$ and 1 Hz) superimposed on a quintet ($J = 12.5$ Hz), both centered at $\delta = 2.81$ ppm [89].

$C_4(CH_3)_2(C_4H_8)Fe_2(CO)_6$ (Table 3, No. 63) has been prepared from $Fe(CO)_5$ and $CH_3C\equiv CCH_2CH_2CH_2C\equiv CCH_3$ (1.2:1 mole) like compound No. 62 but in much better yield (32%). Thin layer chromatography with CH_2Cl_2 separated minor amounts of two other organic products [89].

$C_4(C_4H_8)_2Fe_2(CO)_6$ (Table 3, No. 64) has been prepared from an excess $Fe(CO)_5$ and cyclododeca-1,7-diyne in refluxing toluene for 32 h, 31% yield [62]. ^{57}Fe Mössbauer data were also measured in a glassy matrix of $C_6H_5C_4H_9$-n at 83 K (δ referred to Fe): $\delta = 0.047$, $\Delta = 1.219$ mm·s^{-1} for Fe(2) and $\delta = -0.087$, $\Delta = 0.951$ mm·s^{-1} for Fe(1) [75]. In contrast to the earlier proposed structure [62], an X-ray study revealed that the C_4H_8 chains are bridging the double bonds of the ferrole system. The substance crystallizes in the trigonal system with $a = 9.186(3)$ and $c = 18.604(6)$ Å, space group $P3_1 - C_3^2$. $Z = 3$ gives $D_c = 1.61$, while $D_m = 1.57$ g·cm^{-3}. The molecular structure is shown in **Fig. 14**. Noticeable features of the molecule are the close interaction between Fe(1)

References on p. 57

and C(6) (2.32(2) Å), which causes a bending of the angle Fe(2)–C(6)–O to 162(3)°, and the disorder of the CH$_2$ groups, particularly C(12), C(13), C(16), and C(17). Also noticeable is the finding that a skeletal rearrangement of the starting diyne from a 12-membered ring to a bicyclohexenyl type framework has occurred upon complexation [67].

Fig. 14

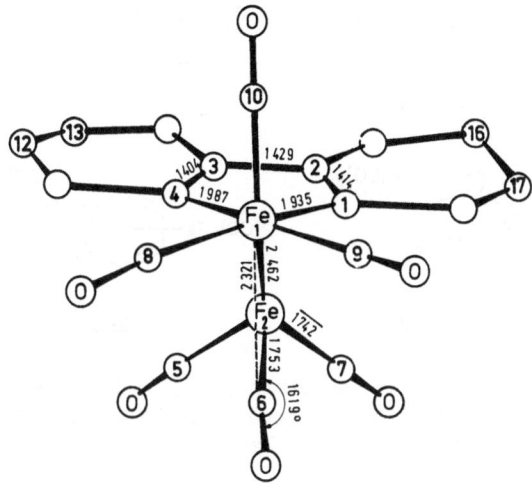

Molecular structure of C$_4$(C$_4$H$_8$)$_2$Fe$_2$(CO)$_6$ [67].

Selected bond lengths (in Å) and angles (°):

Fe(2)–C(1)	2.087(14)	Fe(2)–C(3)	2.189(13)
Fe(2)–C(2)	2.172(12)	Fe(2)–C(4)	2.066(12)

Fe(1)–C(1)–C(2)	116.5(10)	Fe(1)–Fe(2)–C(5)	127.6(6)
C(1)–C(2)–C(3)	114.4(11)	Fe(1)–Fe(2)–C(7)	136.1(8)
C(2)–C(3)–C(4)	110.9(12)	Fe(1)–Fe(2)–C(6)	64.3(6)
C(3)–C(4)–Fe(1)	117.1(11)	Fe(2)–C(6)–O	161.9(28)
C(1)–Fe(1)–C(8)	167.7(9)	C(10)–Fe(1)...C(6)	171.8(9)
C(4)–Fe(1)–C(9)	173.0(7)	Fe(1)...C(6)–Fe(2)	72.8(7)
C(10)–Fe(1)–Fe(2)	129.0(7)		

C$_4$(C$_{11}$H$_{18}$)$_2$Fe$_2$(CO)$_6$ (Table 3, No. 66) can also be obtained in 14% yield from Fe(CO)$_5$ and cyclotrideca-1,7-diyne in boiling toluene for 16 h and is separated from other reaction products by chromatography on Al$_2$O$_3$ with hexane [70]. There are ambiguities concerning the positions of the unequal CH$_2$ chains [69, 70, 75].

C$_4$H$_2$(C$_4$H$_4$)Fe$_2$(CO)$_6$ (Table 3, No. 70) has been prepared from an excess Fe$_2$(CO)$_9$ and 1,4-dibromocyclooctatetraene in heptane at reflux temperature for 20 h, 25% yield. A second product is benzocyclobutadieneirontricarbonyl (XXIV) [84]. It is also formed together with compound XXVII from Fe(CO)$_5$ and 2-bromostyrene under UV irradiation at room temperature and in the UV photolysis of 2-BrC$_6$H$_4$CH=CH$_2$Fe(CO)$_4$ (XXVI) in hexane [57]. The title compound and the isomer No. 71 have been obtained from the reaction of Fe$_3$(CO)$_{12}$ with complex XXIV at 120 °C for 7 h (38% yield each) [73], from the photoreaction of Fe(CO)$_5$ with compound XXV, and the photoreaction of Fe(CO)$_5$

References on p. 57

with compound XXIV [76]. Equimolar amounts of XXVII and XXVIII also give the title compound in moderate yield when treated with an excess $Fe_2(CO)_9$ in hexane at 62 °C for 1.5 h [68].

XXIV   **XXV**   **XXVI**   **XXVII**

XXVIII   **XXIX**

The complex crystallizes from pentane in the monoclinic system with a=8.076, b= 14.188, c=12.408 Å, and β=113.57°; space group $P2_1/c - C_{2h}^5$, Z=4. This X-ray study proved the ferrole structure to be that in Table 3. More details have not been reported [73].

The compound does not decompose in hydrocarbons at 120 °C, and a transformation into the isomer No. 71 is not observed [73]. The reaction with $Fe(CO)_5$ in hexane under UV irradiation leads to complex XXIX, but details have not been reported [57].

$C_4H_2(C_4H_4)Fe_2(CO)_6$ (Table 3, No. **71**) is formed together with No. 70 from $Fe_3(CO)_{12}$ and complex XXIV at 120 °C for 7 h, 38% yield [73]. It occurs also together with No. 70 and several other complexes in the reaction of $Fe(CO)_5$ with compound XXIV under UV irradiation [76].

The complex crystallizes from pentane in the triclinic system with a=8.734, b=14.926, c=12.174 Å, α=99.04°, β=102.06°, and γ=68.96°; space group $P1 - C_i^1$, Z=4. This X-ray study confirmed the expected ferrole structure, but bond lengths and angles have not been reported [73].

The compound is stable in hydrocarbons at 120 °C, and an isomerization to No. 70 does not occur [73].

$C_4H(C_6H_5)(C_4H_4)Fe_2(CO)_6$ (Table 3, No. **72**) has been obtained in small yield from $Fe_3(CO)_{12}$ and $(C_6H_5)_2C=CBr_2$ in petroleum ether at 120 °C for 2 h [38] and from $Fe_3(CO)_{12}$ and diphenylcyclopropenone in petroleum ether at 80 to 100 °C for 2 h [40]. It had earlier also been observed among several organoiron complexes obtained from $Fe_3(CO)_{12}$ and $C_6H_5C≡CC_6H_5$ in petroleum ether at 80 to 100 °C [11]. The empirical formula "$Fe_2(CO)_6(C_6H_5C_2C_6H_5)$" suggested initially a structure [11] which proved to be incorrect, see **Fig. 15**; the ferraindene system is formed by ortho-metallation on one C_6H_5 group of the tolane molecule.

The ^{13}C NMR spectrum (CDCl$_3$) shows the following resonances, compare Fig. 15: δ=113.6 (C-2, J(C,H)=166 Hz), 126.4 to 131.2 (C-11 to C-14, C-16 to C-20), 147.7 (C-15), 149.7 (C-4), 206.6, 207.8, 210.5 (3CO on Fe-1), and 216.5 (2CO on Fe-2) ppm. The assignment for C-4 and C-15 may be reversed. No signal could be assigned to C–3 [81]. The X–ray K absorption edge has a weak absorption at 4.5 eV and the main maximum at 17 eV [11]; however, a diagram of several absorption curves in [96] shows the main absorption closer to 20 eV.

Fig. 15

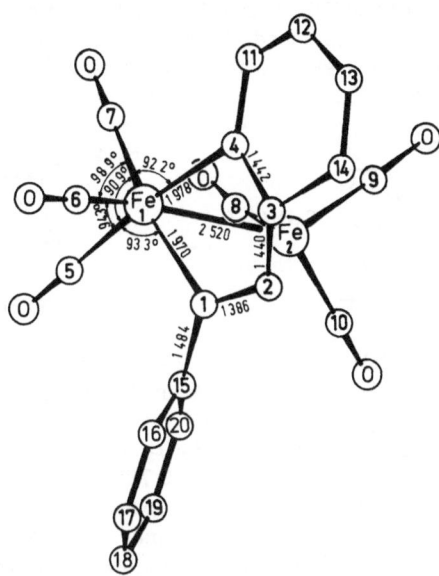

Molecular structure of $C_4H(C_6H_5)(C_4H_4)Fe_2(CO)_6$ [33, 44].

Other selected bond lengths (in Å) and angles (°):

Fe(2)–C(1)	2.132	Fe(2)–C(8)	1.727
Fe(2)–C(2)	2.127	Fe(1)...C(8)	2.472
Fe(2)–C(3)	2.209	Fe–CO (the rest)	1.758 mean
Fe(2)–C(4)	2.188	C–O	1.17 mean
C(1)–Fe(1)–C(4)	82.3	C(1)–Fe(1)–C(6)	92.3
Fe(1)–C(1)–C(2)	115.4	C(1)–Fe(1)–C(7)	167.7
C(1)–C(2)–C(3)	114.6	C(4)–Fe(1)–C(6)	92.3
C(2)–C(3)–C(4)	113.6	C(4)–Fe(1)–C(5)	172.2
C(3)–C(4)–Fe(1)	113.2	C(8)–Fe(2)–C(9)	97.0
Fe(2)–C(8)–O	164.8	C(8)–Fe(2)–C(10)	94.6
Fe–C–O (the rest)	177.4 mean	C(9)–Fe(2)–C(10)	92.2

The complex crystallizes in the triclinic system with a =7.211 ±0.007, b =8.702 ±0.006, c =15.860 ±0.015 Å, α=73.9 ±0.1°, β=103.8 ±0.1°, and γ=101.4 ±0.3°; space group P$\bar{1}$ –C$_i^1$. Z=2 gives D$_c$=1.653 g·cm^{-3}; D$_m$=1.67 g·cm^{-3} [44]. Slightly different lattice parameters are reported in [6, 16, 21, 33]. The molecular structure is shown in Fig. 15. The atomes C(5), C(7), Fe(1), C(1) to C(4) and C(11) to C(15) lie approximately in one plane. Another plane is formed by C(1) and the phenyl carbons. The angle between the two planes is 55.6° if they are defined by C(3), C(11), C(13) and C(15), C(17), C(19). Bond lengths and angles correspond largely to the values found for other ferroles [33, 44], see also [28].

The reductive degradation of the compound (in ether/ethanol) with Na in liquid NH_3 at $-60\ °C$ for 10 min gives exclusively $C_6H_5CH_2CH_2C_6H_5$ in high yield [11]. Even powerful dienophiles do not add to the butadiene part of the condensed six-membered ring [35].

$C_4H(C_6H_4Cl-4)(C_4H_3Cl)Fe_2(CO)_6$ (Table 3, No. 73). Attempts to replace the atom Fe(1) by a ketonic carbonyl group failed: Heating of the compound under 85 atm CO at 250 °C for 2 h yields exclusively the free indenone XXX. No reaction takes place at lower temperatures [35].

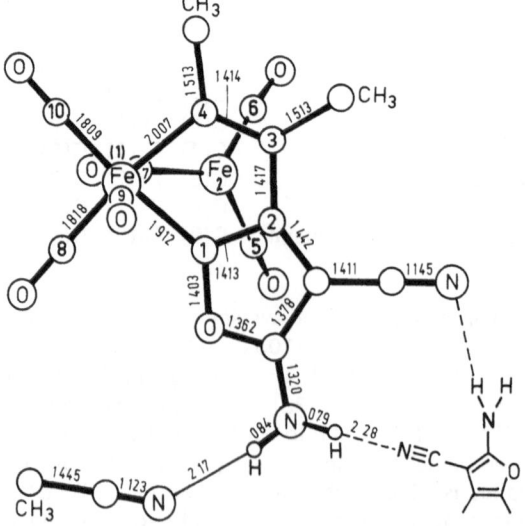

XXX   XXXI

$C_4(CH_3)_2(C_3H_2N_2O)Fe_2(CO)_6$ (Table 3, No. 74) has been obtained in low yield (1.6%) from $Fe_3(CO)_{12}$ and the triafulvene XXXI (about 1:1 mole) in toluene at 100 °C for 30 min. The origin of the hydrogen necessary for the reduction of one CN group is unclear. The complex is recrystallized from CH_2Cl_2. With CH_3CN it forms a 1:1 adduct which crystallizes in the triclinic system with a=9.789(6), b=8.252(5), c=12.371(8) Å, $\alpha=85.59(2)°$, $\beta=103.26(3)°$, and $\gamma=90.45(2)°$; space group $P\bar{1}-C_i^1$. Z=2 gives $D_c=$ 1.65 $g \cdot cm^{-3}$. The two rings are approximately planar and slightly inclined to each other (angle 178.9°), see **Fig. 16**. The bond lengths in the heterocyclic ring correspond to those of the unsubstituted furan. The Fe(1)–C(ring) distances are significantly different. The fractions of multiple bonding in the shorter Fe(1)–C(1) bond are explained by the electronegative O atom which enables C(1) to a greater extend to accept electron density from occupied d orbitals of Fe(1). The NH_2 substituent forms hydrogen bridges to the CH_3CN included in the crystal and to the CN group of another molecule. This is schematically indicated in Fig. 16.

Fig. 16

Molecular structure of $C_4(CH_3)_2(C_3H_2N_2O)Fe_2(CO)_6$ [94].

Other selected bond lengths (in Å) and angles (°):

Fe(1)–Fe(2)	2.510(3)	Fe(2)–CO	1.782 mean
Fe(2)–C(1)	2.099(3)	Fe(1)–C(7)	2.544(3)
Fe(2)–C(2)	2.203(3)	Fe(1)–C(9)	1.762(3)
Fe(2)–C(3)	2.180(4)	C–O	1.136 mean
Fe(2)–C(4)	2.077(3)		

C(1)–Fe(1)–C(4)	79.1(2)	C(5)–Fe(2)–C(6)	95.6(2)
C(1)–Fe(1)–C(8)	92.2(2)	C(5)–Fe(2)–C(7)	96.4(2)
C(1)–Fe(1)–C(9)	93.4(2)	C(6)–Fe(2)–C(7)	97.4(3)
C(1)–Fe(1)–C(10)	166.6(2)	Fe(2)–C(7)–O	167.7(4)
C(4)–Fe(1)–C(8)	165.4(3)	Fe–C–O (the rest)	178.1 mean
C(4)–Fe(1)–C(9)	93.1(2)	Fe(1)–C(1)–C(2)	118.1(4)
C(4)–Fe(1)–C(10)	92.5(2)	C(1)–C(2)–C(3)	113.4(6)
C(8)–Fe(1)–C(9)	99.2(3)	C(2)–C(3)–C(4)	111.5(5)
C(8)–Fe(1)–C(10)	93.6(2)	C(3)–C(4)–Fe(1)	116.0(5)
C(9)–Fe(1)–C(10)	97.6(2)		

The solid is stable in air. It is insoluble in pentane and hexane and dissolves in benzene, toluene, CH_2Cl_2, $CHCl_3$, CH_3CN, ether, and acetone [94].

$C_4H_3(C_2H_5)Fe_2(CO)_6$ and $C_4H_3(OH)Fe_2(CO)_6$ (Table 3, No. **75** and **76**) have been observed together with $C_4H_3(C_2H_5)Fe_3(CO)_8$ and other products in the slow decomposition of $(\pi\text{-}C_3H_5Fe(CO)_3)_2$ in pentane under N_2 at room temperature for a period of 21 d. After laborious chromatographic separations on SiO_2, the complexes have been further purified by vacuum distillation at 40 °C (No. 75) and 60 °C (No. 76) to a 0 °C probe. The OH group of No. 76 can be methylated, see compound No. 77 [99].

References:

[1] W. Reppe, H. Vetter (Lieblgs Ann. Chem. **582** [1953] 133/61). – [2] I. Wender, R.A. Friedel, R. Markby, H.W. Sternberg (J. Am. Chem. Soc. **77** [1955] 4946/7). – [3] R. Clarkson, E.R.H. Jones, P.C. Wailes, M.C. Whiting (J. Am. Chem. Soc. **78** [1956] 6206/7). – [4] H.W. Sternberg, R.A. Friedel, R. Markby, I. Wender (J. Am. Chem. Soc. **78** [1956] 3621/3). – [5] A.A. Hock, O.S. Mills (Proc. Chem. Soc. **1958** 233/4).

[6] K.W. Hübel, Society European Research Associates S.A. (Belg. 567743 [1958]). – [7] E.H. Braye, W. Hübel (Chem. Ind. [London] **1959** 1250/1). – [8] D.A. Brown (J. Inorg. Nucl. Chem. **10** [1959] 39/48). – [9] D.A. Brown (J. Inorg. Nucl. Chem. **11** [1959] 9/12). – [10] J.R. Case, R. Clarkson, E.R.H. Jones, M.C. Whiting (Proc. Chem. Soc. **1959** 150).

[11] W. Hübel, E.H. Braye (J. Inorg. Nucl. Chem. **10** [1959] 250/68). – [12] W. Hübel, E.H. Braye, A. Clauss, E. Weiss, U. Krüerke, D.A. Brown, G.S.D. King, C. Hoogzand (J. Inorg. Nucl. Chem. **9** [1959] 204/10). – [13] W. Hübel, E. Weiss (Chem. Ind. [London] **1959** 703). – [14] G.N. Schrauzer (J. Am. Chem. Soc. **81** [1959] 5307/10). – [15] M.L.H. Green, L. Pratt, G. Wilkinson (J. Chem. Soc. **1960** 989/97).

[16] K.W. Hübel, European Research Associates S.A. (Fr. 1206786 [1958/60]). – [17] W. Hübel, C. Hoogzand (Chem. Ber. **93** [1960] 103/15). – [18] H.D. Kaesz, R.B. King, T.A. Manuel, L.D. Nichols, F.G. Stone (J. Am. Chem. Soc. **82** [1960] 4749/50). – [19] E.H. Braye, C. Hoogzand, W. Hübel, U. Krüerke, R. Merenyi, E. Weiss (in: S.

Kirschner, Advances in Chemistry of the Coordination Compounds, MacMillan, New York 1961, S. 190/7). – [20] A.A. Hock, O.S. Mills (Acta Cryst. **14** [1961] 139/48).

[21] W. Hübel, Union Carbide Corp. (Brit. 885 514 [1961]). – [22] E. Bayer, H.A. Brune, K.L. Hoch (Angew. Chem. **74** [1962] 872/3; Angew. Chem. Intern. Ed. Engl. **1** [1962] 552). – [23] J.R. Case, E.R.H. Jones, N.V. Schwartz, M.C. Whiting (Proc. Chem. Soc. **1962** 256/7). – [24] E. Weiss, W. Hübel, R. Merényi (Chem. Ber. **95** [1962] 1155/69). – [25] E.H. Braye, W. Hübel (Angew. Chem. **75** [1963] 345/6; Angew. Chem. Intern. Ed. Engl. **2** [1963] 217).

[26] K.W. Hübel, E.H. Braye, Union Carbide Corp. (U.S. 3 096 265 [1959/63]). – [27] W. Hübel, R. Merényi (Chem. Ber. **96** [1963] 930/43). – [28] M. Van Meerssche (Ind. Chim. Belge [2] **28** [1963] 507). – [29] A. Nakamura, N. Hagihara (Nippon Kagaku Zasshi **64** [1963] 339/44 according to C.A. **59** [1963] 14869). – [30] M.C. Whiting (Chem. Weekblad **59** [1963] 119/20).

[31] G.F. Emerson, J.E. Mahler, R. Pettit, R. Collins (J. Am. Chem. Soc. **86** [1964] 3590/1). – [32] K.W. Hübel, E.H. Braye, Union Carbide Corp. (U.S. 3 125 594 [1959/64]). – [33] M. Van Meerssche, P. Piret, J. Meunier-Piret, Y. Degréve (Bull. Soc. Chim. Belges **73** [1964] 824/32). – [34] E.H. Braye, W. Hübel (J. Organometal. Chem. **3** [1965] 25/37). – [35] E.H. Braye, W. Hübel (J. Organometal. Chem. **3** [1965] 38/42).

[36] L.F. Dahl, J. Doedens, W. Hübel, J. Nielsen (J. Am. Chem. Soc. **88** [1966] 446/52). .– [37] K.W. Hübel, E.H. Braye, Union Carbide Corp. (U.S. 3 280 017 [1966]). – [38] K.K. Joshi (J. Chem. Soc. A **1966** 594/7). – [39] P.M. Maitlis (Advan. Organometal. Chem. **4** [1966] 95/143, 107). – [40] C.W. Bird, E.M. Briggs, J. Hudec (J. Chem. Soc. C **1967** 1862/4).

[41] J.M. Birchall, F.L. Bowden, R.N. Hazeldine, A.B.P. Lever (J. Chem. Soc. A **1967** 747/53). – [42] E.H. Braye, W. Hübel (J. Organometal. Chem. **9** [1967] 370/2). – [43] G. Cetini, O. Gambino, E. Sappa, M. Valle (Atti Accad. Sci. Torino Classe Sci. Fis. Mat. Nat. **101** [1967] 813/25). – [44] Y. Degréve, J. Meunier-Piret, M. Van Meerssche, P. Piret (Acta Cryst. **23** [1967] 119/26). – [45] E. Bayer, E. Breitmaier, V. Schurig (Chem. Ber. **101** [1968] 1594/600).

[46] F.L. Bowden, A.B.P. Lever (Organometal. Chem. Rev. **3** [1968] 227/79). – [47] R.E. Dessy, R.L. Pohl (J. Am. Chem. Soc. **90** [1968] 1995/2001). – [48] W. Hübel (in: I. Wender, P. Pino, Organic Syntheses via Metal Carbonyls, Vol. 1, Interscience, New York 1968, p. 285/95). – [49] W. Hübel (in: I. Wender, P. Pino, Organic Syntheses via Metal Carbonyls, Vol. 1, Interscience, New York 1968, p. 326/31). – [50] M. Avram, I.G. Dinulescu, E. Avram, C.D. Nenitzescu (Rev. Roumaine Chim. **14** [1969] 1181/9).

[51] M.I. Bruce (Intern. J. Mass Spectrom. Ion Phys. **2** [1969] 349/55). – [52] K.W. Hübel, E.H. Braye, Union Carbide Corp. (U.S. 3 426 052 [1969]). – [53] M. Valle, G. Cetini, O. Gambino, E. Sappa (Atti Accad. Sci. Torino Classe Sci. Fis. Mat. Nat. **103** [1968/69] 913/23). – [54] M.M. Bagga, G. Ferguson, J.A.D. Jeffreys, C.M. Mansell, P.L. Pauson, I.C. Robertson, J.G. Sime (Chem. Commun. **1970** 672/3). – [55] M.I. Bruce, T.A. Kuc (J. Organometal. Chem. **22** [1970] C1/C2).

[56] A.G. MacDiramid, M.A. Nasta (J. Am. Chem. Soc. **93** [1971] 2813/4). – [57] R. Victor, R. Ben-Shosan, S. Sarel (Chem. Commun. **1971** 1241/2). – [58] G.A. Vaglio, O. Gambino, R.P. Ferrari, G. Cetini (Org. Mass Spectrom. **5** [1971] 493/503). – [59] A.G. MacDiarmid, M.A. Nasta, F.E. Saalfeld (J. Am. Chem. Soc. **94** [1972]

2449/55). – [60] O. Gambino, G.A. Vaglio, G. Cetini (Org. Mass Spectrom. **6** [1972] 1297/301).

[61] J.A.D. Jeffreys, C.M. Willis (J. Chem. Soc. Dalton Trans. **1972** 2169/73). – [62] R.B. King, I. Haiduc (J. Am. Chem. Soc. **94** [1972] 4044/6). – [63] K. Oefele, E. Dotzauer (J. Organometal. Chem. **42** [1972] C87/C90). – [64] H.P. Wolf, P. Müller, H.A. Brune (Z. Naturforsch. **27b** [1972] 915/7). – [65] M.J. Bennett, W.A.G. Graham, R.A. Smith, R.P. Stewart (J. Am. Chem. Soc. **95** [1973] 1684/6).

[66] R. Bühler, R. Geist, R. Münich, H. Plieninger (Tetrahedron Letters **1973** 1919/22). – [67] H.B. Chin, R. Bau (J. Am. Chem. Soc. **95** [1973] 5068/70). – [68] D. Ehntholt, A. Rosan, M. Rosenblum (J. Organometal. Chem. **56** [1973] 315/21). – [69] R.B. King, M.N. Ackermann (J. Organometal. Chem. **60** [1973] C57/C59). – [70] R.B. King, I. Haiduc, C.W. Eavenson (J. Am. Chem. Soc. **95** [1973] 2508/16).

[71] A. Messeguer, F. Serratosa, J. Rivera (Tetrahedron Letters **1973** 2895/8). – [72] M.I. Bruce, T.A. Kuc (Australian J. Chem. **27** [1974] 2487/9). – [73] R.E. Davis, B.L. Barnett, R.G. Amiet, W. Merk, J.S. McKennis, R. Pettit (J. Am. Chem. Soc. **96** [1974] 7108/9). – [74] T.J. Devon (Diss. Univ. of Texas, Austin, 1974; Diss. Abstr. Intern. B **35** [1974/75] 3825). – [75] R.H. Herber, R.B. King, M.N. Ackermann (J. Am. Chem. Soc. **96** [1974] 5437/41).

[76] R. Victor, R. Ben-Shosan (J. Chem. Soc. Chem. Commun. **1974** 93/4). – [77] F.W. Grevels, D. Schulz, E. Koerner von Gustorf (J. Organometal. Chem. **91** [1975] 341/6). – [78] P.E. Riley, R.E. Davis (Acta Cryst. B **31** [1975] 2928/30). – [79] E. Sappa, L. Milone, G.D. Andretti (Inorg. Chim. Acta **13** [1975] 67/71). – [80] E. Sappa (Atti Accad. Sci. Torino Classe Sci. Fis. Mat. Nat. **109** [1975] 623/31).

[81] S. Aime, L. Milone, E. Sappa (J. Chem. Soc. Dalton Trans. **1976** 838/40). – [82] T. Chivers, P.L. Timms (J. Organometal. Chem. **118** [1976] C37/C40). – [83] G. Dettlaf, E. Weiss (J. Organometal. Chem. **108** [1976] 213/23). – [84] P.J. Harris, J.A.K. Howard, S.A.R. Knox, R.P. Phillips, F.G.A. Stone, P. Woodward (J. Chem. Soc. Dalton Trans. **1976** 377/82). – [85] R.B. King, C.A. Harmon (Inorg. Chem. **15** [1976] 879/85).

[86] S.R. Prince (Cryst. Struct. Commun. **5** [1976] 451/8). – [87] L.J. Todd, J.P. Hickey, J.R. Wilkinson, J.C. Huffmann, K. Folting (J. Organometal. Chem. **112** [1976] 167/76). – [88] F.R. Young, D.H. O'Brien, R.C. Pettersen, R.A. Levenson, D.L. von Minden (J. Organometal. Chem. **114** [1976] 157/64). – [89] R.S. Dickson, C. Mok, G. Connor (Australian J. Chem. **30** [1977] 2143/51). – [90] P. Hübener, E. Weiss (J. Organometal. Chem. **129** [1977] 105/15).

[91] T. Chivers, P.L. Timms (Can. J. Chem. **55** [1977] 3509/14). – [92] R. Victor, V. Usieli, S. Sarel (J. Organometal. Chem. **129** [1977] 387/99). – [93] D.L. Thorn, R. Hoffmann (Inorg. Chem. **17** [1978] 126/39). – [94] G. Dettlaf, U. Behrens, T. Eicher, E. Weiss (J. Organometal. Chem. **152** [1978] 203/8). – [95] R. Hoffmann, T.A. Albright, D.L. Thorn (Pure Appl. Chem. **50** [1978] 1/9).

[96] A. Schneider (Z. Physik. Chem. [Frankfurt] **31** [1962] 249/73). – [97] R. Bruce, K. Moseley, P.M. Maitlis (Can. J. Chem. **45** [1967] 2011/6). – [98] G.N. Schrauzer, H. Kisch (J. Am. Chem. Soc. **95** [1973] 2501/7). – [99] C.F. Putnik, J.J. Welter, G.D. Stucky, M.J. D'Aniello, B.A. Sosinsky, J.F. Kirner, E.L. Muetterties (J. Am. Chem. Soc. **100** [1978] 4107/16). – [100] S. Aime, L. Milone, D. Osella, E. Sappa, A.M. Manotti Lanfredi, A. Tiripicchio, M. Tiripicchio Camellini (11th Congr. Nazl. Chim. Inorg., Arcavacata di Rende, Italy, 1978, No. 5A).

[101] S. Aime, E. Sappa, A. Tiripicchio, A.M. Manotti Lanfredi (J. Chem. Soc. Dalton Trans. **1979** 1664/70).

2.4.1.2.2.2   Compounds of the $C_4R_4Fe_2(CO)_5{}^2D$ Type

The following substances are monosubstitution products of the parent compounds in 2.4.1.2.2.1. An X-ray study of $C_4(C_6H_5)_4Fe_2(CO)_5P(C_6H_5)_3$ demonstrated that the phosphine ligand is attached to the ferrole iron Fe(1) [4] contrary to previous belief [1, 2], see Formula I. This is consistent with ^{13}C NMR observations as phosphorus–carbon coupling only occurs to the adjacent C(1,4) ring atoms and the two equivalent carbonyl groups on Fe(1) [4].

I

$C_4H_4Fe_2(CO)_5P(C_6H_5)_3$ has been obtained in 37% yield from $C_4H_4Fe_2(CO)_6$ and $P(C_6H_5)_3$ (about 2:5 mole) in petroleum ether at 120 °C for 12 h in a sealed tube. It precipitates from ether/methanol as orange crystals melting with decomposition at 173 to 175 °C. IR spectrum (KBr): $v(CO)$ bands at 1931, 1980, and 2037 cm^{-1}. The compound dissolves readily in benzene, moderately in ether, and slightly in methanol and petroleum ether [1].

$C_4(C_6H_5)_4Fe_2(CO)_5P(C_4H_9\text{-}n)_3$ has been prepared in 55% yield by the same procedure employed for the following complex [4]. The substance melts at 142 to 144 °C. ^{13}C NMR spectrum (CD$_2$Cl$_2$): $\delta=148.3$ (C-2,3), 172.1 (C-1,4, $J(C,P)=15.8$ Hz), 208.3 (2 CO on Fe-1, $J(C,P)=15.5$ Hz), 220.9 (3 CO on Fe-2) ppm [4].

$C_4(C_6H_5)_4Fe_2(CO)_5P(C_6H_5)_3$ has been prepared in 80% yield from $C_4(C_6H_5)_4Fe_2(CO)_6$ and $P(C_6H_5)_3$ in boiling toluene in 20 h and can be separated from the starting material by chromatography on Al_2O_3 with benzene. It precipitates from benzene/petroleum ether as red needles and prisms, melting at 180 to 182 °C with decomposition [1].

^{13}C NMR spectrum (CD$_2$Cl$_2$): $\delta=149.3$ (C-2,3), 176.6 (C-1,4, $J(C,P)=12.4$ Hz), 209.5 (2 CO on Fe-1, $J(C,P)=13.1$ Hz), 217.7 (3 CO on Fe-2) ppm [4]. The IR spectrum (KBr) shows $v(CO)$ bands at 1894, 1957, 1969, 2000, and 2033 cm^{-1} [1]. The complex crystallizes in the triclinic system with a$=12.868(9)$, b$=11.667(8)$, c$=13.867(9)$ Å, $\alpha=85.98(3)°$, $\beta=90.35(3)°$, and $\gamma=83.36(3)°$; space group $P\bar{1}-C_i^1$. Z$=2$ gives $D_c=1.401$ g·cm^{-3}. The molecular structure is presented in **Fig. 17**. In general, the bond distances and angles are comparable to those found for unsubstituted ferroles. The Fe(1)–C(7) distance is somewhat longer than the corresponding distance in the parent compound which may be due to the electron withdrawing capability of phosphine being weaker than CO. The two carbonyls C(5)–O and C(6)–O are bent out of the ferracyclopentadiene plane by 15.4° and 17.7°, respectively, probably because of the steric requirements of the phosphine [4].

Fig. 17

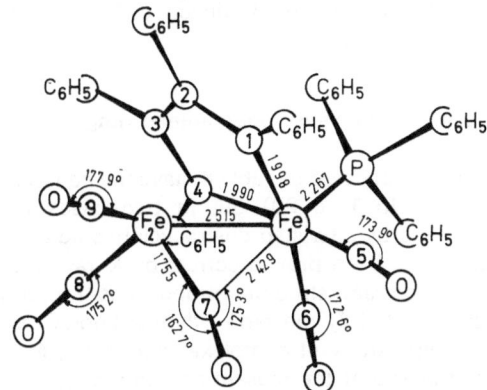

Molecular structure of $C_4(C_6H_5)_4Fe_2(CO)_5P(C_6H_5)_3$ [4].

Other selected bond lengths (in Å) and angles (°):

Fe(2)–C(1)	2.131(9)	C(8,9)–O	1.182 mean
Fe(2)–C(2)	2.169(9)	C(1)–C(2)	1.422(12)
Fe(2)–C(3)	2.132(9)	C(2)–C(3)	1.436(12)
Fe(2)–C(4)	2.109(9)	C(3)–C(4)	1.430(13)
Fe(1)–C(5,6)	1.771 mean	C–C$_6$H$_5$	1.509 mean
Fe(2)–C(8,9)	1.730 mean	P–C$_6$H$_5$	1.840 mean
C(5,6)–O	1.167 mean		

Fe(2)–Fe(1)–C(1,4)	54.7 mean	Fe(1)–C(1)–C(2)	116.6(6)
Fe(2)–Fe(1)–C(5)	111.6(3)	C(1)–C(2)–C(3)	113.3(8)
Fe(2)–Fe(1)–C(6)	109.9(3)	C(2)–C(3)–C(4)	113.3(8)
Fe(2)–Fe(1)–P	141.1(1)	C(3)–C(4)–Fe(1)	116.5(7)
C(1)–Fe(1)–C(4)	79.7(4)	Fe(1)–Fe(2)–C(7)	66.6(4)
C(1)–Fe(1)–C(5)	92.1(4)	Fe(1)–Fe(2)–C(8)	123.8(4)
C(1)–Fe(1)–C(6)	163.5(4)	Fe(1)–Fe(2)–C(9)	142.3(4)

The degradation of the complex with Br_2 in CCl_4 yields $C_4(C_6H_5)_4COFe$-$(CO)_2P(C_6H_5)_3$ ($C_4(C_6H_5)_4CO$ =tetraphenylcyclopentadienone) [1]; in [3] only the free cyclic ketone is mentioned as reaction product. Pentaphenylarsole, cyclo-$C_4(C_6H_5)_4AsC_6H_5$, results from the reaction with $C_6H_5AsCl_2$ in CH_2Cl_2 under UV irradiation [3].

$C_4H_2(OCH_3)_2Fe_2(CO)_5SO_2$ has been isolated from a reaction of $C_4H_2(OCH_3)_2Fe_2(CO)_6$ with boiling SO_2 under UV irradiation for 24 h, 55% yield. The orange–red compound (from CH_2Cl_2) melts with decomposition at 155 to 156 °C. The CO stretching vibrations are shifted to higher frequencies, but spectral data have not been reported. It has been assumed that substitution occurs in the π–bonded $Fe(CO)_3$ moiety; however, this is doubtful in view of the structure of the previous complex. The substance is less stable than the parent complex. It is insoluble in ether and decomposes in acetonitrile and tetrahydrofuran [2].

References:

[1] E. Weiss, W. Hübel, R. Merényi (Chem. Ber. **95** [1962] 1155/69). – [2] E.H. Braye, W. Hübel (Angew. Chem. **75** [1963] 345/6; Angew. Chem. Intern. Ed. Engl. **2** [1963] 217). – [3] K.W. Hübel, E.H. Braye, Union Carbide Corp. (U.S. 3280017

[1966]. – [4] L.J. Todd, J.P. Hickey, J.R. Wilkinson, J.C. Huffmann, K. Folting (J. Orga-nometal. Chem. **112** [1976] 167/76).

2.4.1.2.3 Compounds with a Diferracycloheptadiene Ring

The following compounds, collected in Table 4, have the composition $C_4R_4CXFe_2(CO)_6$ where X is usually oxygen (No. 1 to 20) but in two cases CH_2 and CHR (No. 21 and 22). They are characterized by a twisted diferracycloheptadiene system as represented by Formula I. This structure has been proved correct by X-ray studies of three complexes [15, 24, 28]. A previous incorrect structure proposal is found in [1, 3 to 11, 17]. In the more recent literature [31, 32] the term "flyover bridge" is used for the bridging ^4L ligand. In a molecular orbital study of complexes containing $M_2(CO)_6$ transition metal fragments, the electronic structure of the present compound type has also been discussed [32].

I

For symmetrically substituted complexes ^{57}Fe Mössbauer spectra revealed equivalent Fe atoms, thus supporting structure I and not the earlier proposal [21]. In general, the IR spectra show three bands of the terminal CO ligands; the ketonic ring CO occurs close to 1670 cm^{-1} [23]. With the exception of compound No. 16 ^1H NMR spectra have not been measured [30]. Variable-temperature ^{13}C NMR spectra indicate scrambling of the CO ligands at each Fe atom [27, 28, 29], see compounds No. 3, 4, and 8.

The $C_4R_4CXFe_2(CO)_6$ complex type is diamagnetic [3, 8]. Compounds with X=O usually decompose already below the melting point. The thermal and photolytic decomposition leads frequently to cyclopentadienone derivatives. Quinone derivatives can be formed by CO insertion under alkaline reducing conditions. Six-membered heterocycles become accessible by insertion of hetero atoms like O, S, and Se. Typical organic reactions of the ketonic CO have not been observed, but it seems to be possible to obtain from $C_4(C_6H_5)_4COFe_2(CO)_6$ an adduct with $FeCl_3$ [23]. The chemical behavior of $C_4(C_6H_5)_4COFe_2(CO)_6$ has been investigated in detail and is summarized in Scheme 3.

Similar to ferroles, see 2.4.1.2.2.1, the diferracycloheptadiene complexes are built up from iron carbonyls and alkynes. These reactions usually afford a variety of other complex types which must be separated by chromatography (for formulas see p. 64).

Method I:　Fe$_3$(CO)$_{12}$ is reacted with alkynes, RC≡CR or RC≡CR', about 1:3 moles, in an inert solvent like petroleum ether, n-heptane, or toluene at about 80 to 100 °C. The reaction mixture frequently contains also the complex types II to V as well as substituted benzenes and cyclopentadienones. The product distribution depends on the kind of alkyne [22] and the reaction conditions [30]. Isomers like formulas VI, VII, and VIII have to be expected from reactions with unsymmetrical alkynes, RC≡CR'. Their existence has been proved by spectroscopic methods and in some cases they could be separated.

Scheme 3

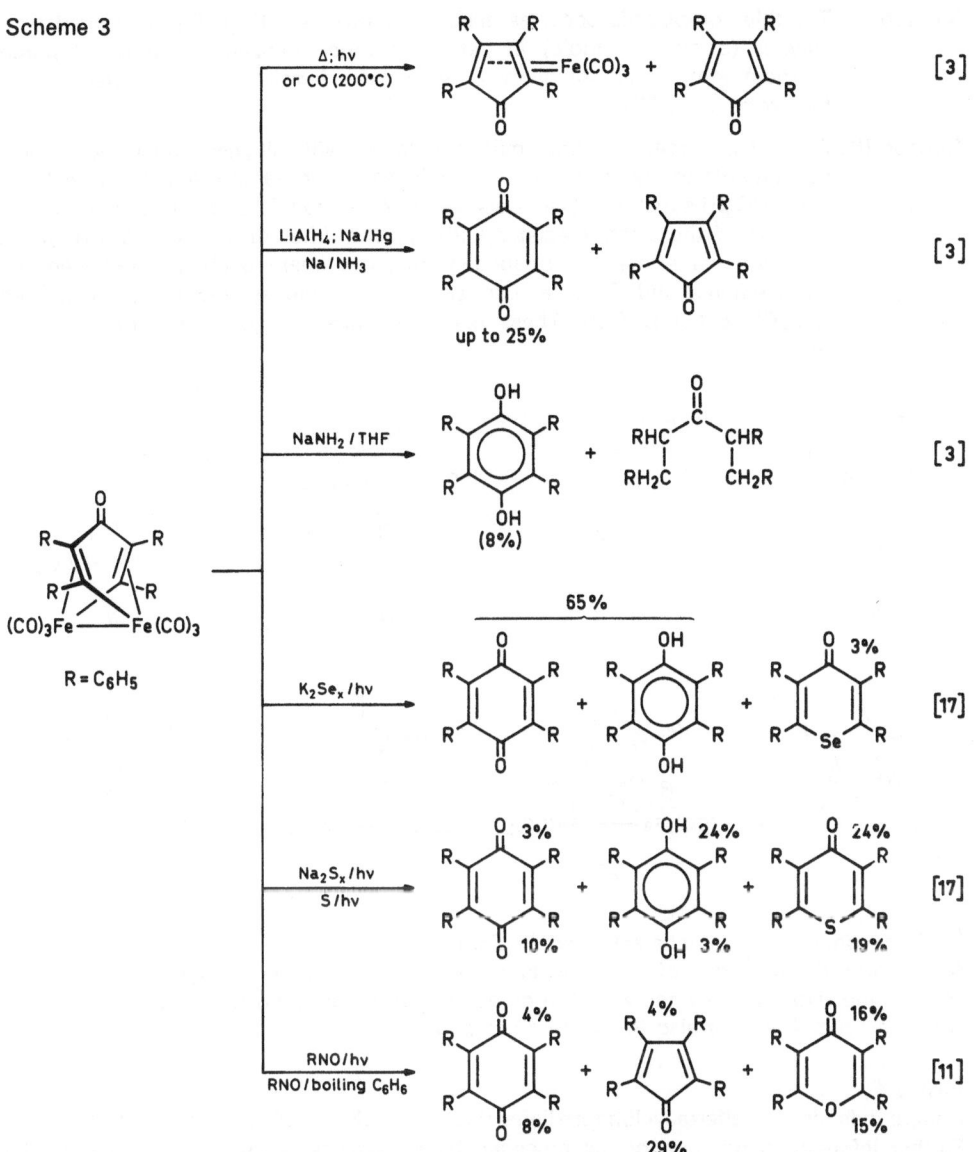

Typical reactions of the $C_4R_4COFe_2(CO)_6$ complex type [23].

Method II: The title compounds occur as major components if $Fe_2(CO)_9$ is treated with alkynes (about 1:1 mole) in petroleum ether, benzene, or tetrahydrofuran at room temperature or slightly higher; the ferrole type IV is generally a by-product [2, 22].

Method III: $Fe_2(CO)_9$ is reacted under mild conditions with alkynes which bear bulky substituents to give unstable intermediates of the composition $RC_2RFe(CO)_4$ and "$RC_2RFe_2(CO)_7$" (see, however, C 2, section 2.2.1.2.1.1, p. 72). Further reaction of these intermediates with another different alkyne $R'C \equiv CR'$ yields a fourth isomer type (formula IX). But a primary exchange between the intermediates and $R'C \equiv CR'$ can also lead to the symmetrically substituted $C_4R_4'COFe_2(CO)_6$ [22]. These reactions have not been described in more detail.

The compounds are characterized in Table 4 by the "flyover" bridging ligand. In some cases the positions of the substituents on the five-carbon bridge are not known (compounds No. 17 to 20 and 22). The numbering of the bridging C atoms for compound No. 1 in Table 4 applies also to all other complexes.

Table 4
Compounds with a diferracycloheptadiene ring, $C_4R_4CXFe_2(CO)_6$ with $X=O$ and CR_2. Further information on compounds preceded by an asterisk is given at the end of the table. For abbreviations and dimensions see p. 170.

No. C_4R_4CX ligand	Method of preparation, reaction conditions (yield in %)	Properties and further remarks	Ref.
*1	I, – (–)	^{13}C NMR ($CDCl_3$): 20.8 (CH_3), 100.7 (C-2,4), 171.1 (C-1,5, J(C,H) =161), 199.6 (C-3)	[28]

Table 4 [continued]

No. C_4R_4CX ligand	Method of preparation, reaction conditions (yield in %)	Properties and further remarks	Ref.
*2 [structure: CH_3, CH_3]	I, — (—)	^{13}C NMR ($CDCl_3$): 20.2 (CH_3-4), 37.7 (CH_3-1), 83.5 (C-2, J(C,H) =161), 101.4 (C-4), 169.7 (C-5, J(C,H) =166), 200.1 and 200.9 (C-3 and C-1, assignment could be reversed)	[28]
*3 [structure: $(CH_3)_3C$, $\overset{2'}{}$, $\overset{4'}{}\dot{C}(CH_3)_3$]	I, — (—)	^{13}C NMR ($CDCl_3$): 30.5 (CH_3), 37.8 (C-2',4'), 122.3 (C-2,4), 163.1 (C-1,5, J(C,H) =160), 194.2 (C-3); for CO signals see further information	[28]
*4 [structure: $\overset{4'}{}\dot{C}(CH_3)_3$, $(CH_3)_3\overset{1'}{C}$]	I, — (—)	^{13}C NMR ($CDCl_3$): 29.2 (CH_3 on C-4'), 32.6 (CH_3 on C-1'), 37.1 (C-4'), 48.0 (C-1'), 78.2 (C-2, J(C,H) =166), 123.9 (C-4), 161.9 (C-5, J(C,H) = 161), 197.3 (C-3), 216.3 (C-1); for CO signals see further information	[28]
*5 [structure: CH_3, CH_3, CH_3, CH_3]	II, from $Fe_2(CO)_9$ and $CH_3C≡CCH_3$ in petroleum ether at 25°/44 h (—)	red orange; dec. at 150°, 150 to 170° IR (CCl_4): 2005, 2007, 2041, 2074; v(C=O) at 1667	[23, 24, 25]
*6 [structure: C_2H_5, C_2H_5, C_2H_5, C_2H_5]	III, see further information	orange; m.p. 155 to 175° (dec.), 158 to 171° (dec.) IR (CCl_4): 1990, 2005, 2040, 2072; v(C=O) at 1673	[22, 23, 25]
*7 [structure: CH_3O, OCH_3, CH_3O, OCH_3]	II, from $Fe_2(CO)_9$ and $CH_3OC≡COCH_3$ at −60 to +20° in pentane (very low)	IR (CCl_4): 1985, 2000, 2010, 2050, 2085; other bands at 1640, 1685, 1730 UV (CCl_4): λ_{max} =273, 494	[26]
*8 [structure: phenyl groups, $\overset{2'}{}$, $\overset{4'}{}$, $\overset{1'}{}$, $\overset{5'}{}$]	I, in petroleum ether at ≈90°/2 h (24) II, in THF at 23°/4 h (80)	red; m.p. 154 to 155° (dec.), 160° (dec.) ^{13}C NMR ($CDCl_3$): 94.2 (C-2,4), 129.0 to 130.4 (C_6H_5), 137.2 (C-2',4'), 148.7 (C-1',5'), 192.4 and 195.5 (C-1,5 and C-3, assignment could be reversed), 206.3, 206.9 and 210.0 (CO on Fe, at −25°)	[3, 4, 21, 28]

[continued on p. 66]

Table 4 [continued]

No. C_4R_4CX ligand	Method of preparation, reaction conditions (yield in %)	Properties and further remarks	Ref.
*8 [continued]		$^{57}Fe-\gamma$ (80 K): $\delta=0.246$, $\Delta=0.995$ IR (methylcyclohexane): 2010, 2043, 2065; $\nu(C=O)$ at 1665	
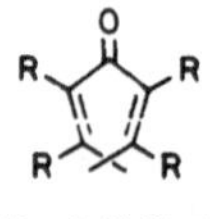 *9 R = C_6H_4Cl-4	I, from $Fe_3(CO)_{12}$ and 4-$ClC_6H_4C\equiv C$-C_6H_4Cl-4 in petroleum ether at 80°/3 h (5 to 10)	red; m.p. 200 to 220° (dec.) IR (KBr): 2016, 2049, 2070; $\nu(C=O)$ at 1667	[1, 5, 8, 14]
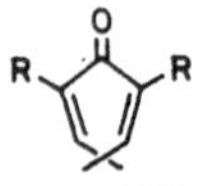 10 R = $COOCH_3$	I, no details available (−)	dark red; m.p. 133 to 135° (dec.)	[23]
*11	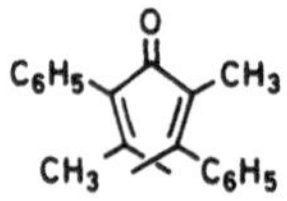 I, no details available (−)	red; two crystal modifications with m.p. 151 to 164° (dec.) and m.p. 161 to 170° (dec.)	[23]
12	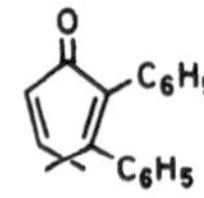 III, no details available (−)	orange red; m.p. 143 to 145° (dec.)	[23]
*13	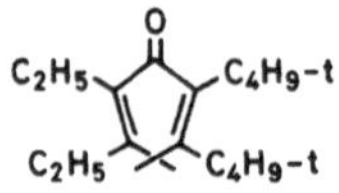 III, at 20° (23)	orange red; m.p. 127 to 130° (dec.)	[22, 23]
*14	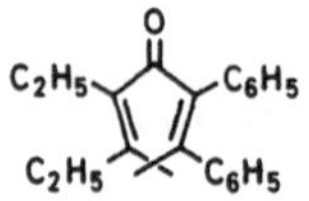 III, at 20° (55)	red; m.p. 147 to 152° (dec.)	[22, 23]
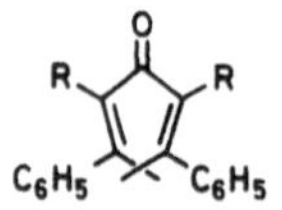 *15 R = $COOCH_3$	I, from $Fe_3(CO)_{12}$ and $C_6H_5C\equiv CCOOCH_3$ in petroleum ether at 75°/2 h (−)	red; dec. at 170 to 180° IR (CH_2Cl_2): 2033, 2066, 2092; $\nu(C=O$, ring) at 1684, $\nu(C=O$, acyle) at 1718	[16, 23]
*16	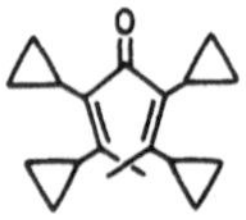 see further information	red−orange prisms; m.p. 132° (dec.) ^1H NMR ($CDCl_3$): 0.53 to 1.67 (9 H), 2.53 (1 H) IR (C_6H_{14}): 2001, 2007, 2040, 2071; (KBr): $\nu(C=O)$ at 1668	[30]

References on p. 74

Table 4 [continued]

No. C₄R₄CX ligand	Method of preparation, reaction conditions (yield in %)	Properties and further remarks	Ref.
17 2R=CH₃ 2R=C₆H₅ 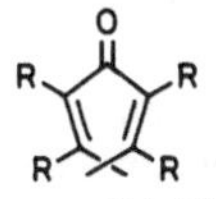	I, no details available (−)	orange; m.p. 178 to 179° (dec.) see further information to No. 11	[23]
18 2R=CH₃ 2R=C₆H₅ 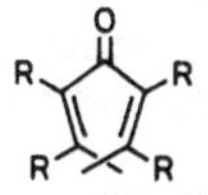	I, no details available (−)	dark red; m.p. ≈170° (dec.) see further information to No. 11	[23]
19 2R=C₆H₅ 2R=Si(CH₃)₃ 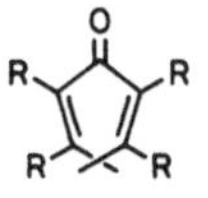	I, no details available (−)	red; m.p. 167° (dec.) the positions of the substituents are not known	[23]
*20 2R=C₆H₅ 2R=C≡CC₆H₅ 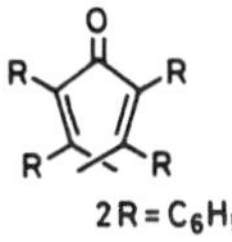	I, in petroleum ether at 80 to 100°/1 h (15.5) II, in C₆H₆ at 40 to 45°/1 h (10.5)	dark red; dec. at 185 to 195° IR (KBr): 2024, 2062, 2088; ν(C=O) at 1672, ν(C≡C) at 2183	[10]
*21 CH₂	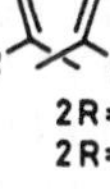 I, see further information	orange red; m.p. 62 to 63°; subl. at 30 to 40°/10⁻² μ_D=1.70±0.11 D (in C₆H₆) χ_{mol} = −(125.3±3.6)×10⁻⁶ (293 K) IR (KBr): ≈2000, 2045, 2092; ν(C=CH₂) at 1621	[9]
*22 CH–C₆H₄Br-4 2R=H 2R=C₆H₄Br-4 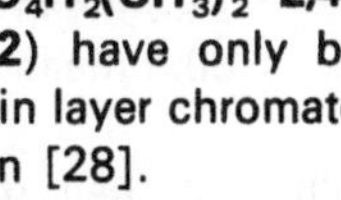	I, in petroleum ether at 65°/3 h (3)	red violet; dec. at ≈200° IR (KBr): 2012, 2049, 2075	[13]

*Further information:

(C₄H₂(CH₃)₂-2,4)COFe₂(CO)₆ and (C₄H₂(CH₃)₂-1,4)COFe₂(CO)₆ (Table 4, No. 1 and 2) have only been obtained as a binary mixture which could not be separated by thin layer chromatography. Spectral assignments were possible except for the carbonyl region [28].

(C₄H₂(C₄H₉-t)₂-2,4)COFe₂(CO)₆ and (C₄H₂(C₄H₉-t)₂-1,4)COFe₂(CO)₆ (Table 4, No. 3 and 4) are formed both together by method I and are separated by thin layer chromatography.

References on p. 74

The ^{13}C NMR spectrum of No. 3 shows at room temperature two broad resonances which are resolved at -25 °C into three sharp peaks at $\delta=205.8$, 209.5, and 209.9 ppm. These signals coalesce at 65 °C to a single resonance at $\delta=208.4$ ppm [28].

The ^{13}C NMR variable-temperature spectra of No. 4 are given as a figure in [28]. The limiting spectrum at -82 °C has five resonances of the intensities $1:2:1:1:1$. The peak with intensity 2 due to a superposition. At -64 °C, three of these coalesce, likely because of CO exchange at one Fe atom. At 25 °C, the remaining four signals are found at $\delta=206.9$, 208.6, 210.5, and 212.2 ppm with the intensities $1:3:1:1$. A second coalescence is observed at 63 °C leaving a single peak which may be a either superposition of the two averaged CO resonances of each $Fe(CO)_3$ unit or a really unique peak due to CO scrambling between the Fe atoms [28].

$C_4(CH_3)_4COFe_2(CO)_6$ (Table **4**, No. **5**) crystallizes in the monoclinic system with $a=8.336\pm0.010$, $b=26.164\pm0.040$, $c=8.308\pm0.010$ Å, and $\beta=112°24'\pm6'$; space group $P2_1/c-C_{2h}^5$. $Z=4$ gives $D_c=1.63$ g·cm^{-3}, $D_m=1.62$ g·cm^{-3}. Five C atoms and the two bonded Fe atoms form a twisted seven-membered ring, see **Fig. 18**.

Fig. 18

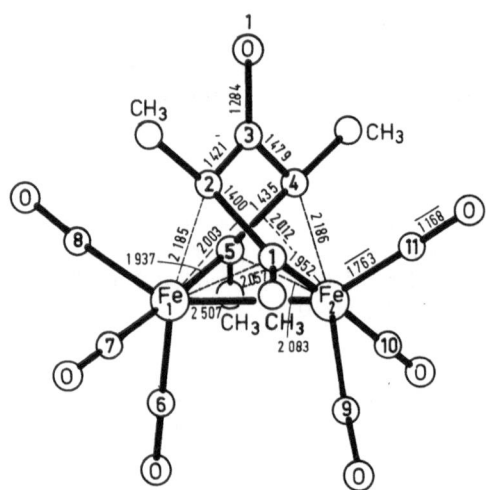

Molecular structure of $C_4(CH_3)_4COFe_2(CO)_6$ [24].

Selected angles (°):

Fe(1)–C(1)–C(2)	117.6±1.4	Fe(2)–Fe(1)–C(6)	97.9±0.9
C(1)–C(2)–C(3)	107.8±2.2	Fe(2)–Fe(1)–C(7)	119.4±1.0
C(2)–C(3)–C(4)	119.2±1.8	Fe(2)–Fe(1)–C(8)	141.6±1.0
C(3)–C(4)–C(5)	111.3±2.1	Fe(1)–Fe(2)–C(9)	95.3±1.0
C(4)–C(5)–Fe(1)	113.1±1.3	Fe(1)–Fe(2)–C(10)	120.8±0.8
Fe(1)–C(1,5)–Fe(2)	77.5 mean	Fe(1)–Fe(2)–C(11)	142.2±0.8
C(1)–Fe(1,2)–C(5)	89.3 mean	Fe(1)–Fe(2)–C(5)	48.9 mean
C(5)–Fe(1)–C(2)	85.7±0.8	Fe(1)–Fe(2)–C(4)	74.4 mean
C(1)–Fe(2)–C(4)	82.9±0.8	Fe(1)–Fe(2)–C(1)	60.9 mean

It appears that the bonds C(1)–C(2) and C(4)–C(5) have partially lost their double bond character. Each Fe atom acquires the rare gas configuration by σ- and π-bonding to three C atoms of the bridging ligand. The carbonyl groups are approximately linear (angles Fe–C–O between 173.3 and 179.5°) [24].

References on p. 74

$C_4(C_2H_5)_4COFe_2CO)_6$ (Table **4**, No. **6**) has been obtained in small amounts (1% yield) from the unstable $(t-C_4H_9C_2C_4H_9-t)Fe(CO)_4$ and $C_2H_5C\equiv CC_2H_5$ by alkyne exchange and further reaction of a dinuclear intermediate with $C_2H_5C\equiv CC_2H_5$. Major products are tetraethyl-p-quinone (33%) and a complex of the composition $(C_2H_5C_2C_2H_5)_2CO-Fe_2(CO)_5$. More details have not been reported [22, p. 287].

$C_4(OCH_3)_4COFe_2(CO)_6$ (Table **4**, No. **7**) is formed by method II in very low yield since $CH_3OC\equiv COCH_3$ rapidly decomposes at the temperature at which the reaction with $Fe_2(CO)_9$ takes place. Thin layer chromatography on Al_2O_3 only gives the complex as an oily mixture with hexamethoxybenzene [26].

$C_4(C_6H_5)_4COFe_2(CO)_6$ (Table **4**, No. **8**) is best prepared by method II in tetrahydrofuran [29]. The reaction of $Fe_2(CO)_9$ with $C_6H_5C\equiv CC_6H_5$ in hydrocarbon solvents at room temperature [2, 22], particularly in an atmosphere of CO [22, p. 285], first gives a dark-green intermediate $C_6H_5C_2C_6H_5Fe_2(CO)_6$ (see C 2, section 2.2.1.2.1.1, p. 72), which on standing or slight heating decomposes into the title compound, $C_4(C_6H_5)_4Fe_2(CO)_6$ (mole ratio 2:1), and exactly one mole $Fe(CO)_5$ [2]. Further reaction of the intermediate with $C_6H_5C\equiv CC_6H_5$ at 20 °C (method III) yields the title compound as main product (75%), the ferrole (5 to 10%), and tetraphenyl-p-quinone iron tricarbonyl (Formula XI, 7%) [22, p. 286], see also [12]. A yield of 77% is obtained by starting from $(t-C_4H_9C_2C_4H_9-t)Fe(CO)_4$ and $C_6H_5C\equiv CC_6H_5$ as the mononuclear t-butylacetylene complex apparently readily exchanges with other alkynes [22, p. 287]. The mixture of products formed in method I is separated by chromatography on Al_2O_3. The complexes II, IV, V (see p. 64), and X are first removed from the column with petroleum ether and benzene, while the title compound is eluted with ether, followed by the cyclopentadienone complex III eluted with methanol [3]. Small amounts of $C_4(C_6H_5)_4COFe_2(CO)_6$ have also been obtained from $Fe(CO)_5$ and $C_6H_5C\equiv CC_6H_5$ by short irradiation (1 to 2 h) in refluxing benzene; long irradiation under the same conditions gives a mixture of complexes like method I. Short irradiation at lower temperature, e.g., boiling ether, increases the yield of the title compound [4]. $C_4(C_6H_5)_4COFe_2(CO)_6$ has been isolated among other complexes from the reaction of $Fe_2(CO)_9$ with diphenylcyclopropenone in benzene at room temperature for 24 h [18].

X   XI

^{13}C NMR shifts very similar to those listed in Table 4 have been measured in CD_2Cl_2 at −68 °C [27]. Variable-temperature spectra from −30 to +85 °C are shown as diagrams in [27, 29]. The three CO resonances begin to broaden at about 10 °C [27] or a little lower [29] and coalesce above about 50 °C to one single signal [27, 28] with $\delta = 207.7$ ppm [28]. The observed pattern of broadening and collapse has been reproduced via computer simulation based on the assumption that the lifetime of CO in each of the three different sites is the same [29]. Scrambling rates of $k = 24.6$ (10 °C) and 275 (41 °C) s^{-1} have been estimated, giving $E_a \approx 13.4 \pm 1.0$ kcal·mol^{-1} provided that $\Delta S^* = 0$ [27]. An Arrhenius plot gave $E_a = 14.7(2)$ kcal·mol^{-1} and $A = 3.4(8) \times 10^{12}$ s^{-1} corresponding to $\Delta G^* = 15.1(2)$ and $\Delta H^* = 14.1(1)$ kcal·mol^{-1} and $\Delta S^* = -3.2(5)$ cal·K^{-1}·mol^{-1} [29].

References on p. 74

The ν(CO) region of the IR spectrum of the solid (Nujol, KBr) has been reported in [3, 4]. The complex crystallizes in the monoclinic system with a$=$13.167(5), b$=$ 11.535(3), c$=$21.787(8) Å, and $\beta=$95.11(3)°; space group P2$_1$/n$-$C$_{2h}^5$ and Z$=$4 [29]. Slightly different lattice parameters are given in [1, 5, 8]; D$_c=$1.43 g$\cdot$cm^{-3} [1, 3, 5, 8]. The molecular structure is analogous to that of compound No. 5. It possesses no rigorous element of symmetry, but it comes very close to having a twofold axis passing through the keto group and bisecting the Fe–Fe bond. Fe–C–O angles lie between 174° and 179°. A stereoscopic view of the entire molecule is given [29]. The numbering of the atoms for the selected bond lengths (in Å) and angles (°) refers to Fig. 18 on p. 68.

Fe(1)–Fe(2)	2.536(1)	C(1)–C(2)	1.414(6)
Fe(1)–C(1)	2.071(4)	C(2)–C(3)	1.493(6)
Fe(1)–C(2)	2.179(4)	C(3)–C(4)	1.516(6)
Fe(1)–C(5)	2.050(4)	C(4)–C(5)	1.410(6)
Fe(2)–C(4)	2.159(4)	C(3)–O(1)	1.213(5)
Fe(2)–C(5)	2.033(4)	Fe–CO	1.812 mean
Fe(2)–C(1)	2.018(4)	C–O	1.128 mean
Fe(2)–C(1)–C(2)	114.1(3)	C(1)–Fe(1,2)–C(5)	90.1 mean
C(1)–C(2)–C(3)	115.3(4)	Fe(2)–Fe(1)–C(6)	94.1(2)
C(2)–C(3)–C(4)	114.6(4)	Fe(2)–Fe(1)–C(7)	124.4(2)
C(3)–C(4)–C(5)	108.4(4)	Fe(2)–Fe(1)–C(8)	138.4(2)
C(4)–C(5)–Fe(1)	111.7(3)	Fe(1)–Fe(2)–C(9)	91.8(2)
C(1)–Fe(1)–C(2)	38.8(2)	Fe(1)–Fe(2)–C(10)	127.1(2)
C(4)–Fe(2)–C(5)	39.2(2)	Fe(1)–Fe(2)–C(11)	135.8(2)
Fe(1)–C(1,5)–Fe(2)	76.8 mean		

C$_4$(C$_6$H$_5$)$_4$COFe$_2$(CO)$_6$ is soluble in benzene and acetone and is less soluble in petroleum ether and alcohols. It crystallizes from benzene/petroleum ether with 0.5 moles C$_6$H$_6$. These diamagnetic solvate crystals melt at 160 °C with decomposition. Solvate–free crystals can be obtained from acetone or CH$_3$OH [3]. Heating in a sealed tube in toluene solution at 130 °C/3 h readily converts the compound into tetraphenylcyclopenta-dienone iron tricarbonyl (Formula III, p. 64) in 82% yield and the free cyclopentadienone (13%). The same conversion can be accomplished in boiling benzene [4]. The thermal decomposition under 200 atm CO at 170 to 200 °C also gives the complex III in 55% yield. Complex III (69% yield) and the free cyclopentadienone (2% yield) are formed by photolysis in toluene [3].

The polarographic reduction in 1,2–dimethoxyethane at 22 °C proceeds irreversibly in two one–electron steps at E$_{1/2}=-$1.27 and $-$2.01 V (referred to 10^{-3} M Ag$^+$/Ag) [19]. Tetraphenylcyclopentadienone is the major degradation product when the complex is treated with Br$_2$ in warm acetic acid or with dilute NaOH in CH$_3$OH/H$_2$O [4]. The reaction with P(C$_6$H$_5$)$_3$ gives no substitution product except the mononuclear C$_4$(C$_6$H$_5$)$_4$COFe(CO)$_2$P(C$_6$H$_5$)$_3$ [23, p. 334]. For other reactions refer to Scheme 3 on p. 63.

C$_4$(C$_6$H$_4$Cl–4)$_4$COFe$_2$(CO)$_6$ (Table 4, No. 9) is formed by method I together with the other complex types described for compound No. 8. The formation from Fe$_2$(CO)$_9$ and 4–ClC$_6$H$_4$C$\equiv$CC$_6$H$_4$Cl–4 in benzene at 70 °C is complete within 5 to 10 min. A second product is the corresponding ferrole. This reaction also takes place at room temperature over a period of several hours giving first an unstable black intermediate,

References on p. 74

which on standing overnight or heating for several minutes to 60 to 70 °C produces quantitatively the titel compound and the ferrole in a 2:1 ratio [1, 5, 8].

The compound is highly soluble in benzene and other organic solvents. It crystallizes in the triclinic system. In boiling benzene it decomposes into tetra-4-chlorophenylcyclo-pentadienone and the corresponding iron tricarbonyl complex (Formula III, p. 64) [1, 5, 8]. In the polarographic reduction in dimethoxyethane the first reversible step at $E_{1/2} = -1.24$ V (referred to 10^{-3} M Ag$^+$/Ag) produces a yellow radical anion showing an one-line ESR spectrum with g=2.0619; it can be reoxidized at -0.4 V to the parent compound. The further reduction of the radical anion at $E_{1/2} = -1.87$ V proceeds irreversibly [19]. One-electron exchange processes of the radical anion are discussed in [20].

$C_4(CH_3)_2(C_6H_5)_2COFe_2(CO)_6$ (Table 4, No. 11). Three isomeric complexes have been obtained from $Fe_3(CO)_{12}$ and $CH_3C \equiv CC_6H_5$ (method I), see compounds No. 17 and 18. The title compound is the major component. The unsymmetrical substitution is con-cluded from the fact that degradation in boiling SO_2Cl_2 gives 2,5-dimethyl-3,6-diphenyl-quinone [23, p. 332]. The positions of the substituents in the isomers No. 17 and 18 are not known.

The polarographic reduction (for conditions see No. 9) gives at $E_{1/2} = -1.50$ V revers-ibly a dark red-brown radical anion which has two lines in the ESR spectrum separated by 104 G, g=2.0322 at half the distance between the signals. The regeneration of the parent complex occurs in two steps at -0.7 and -1.2 V. A second reduction step at $E_{1/2} = -2.12$ V is irreversible [19], see also [20].

$C_4(C_2H_5)_2(C_4H_9-t)_2COFe_2(CO)_6$ and $C_4(C_2H_5)_2(C_6H_5)_2COFe_2(CO)_6$ (Table 4, No. 13 and 14). Starting materials in method III are the unstable intermediates $(t-C_4H_9C_2C_4H_9-t)Fe_2(CO)_6$ for No. 13 and $C_6H_5C_2C_6H_5Fe_2(CO)_6$ for No. 14 which are reacted with $C_2H_5C \equiv CC_2H_5$. Some tetraethylquinone is formed with No. 13 and the ferrole XII in 25% yield with No. 14 [22, p. 286].

XII   XIII   XIV

$C_4(C_6H_5)_2(COOCH_3)_2COFe_2(CO)_6$ (Table 4, No. 15). The preparation by method I also gives the complex types III and IV, p. 64. The compound has not been obtained by method II; only mononuclear complexes were formed from $Fe_2(CO)_9$ and $C_6H_5C \equiv CCOOCH_3$ in benzene at room temperature [16]. The positions of the substituents were located by the reaction of the complex with $CH_3OOCC \equiv CCOOCH_3$. This gives appreciable amounts of compound XIII which can only be formed from an intermediate cyclopentadienone XIV [23, p. 332].

$C_4(C_3H_5-cyclo)_4COFe_2(CO)_6$ (Table 4, No. 16) is formed in 47% yield from $Fe(CO)_5$ (≈ 0.1 M) and cyclo-$C_3H_5C \equiv CC_3H_5$-cyclo (≈ 0.05 M) in hexane on irradiation for 45 min. Seven other iron complexes have been isolated as minor components. A similar distribution of major products was obtained from $Fe_2(CO)_9$ and the alkyne (about 1.4:1 mole) in refluxing pentane for 1 h, 42% yield of the title compound. Only traces of the complex are formed in the thermal reaction of $Fe_3(CO)_{12}$ with the alkyne in refluxing benzene [30]. For the formation from a $(CO)_3Fe(C_8H_{10}CO)Fe(CO)_3$ complex see 2.4.1.2.4.

The title compound decomposes in refluxing toluene (8 h) giving the complex type III (24%) and traces of IV (see p. 64, R=cyclo-C_3H_5) in addition to 30% recovered starting material. It remains unchanged on UV irradiation in hexane for 2 h. The reaction with an excess of cyclo-$C_3H_5C{\equiv}CC_3H_5$-cyclo in hexane under irradiation for about 5 h leads to compound XV in 54% yield and the ferrole IV [30].

XV   XVI

$C_4(C_6H_5)_2(C{\equiv}CC_6H_5)_2COFe_2(CO)_6$ (Table **4**, No. **20**). Other products, in comparable yields, of the preparations by methods I and II are the cyclopentadienone type III and the ferrole IV (see p. 64, R=C_6H_5 and C≡CC_6H_5). A similar product distribution is obtained from $Fe(CO)_5$ and $C_6H_5C{\equiv}CC{\equiv}CC_6H_5$ in petroleum ether in an autoclave at 150 °C for 20 h, 10% yield of the title compound and small amounts of the symmetrical trimerization product XVI. In chromatographic separation on Al_2O_3, the title complex is eluted the last with benzene/ether.

Several crystal fractions, obtained from benzene/petroleum ether (dark-red prisms) showed slight differences in the 750 to 1200 cm^{-1} region of the IR spectrum. Thus the product appeared to consist of a mixture of isomers [10].

$C_4H_4C{=}CH_2Fe_2(CO)_6$ (Table **4**, No. **21**) is one of a large number of iron carbonyl complexes formed from $Fe_3(CO)_{12}$ and HC≡CH in an autoclave. The best yields of the title compound (up to 9%) are obtained with an initial acetylene pressure of 22 atm and heating at 45 to 50 °C in petroleum ether for about 16 h. In chromatographic separation on neutral or acidic Al_2O_3 the complex leaves the column at the top with petroleum ether, but the separation from the following ferrole type IV is not complete and can only be achieved by repeated chromatography.

The compound crystallizes in orange-red prisms from a very concentrated benzene solution after addition of methanol or petroleum ether and long storage at 4 °C [9]. An incorrect molecular structure is assumed in [9].

The compound crystallizes in the monoclinic system with a=14.75±0.02, b= 13.26±0.02, c=7.036±0.01 Å, and β=94.6°±0.2°; space group Cc−C_s^4. Z=4 gives D_c= 1.733 g·cm^{-3}, D_m=1.724 g·cm^{-3}. The molecule, see **Fig. 19**, shows the typical twisted diferracycloheptadiene system with an exocyclic methylene group. The bonds within the bridging ligand preserve their double and single bond character. The bond lengths between the Fe atoms and the C atoms of the ligand are not nearly equivalent, but the distances to the middle of each double bond are equal (2.06 Å). There is also

References on p. 74

Fig. 19

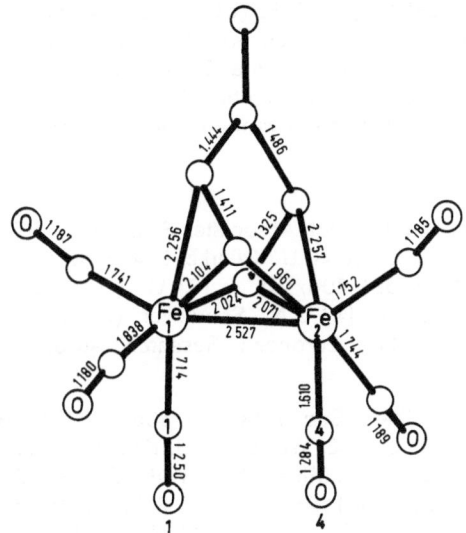

Molecular structure of $C_4H_4C=CH_2Fe_2(CO)_6$ [15].

Bond angles (°), mean values of corresponding angles:

Fe(1)–Fe(2)–C(4)	91.7	C(7)–Fe(2)–C(8)	36.5
Fe(1)–Fe(2)–C(5)	115.9	C(7)–Fe(2)–C(11)	86.6
Fe(1)–Fe(2)–C(7)	50.0	C(8)–Fe(2)–C(11)	79.7
Fe(1)–Fe(2)–C(8)	73.6	C(4)–Fe(2)–C(11)	88.2
Fe(1)–Fe(2)–C(11)	53.5	C(6)–Fe(2)–C(11)	92.8
Fe(1)–C(11)–Fe(2)	76.5	C(5)–Fe(2)–C(8)	98.2
C(4)–Fe(2)–C(5)	91.8	C(6)–Fe(2)–C(8)	88.6
C(4)–Fe(2)–C(6)	101.7	C(7)–C(8)–C(9)	115.4
C(5)–Fe(2)–C(6)	97.7	C(8)–C(9)–C(12)	124.3
Fe–C–O	174.6	C(8)–C(9)–C(10)	111.4

a difference in the bond lengths within Fe(1)–C(1)–O(1) and Fe(2)–C(4)–O(4) (average C–O 1.27 and average Fe–C 1.66 Å) as compared to all other CO groups (average C–O 1.18 and average Fe–C 1.77 Å) [15].

The compound is very soluble in the common organic solvents. The thermal decomposition of the solid in a sealed tube at 110 °C for 13 h gives the complexes XVII (45% yield), XVIII (3%), and Fe(CO)$_5$. Compound XVII is formed exclusively (89% yield) if the decomposition is carried out under a pressure of 50 atm CO [9].

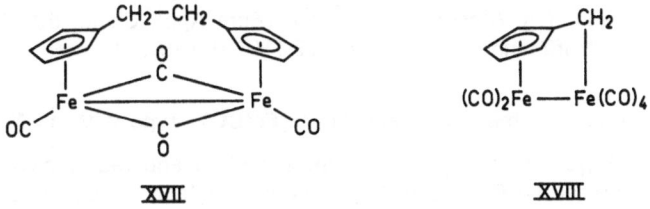

XVII   XVIII

$(C_4H_2(C_6H_4Br-4)_2C=CHC_6H_4Br-4)Fe_2(CO)_6$ (Table **4**, No. **22**) is formed by method I together with several other complexes and 1,2,4-tris(4-bromophenyl)benzene. It is eluted from Al_2O_3 with CS_2, usually contaminated with the benzene derivative. Their difficult separation can be effected by fractional crystallization from benzene/methanol. The compound is readily soluble in conventional solvents, even in light petroleum [13].

References:

[1] K.W. Hübel, European Research Associates S.A., (Belg. 567743 [1958]). – [2] W. Hübel, E.H. Braye, A. Clauss, E. Weiss, U. Krüerke, D.A. Brown, G.S.D. King, C. Hoogzand (J. Inorg. Nucl. Chem. **9** [1959] 204/10). – [3] W. Hübel, E.H. Braye (J. Inorg. Nucl. Chem. **10** [1959] 250/68). – [4] G.N. Schrauzer (J. Am. Chem. Soc. **81** [1959] 5307/10). – [5] K.W. Hübel, European Research Associates S.A. (Fr. 1206786 [1958/60]).

[6] P.L. Pauson (Proc. Chem. Soc. **1960** 297/305). – [7] E.H. Braye, C. Hoogzand, W. Hübel, U. Krüerke, R. Merényi, E. Weiss (in: S. Kirschner, Advances in the Chemistry of the Coordination Compounds, McMillan, New York 1961, p. 190/8). – [8] K.W. Hübel, Union Carbide Corp. (Brit. 885514 [1961]). – [9] E. Weiss, W. Hübel, R. Merényi (Chem. Ber. **95** [1962] 1155/69). – [10] W. Hübel, R. Merényi (Chem. Ber. **96** [1963] 930/43).

[11] K.W. Hübel, E.H. Braye, Union Carbide Corp. (U.S. 3096265 [1959/63]). – [12] C. Hoogzand, W. Hübel (Proc. 8th Intern. Conf. Coord. Chem., Vienna 1964, p. 258/9). – [13] E.H. Braye, W. Hübel (J. Organometal. Chem. **3** [1965] 25/37). – [14] E.H. Braye, W. Hübel (J. Organometal. Chem. **3** [1965] 38/42). – [15] P. Piret, J. Meunier-Piret, M. Van Meerssche, G.S.D. King (Acta Cryst. **19** [1965] 78/84).

[16] L.F. Dahl, R.J. Doedens, W. Hübel, J. Nielsen (J. Am. Chem. Soc. **88** [1966] 446/52). – [17] K.W. Hübel, E.H. Braye, Union Carbide Corp. (U.S. 3280017 [1959/66]). – [18] C.W. Bird, E.M. Briggs, J. Hudec (J. Chem. Soc. C **1967** 1862/4). – [19] R.E. Dessy, R.L. Pohl (J. Am. Chem. Soc. **90** [1968] 1995/2001). – [20] R.E. Dessy, R.L. Pohl (J. Am. Chem. Soc. **90** [1968] 2005/8).

[21] R. Greatrex, N.N. Greenwood, P.L. Pauson (J. Organometal. Chem. **13** [1968] 533/4). – [22] W. Hübel (in: I. Wender, P. Pino, Organic Syntheses via Metal Carbonyls, Vol. 1, Interscience, New York 1968, p. 284/92). – [23] W. Hübel (in: I. Wender, P. Pino, Organic Syntheses via Metal Carbonyls, Vol. 1, Interscience, New York 1968, p. 328/34). – [24] J. Piron, P. Piret, J. Meunier-Piret, M. Van Meerssche (Bull. Soc. Chim. Belges **78** [1969] 121/30). – [25] M. Valle, G. Cetini, O. Gambino, E. Sappa (Atti Accad. Sci. Torino **103** [1968/69] 913/23).

[26] A. Messeguer, F. Serratosa, J. Rivera (Tetrahedron Letters **1973** 2895/8). – [27] J.P. Hickey, J.R. Wilkinson, L.J. Todd (J. Organometal. Chem. **99** [1975] 281/6). – [28] S. Aime, L. Milone, E. Sappa (J. Chem. Soc. Dalton Trans. **1976** 838/40). – [29] F.A. Cotton, D.L. Hunter, J.M. Troup (Inorg. Chem. **15** [1976] 63/7). – [30] R. Victor, V. Usieli, S. Sarel (J. Organometal. Chem. **129** [1977] 387/99).

[31] R. Hoffmann, T.A. Albright, D.L. Thorn (Pure Appl. Chem. **50** [1978] 1/9). – [32] D.L. Thorn, R. Hoffmann (Inorg. Chem. **17** [1978] 126/39).

2.4.1.2.4 Compounds of the $(CO)_3Fe(\mu-^3L-^1L)Fe(CO)_n$ Type with n=3 and 4

$(CO)_3Fe(\mu-C_8H_{10}CO)Fe(CO)_3$. Two isomers, violet I and red II, have been isolated from a reaction between $Fe(CO)_5$ and dicyclopropylacetylene in pentane or hexane under

UV irradiation for 45 min. These products are only formed with an excess of $Fe(CO)_5$ at low concentrations of the starting materials, about 0.01 M of the alkyne and about 0.05 M of the carbonyl. A similar distribution of products resulted from a thermal reaction of $Fe_2(CO)_9$ with the alkyne in low concentration in refluxing pentane. For compounds formed at higher concentrations see 2.4.1.2.2.1 and 2.4.1.2.3.

IR and NMR data indicate that both cyclopropane rings of the original alkyne open in the course of the reaction. Structures I and II have been proposed on the basis of spectral analyses [2]. The proposed coordination of one Fe atom to the cyclopentenone carbonyl group is substantiated by the lowered carbonyl absorption in the IR spectrum; however, this could be also a donor type interaction $(C=O \rightarrow Fe)$, in which case the ligand would belong to the $^3L-^2D$ type.

$(CO)_3Fe(\mu-C_8H_{10}CO)Fe(CO)_3$, isomer I, forms violet–black crystals from pentane which melt above 120 °C with decomposition. ^1H NMR spectrum $(CDCl_3)$: $\delta=1.84$ (CH_3, $J\approx6.4\,Hz$), 2.43 to 2.63 (2 H), 2.86 to 3.23 (2 H), 5.84 (split d, H-7, $J\approx9.4\,Hz$), and 7.53 (narrow m, 1 H) ppm. IR spectrum (hexane): $\nu(CO)$ bands at 1954, 1974, 1992, 2015, and 2060 cm^{-1}, $\nu(C=O)$ at 1631 cm^{-1}.

Short UV irradiation (5 min) in pentane does not affect the complex. In the presence of dicyclopropylacetylene the same reaction gives detectable amounts of $C_4(C_3H_5-cyclo)_4-Fe_2(CO)_6$ and $C_4(C_3H_5-cyclo)_4COFe_2(CO)_6$, see 2.4.1.2.2.1 and 2.4.1.2.3.

$(CO)_3Fe(\mu-C_8H_{10}CO)Fe(CO)_3$, isomer II, crystallizes from pentane in red needles, melting point above 140 °C with decomposition. ^1H NMR spectrum $(CDCl_3)$: $\delta=1.52$ (d, CH_3, $J\approx5.8\,Hz$), 2.48 (broad, 4 H), 4.38 (q, 1 H, $J\approx5.8\,Hz$), 4.44 (s, 1 H), and 7.68 (dd, 1 H) ppm. ^{13}C NMR spectrum $(CDCl_3)$: $\delta=19.30$ (CH_3), 28.54 and 34.96 (C-2,3), 61.02 and 81.94 (C-6,8), 160.83 (C-4), 147.54 and 182.58 (C-5,7), 212.61 (C-1), 209.56 and 214.05 (CO ligands) ppm. IR spectrum (hexane): $\nu(CO)$ bands at 1948, 1957, 1975, 1991, 2015, and 2061 cm^{-1}, $\nu(C=O)$ at 1656 cm^{-1}.

No significant product was detected after UV irradiation in the presence of dicyclopropylacetylene in pentane for 1 h [2].

$(CO)_3Fe(\mu-C_4H_4(COOCH_3)_2)Fe(CO)_4$ (Formula III) occurs as a by-product (1.5 and 3% yield) when the cis or trans isomer of compound IV is transformed by $Fe_2(CO)_9$ in hexane at 40 °C/4 h into the anti or syn form of the $^4LFe(CO)_4$ type V or VI, respectively. The dinuclear complex is eluted first from SiO_2 with hexane.

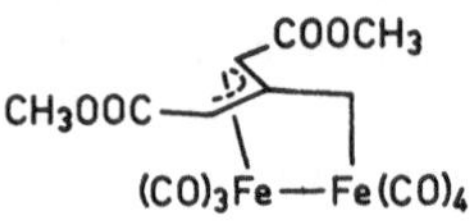

III 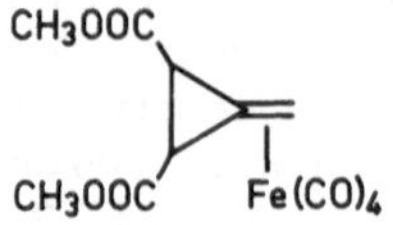   IV 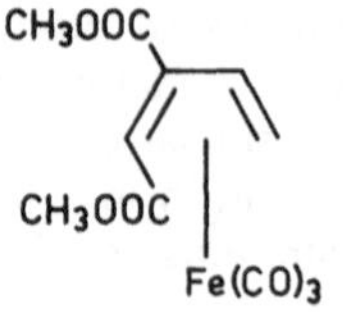   V 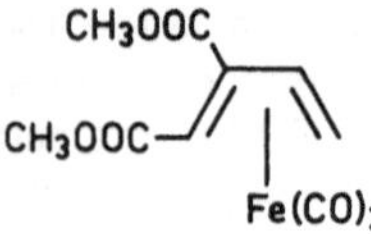   VI

The title compound is a deep red solid, melting at 71 to 72 °C. ^{1}H NMR spectrum (CS$_2$): δ=2.40 (AB quartet, 2 H, J=17.5 Hz), 3.55 and 3.68 (s's, 2 CH$_3$), 3.70 and 4.10 (s's, 2 allyl H) ppm. IR spectrum (CS$_2$): ν(CO) bands at 1972, 1992, 2008, 2032, and 2079 cm^{-1}, ν(C=O) at 1739 cm^{-1} [1].

References:

[1] T.H. Whitesides, R.W. Slaven (J. Organometal. Chem. **67** [1974] 99/108). –
[2] R. Victor, V. Usieli, S. Sarel (J. Organometal. Chem. **129** [1977] 387/99).

2.4.1.2.5   Compounds of the (CO)$_3$Fe(μ-^{4}L-^{2}D)Fe(CO)$_2$ Type

Reactions of the complex I (see C 2, 2.2.1.2.3.3, p. 86) with alkynes RC≡CR' leads to coupling and insertion generating the type II of the general composition **(CO)$_3$Fe(μ-CR'=CR-C=C(C$_4$H$_9$-t)-CO-P(C$_6$H$_5$)$_2$)Fe(CO)$_2$**. These reactions have been carried out at or slightly above 25 °C with CF$_3$C≡CCF$_3$, C$_2$H$_5$C≡CC$_2$H$_5$, C$_6$H$_5$C≡CC$_6$H$_5$, C$_2$H$_5$OOCC≡COOC$_2$H$_5$, and C$_6$H$_5$C≡CCOOC$_2$H$_5$ giving high yields of purple, crystalline, air–stable complexes which are described only in general terms: ^{57}Fe Mössbauer spectra show nonequivalent iron sites, the ketonic carbonyl appears at 1699 cm^{-1} (for R=R'=CF$_3$), and the mass spectrum establishes the presence of six carbonyl groups, one being ketonic. In refluxing benzene, compounds II are converted into the derivatives III, which belong to a (CO)$_3$Fe(μ-^{5}L-^{2}D)Fe(CO)$_2$ type. Obviously in error, compound III is formulated in the literature with two Fe(CO)$_3$ groups.

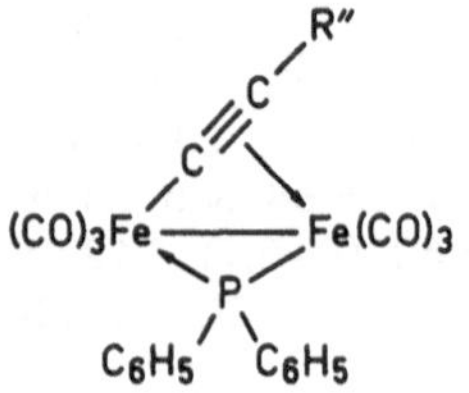

I 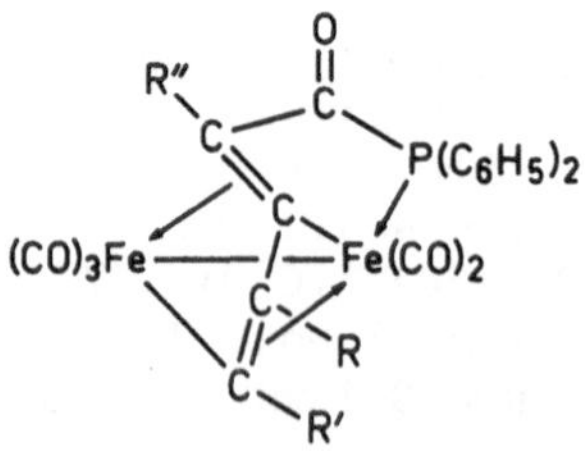   II 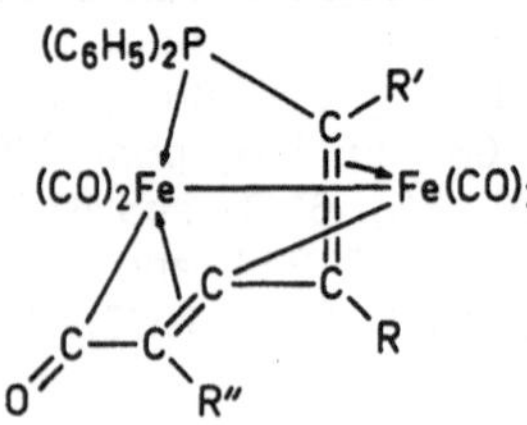   III

(CO)$_3$Fe(μ-C(COOC$_2$H$_5$)=C(COOC$_2$H$_5$)-C=C(C$_4$H$_9$-t)-CO-P(C$_6$H$_5$)$_2$)Fe(CO)$_2$ has been more fully characterized by an X-ray study. It crystallizes in the monoclinic system with a=11.474(4), b=16.779(5), c=9.926(3) Å, and β=97.35(4)°; space group P2$_1$–C$_2^2$, Z=2. The molecular structure is shown in **Fig. 20**.

Fig. 20

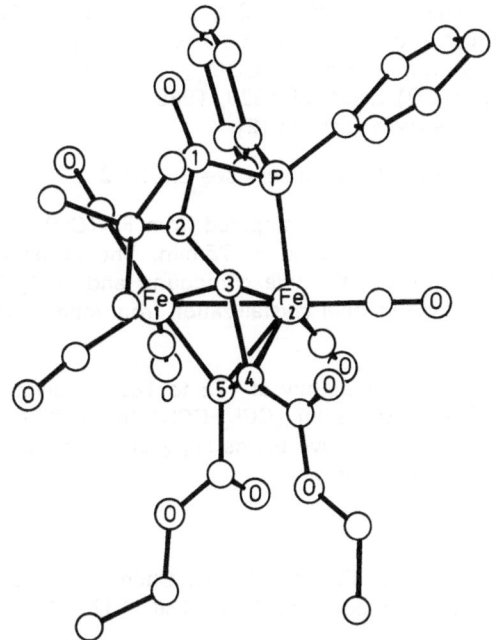

Molecular structure of
$(CO)_3Fe(\mu-C(COOC_2H_5)=C(COOC_2H_5)-C=C(C_4H_9-t)-CO-P(C_6H_5)_2)Fe(CO)_2$.

Important bond lengths (in Å):

Fe(1)–Fe(2)	2.524(1)	Fe(1)–C(2)	2.405(6)
Fe(2)–P	2.193(2)	Fe(1)–C(3)	2.159(5)
P–C(1)	1.918(6)	Fe(1)–C(5)	2.022(6)
C(1)–C(2)	1.530(8)	Fe(2)–C(3)	1.931(6)
C(2)–C(3)	1.341(8)	Fe(2)–C(4)	2.090(6)
C(3)–C(4)	1.455(8)	Fe(2)–C(5)	2.089(6)
C(4)–C(5)	1.400(9)		

Reference:

W.F. Smith, N.J. Taylor, A.J. Carty (J. Chem. Soc. Chem. Commun. **1976** 896/9).

2.4.1.2.6   Compounds of the $(CO)_3Fe(\mu-{}^4L-{}^2D)Fe(CO)_3$ Type

$C_4H_2(CH_3)_2P(C_4H_9-t)Fe_2(CO)_6$ (Formula II) has been obtained from the reaction of $Fe_3(CO)_{12}$ with the phosphole I (molar ratio 2:1) in refluxing toluene for 4 h. The compound is eluted from SiO_2 with hexane/benzene (8:2), 71% yield.

CH₃   CH₃

P

C₄H₉-t

I

t-C₄H₉ — P ... CH₃ / CH₃

(CO)₃Fe ⟵— Fe(CO)₃

II

The red complex, crystallized from CH_3OH, melts at 80 °C. The ^1H NMR spectrum $(CDCl_3)$ shows resonances at $\delta=0.89$ (t-C_4H_9, J(H,P) =20.3 Hz), 2.12 (H-2,5, J(H,P) = 32.7 Hz), and 2.22 (CH_3) ppm. ^{31}P NMR $(CDCl_3)$: $\delta=73$ ppm relative to external P_4O_6. IR spectrum (decaline): $\nu(CO)$ bands at 1929, 1969, 1986, 1998, and 2067 cm^{-1}. The mass spectrum contains the molecular ion [2].

For another type of phosphole complexes see 2.4.1.1.2, p. 8.

$C_4(C_6H_5)_4PC_6H_5Fe_2(CO)_6$ has been prepared from $Fe_3(CO)_{12}$ and pentaphenylphosphole (3:1 mole) in refluxing isooctane for 75 min. The chromatographic purification $(SiO_2/benzene)$ gave a mixture of the title compound and $(CO)_4FeP(C_6H_5)C_4(C_6H_5)_4$, which could be separated by fractional crystallization from ethanol/petroleum ether, about 24% yield.

The orange plates decompose slightly at 178 to 183 °C, giving a viscous oil, which evolves gas at 250 °C. The IR spectrum $(CCl_2=CCl_2)$ has $\nu(CO)$ bands at 1938, 1996, 2020, 2062, and 2088 cm^{-1}. These five bands suggest a structure analogous to II and with pentaphenylphosphole as ligand [1].

References:

[1] E.H. Braye, W. Hübel, J. Caplier (J. Am. Chem. Soc. 83 [1961] 4406/13). – [2] F. Mathey, G. Muller (J. Organometal. Chem. 136 [1977] 241/9).

2.4.1.2.7 $(CO)_3Fe(\mu-SCH_3)_2Fe(CO)C_7H_8$

The title compound represents a unique example of a complex with one diolefine as a terminal ^4L ligand, see Formula I. Attempts to prepare analogous complexes with cycloocta-1,5-diene, cycloocta-1,3,5-triene, cyclooctatetraene, cycloheptatriene, or butadiene were unsuccessful.

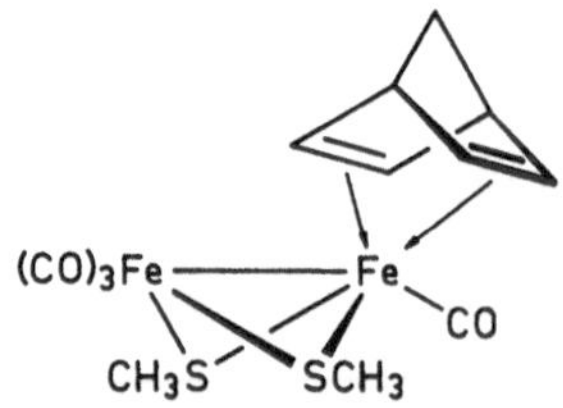

I

The complex is formed from $(CO)_3Fe(\mu-SCH_3)_2Fe(CO)_3$ (about 80% anti and 20% syn, see C 1, 2.1.3.2.2.2, p. 77) and bicyclo[2.2.1]heptadiene (1:5 mole) in refluxing benzene for 70 h. In the chromatographic separation on Al_2O_3, the complex is eluted with pentane, following the starting materials (38% recovery). It slowly crystallizes from pentane at −78 °C, 24% yield (15% conversion). Reactions in boiling toluene or xylene gave little or none of the desired complex.

The dark brown crystals melt at 84 to 86 °C. The ^1H NMR spectrum has been recorded in CS_2, C_6H_6, and C_6F_6. In CS_2, one SCH_3 group and the bridge CH protons of C_7H_8 show identical chemical shifts. In order to separate them the spectrum has been recorded in both C_6H_6 and C_6F_6 which exhibit opposite solvent effects. In C_6H_6 (multiplicity, relative intensity) $\delta=0.60$ (s, 3), 0.74 (s, 2), 1.11 (s, 3), 2.86 (m, 2), and 3.20 (m, 4);

in C_6F_6: $\delta = 1.11$ (s, 3), 1.18 (s, 3), 1.27 (s, 2), 3.42 (m, 2), and 3.70 (m, 4) ppm. The IR spectrum (in KBr) is completely given from 685 to 2910 cm^{-1}; $\nu(CO)$ bands (in halocarbon oil) lie at 1943, 1963, 1974, 2029, and 2040 cm^{-1}.

Reference:

R.B. King, M.B. Bisnette (Inorg. Chem. **4** [1965] 1663/5).

2.4.2 Compounds with Two ^4L Ligands

The following compounds with the common composition $(^4L)_2Fe_2(CO)_3$ bear two terminal cyclobutadiene ligands. However, two types must be distinguished. The first contains only one bridging CO (Formula II). This structure is based on spectroscopic data [4]. The second has all CO ligands as bridging groups (Formula III). This structure could be proved by an X-ray study [3].

II   III

cyclo-$C_4H_4Fe(CO)(\mu$-CO)(CO)FeC_4H_4-cyclo (Formula II) is formed on irradiation ($\lambda \geq 280$ nm) of cyclo-$C_4H_4Fe(CO)_3$ in tetrahydrofuran [2, 4] or 2,3-dimethylbut-2-ene [2] at -40 °C for 6 h while bubbling N_2 through the solution. After removal of the solvent at -20 °C and the excess of starting material in a high vacuum, the residue is extracted with pentane at -20 °C. Cooling of the extract to -70 °C gives the deep red compound in about 8% yield [4], see also [1, 2].

The diamagnetic complex shows in the ^1H NMR spectrum ($C_6D_5CD_3$ at -30 °C) only one singlet of the cyclobutadiene protons at $\delta - 4.22$ ppm. The C atoms of the ^4L ligand exhibit in the ^{13}C NMR spectrum also one chemical shift at $\delta = 77.8$ ppm with $^1J(C,H) = 180$ Hz. The other coupling (^2J and ^3J) could not be resolved. One single peak of the CO ligands at $\delta = 238.8$ ppm can be explained by a rapid fluctuation. The IR spectrum (C_5H_{12} at -20 °C) indicates two terminal and one bridging CO by bands at 1980, 2051, and at 1861 cm^{-1}, respectively.

The mass spectrum contains the molecular ion $[M]^+$ and the fragments $[M-2CO]^+$, $[M-3CO]^+$, and $[M-3CO - C_4H_4]^+$. The complex reacts at -20 °C with CO giving quantitatively cyclo-$C_4H_4Fe(CO)_3$. An 1:1 mixture of cyclo-$C_4H_4Fe(CO)_2P(OCH_3)_3$ and cyclo-$C_4H_4Fe(CO)(P(OCH_3)_3)_2$ is formed with an excess of $P(OCH_3)_3$ [4].

cyclo-$C_4(C_4H_9$-t)$_2(C_6H_5)_2Fe(\mu$-CO)$_3FeC_4(C_4H_9$-t)$_2(C_6H_5)_2$-cyclo (Formula III, R = t-C_4H_9, R'=C_6H_5) has been obtained by UV irradiation of cyclo-$C_4(C_4H_9$-t)$_2(C_6H_5)_2$-Fe(CO)$_3$ in hexane followed by rapid thin layer chromatography on SiO_2 in Ar atmosphere. More details are not reported.

The diamagnetic violet crystals melt at 204 to 205 °C with decomposition. ^1H NMR spectrum (solvent not given): $\delta = 1.35$ (t, t-C_4H_9) and 7.13 (m, C_6H_5) ppm. The IR spectrum (no details) shows only bands of bridging carbonyls at 1830 and 1837 cm^{-1}.

The complex crystallizes in the tetragonal system with a=b=15.093(5) and c= 18.641(4) Å; space group $I\bar{4}$–S_4^2. Z=4 gives $D_c = 1.296$ g·cm^{-3}, $D_m = 1.305$ g·cm^{-1}.

The molecule, see **Fig. 21**, is located on the crystallographic two-fold axis passing through the center of the Fe-Fe bond and the carbonyl group C(1)-O. The extremely short Fe-Fe distance (2.177(3) Å) suggests a sharing of three electron pairs between the Fe atoms. The average Fe-CO distance is 1.974 Å. The average bond lengths within the essentially squareplanar cyclobutadiene rings are 1.468 Å for C-C and 2.062 Å for Fe-C.

Fig. 21

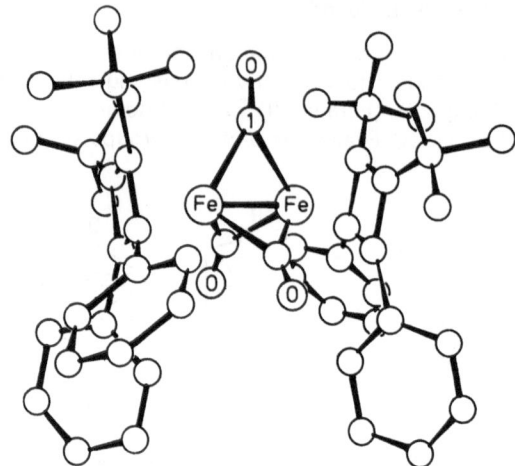

Molecular structure of
cyclo-$C_4(C_4H_9$-t$)_2(C_6H_5)_2$Fe$(\mu$-CO$)_3$FeC$_4(C_4H_9$-t$)_2(C_6H_5)_2$-cyclo [3].

Under 140 atm CO at 80 °C the complex is quantitatively converted into the starting material [3].

cyclo-$C_4(C_6H_5)_4$Fe(μ-CO$)_3$FeC$_4(C_6H_5)_4$-cyclo (Formula III, R=R'=C_6H_5) could be prepared like the previous complex from cyclo-$C_4(C_6H_5)_4$Fe(CO)$_3$. The dark violet crystals melt at 132 to 134 °C with decomposition. The v(CO) bands of the bridging carbonyls are found at 1849 and 1858 cm^{-1}. A similar treatment with CO leads to the starting material [3].

References:

[1] E.A. Koerner von Gustorf, I. Fischler, F.-W. Grevels, D. Schulz, R. Wagner (6th Intern. Conf. Organometal. Chem., Amherst, Mass., 1973, Abstr. No. 46). – [2] E.A. Koerner von Gustorf, I. Fischler, R. Wagner (Proc. 16th Intern. Conf. Coord. Chem., Dublin 1974, Abstr. 4.20). – [3] S.-I. Murahashi, T. Mizoguchi, T. Hosokawa, I. Moritani, Y. Kai, M. Kohara, N. Yasuoka, N. Kasai (J. Chem. Soc. Chem. Commun. **1974** 563/4). – [4] I. Fischler, K. Hildenbrand, E.A. Koerner von Gustof (Angew. Chem. **87** [1975] 35/7; Angew. Chem. Intern. Ed. Engl. **14** [1975] 54/6).

2.4.3   Other Compounds with an Uncertain Structure

The following two compounds are derived from the cluster molecule I (R=C_6H_5 and $C_6H_4CH_3$-4), whose structure has been elucidated by an X-ray study. The spectral data suggest a structure in which the Ir$_2$Cu$_4$ cluster is preserved, with two Fe(CO)$_4$

moieties bonded to two acetylenic groups to form η^2-alkyne systems as shown by Formula II for one unit. Thus the entire cluster molecule has formally to be regarded as a ^{4}L ligand connecting two Fe(CO)$_4$ units. The mutual positions of these units are unknown. The observed instability of the complexes also supports structure II.

I   II

Ir$_2$Cu$_4$(P(C$_6$H$_5$)$_3$)$_2$(C≡CC$_6$H$_5$)$_8$Fe$_2$(CO)$_8$ has been prepared from compound I (R= C$_6$H$_5$) and an eightfold excess of Fe$_2$(CO)$_9$ in benzene at room temperature for 2.5 h. Chromatography on SiO$_2$ with benzene/petroleum ether (1:1) gives the complex in 63% yield.

The purple crystals melt at 154 to 156 °C with decomposition. ^{1}H NMR spectrum (CDCl$_3$): δ=6.95 (m, C≡CC$_6$H$_5$), 7.24 and 7.86 (m's, PC$_6$H$_5$) ppm. IR spectrum (CDCl$_3$): ν(CO) bands at 1970, 1982, 2008, and 2048 cm^{-1}. The solid-state complex does not have the characteristic iridescence of its cluster precursor.

The slightly air-sensitive compound is soluble in most organic solvents. Storage of a solution for a prolonged period of time results in decomposition to the starting material and presumably iron oxides.

Ir$_2$Cu$_4$(P(C$_6$H$_5$)$_3$)$_2$(C≡CC$_6$H$_4$CH$_3$-4)$_8$Fe$_2$(CO)$_8$ has been obtained like the previous complex from compound I with R=C$_6$H$_4$CH$_3$-4. The chromatographic isolation on SiO$_2$ with benzene/petroleum ether (3:7) gives the product in 53% yield.

The purple crystals melt at 148 to 152 °C with decomposition. ^{1}H NMR spectrum (CDCl$_3$): δ=2.09 (unsymmetric multiplet, CH$_3$), 6.63 (m, C≡CC$_6$H$_4$), and 7.18 and 7.84 (m's, PC$_6$H$_5$) ppm. IR spectrum (CDCl$_3$): ν(CO) bands at 1967, 1980, 2006, and 2046 cm^{-1}.

All other properties correspond to those of the previous compound.

Reference:

A.M. Abu Salah, M.I. Bruce (Australien J. Chem. 29 [1976] 531/41).

2.5   Compounds with Ligands Bonded by Five Carbon Atoms

2.5.1   Compounds with One ^{5}L Ligand

2.5.1.1   Carbonyl Complexes with One Terminal C_5H_5 Ligand

2.5.1.1.1   Compounds of the $C_5H_5(CO)_2Fe(\mu-ER_n)Fe(CO)_4$ Type (E=S, P, and As) and Donor Substitution Products

The compounds listed in Table 5 contain PR_2 and AsR_2 bridging groups. Two compounds with SR bridges have been obtained from $Fe_2(CO)_9$ and $C_5H_5Fe(CO)_2SR$ in benzene. Because of their instability they have only been characterized by the IR spectra in solution:

$C_5H_5(CO)_2Fe(\mu-SCH_3)Fe(CO)_4$, IR spectrum (CH_2Cl_2): $\nu(CO)$ bands at 1918, 1947, 1991, 2028, and 2048 cm^{-1} [3].

$C_5H_5(CO)_2Fe(\mu-SC_6H_5)Fe(CO)_4$, IR spectrum (CH_2Cl_2): $\nu(CO)$ bands at 1920, 1952, 1992, 2029, and 2050 cm^{-1} [3].

UV irradiation of the above compounds yields carbon monoxide, $(C_5H_5Fe(CO)_2)_2$, and $(CO)_3Fe(\mu-SR)_2Fe(CO)_3$. The expected products $C_5H_5(CO)Fe(\mu-CO)(\mu-SR)Fe(CO)_3$ could not be detected in the reaction mixtures probably due to their rearrangement in solution [3], see also [1].

The compounds $C_5H_5(CO)_2Fe(\mu-ER_2)Fe(CO)_4$ (E=P and As) and their donor substitution products are subdivided in Table 5 into the types I to IV. Type IV with a bidentate diphosphine as a second bridging ligand is represented only by compound No. 15.

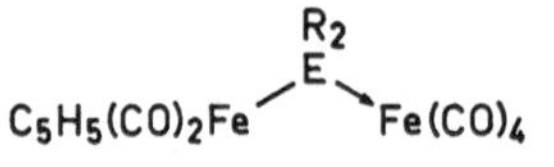

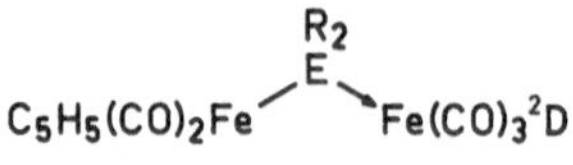

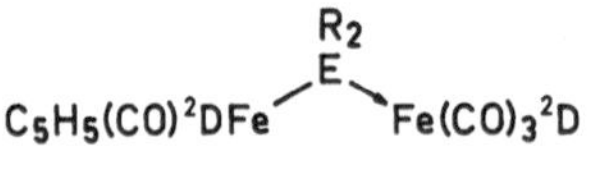

The following preparative methods are used:

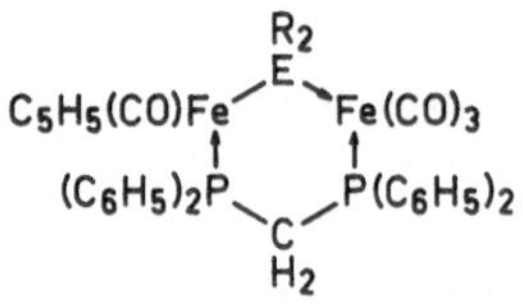

Method I: Reaction of $C_5H_5Fe(CO)_2ER_2$ (E=P and As) with $Fe_2(CO)_9$ (2:1 mole) in C_6H_6 [3] or CH_2Cl_2 [10] at room temperature for 1.5 to 24 h [3, 10] and purification by chromatography on Al_2O_3 [3] or sublimation [10].

Method II: Reaction of $Na[C_5H_5Fe(CO)_2]$ with $(CO)_4FeER_2Cl$ (E=P and As) in equimolar amounts in tetrahydrofuran at room temperature [5, 7] or between $-80\,°C$ and room temperature [3] and purification by chromatography on Al_2O_3 [3] or by recrystallization from hexane [5, 7].

Method III: Irradiation of $C_5H_5(CO)Fe(\mu-CO)(\mu-P(C_6H_5)_2)Fe(CO)_3$ in the presence of a phosphine in benzene at 25 °C for 40 min [3].

References on p. 86

Method IV: Heating of $C_5H_5(CO)Fe(\mu\text{-}CO)(\mu\text{-}P(C_6H_5)_2)Fe(CO)_3$ in the presence of a ^2D compound in benzene at reflux temperature for one to several hours. Separation of the derivatives II and III from the same reaction has been achieved by fractional crystallization [3].

There is experimental evidence that the disubstituted derivatives III are only formed via the type $C_5H_5(CO)Fe(\mu\text{-}CO)(\mu\text{-}PR_2)Fe(CO)_3$ as intermediates. Thus the overall reaction scheme is:

$$C_5H_5(CO)_2Fe(\mu\text{-}PR_2)Fe(CO)_4 + {}^2D$$

$$C_5H_5(CO)_2Fe(\mu\text{-}PR_2)Fe(CO)_3^2D \xrightarrow{\;h\nu\;} C_5H_5(CO)Fe(\mu\text{-}CO)(\mu\text{-}PR_2)Fe(CO)_3$$

$$\downarrow \substack{\text{heating with} \\ \text{excess of } {}^2D} \Big| h\nu/{}^2D$$

$$C_5H_5(CO)({}^2D)Fe(\mu\text{-}PR_2)Fe(CO)_3{}^2D$$

All compounds are diamagnetic and most have a poorly defined melting point. Compounds No. 7 and 10 have not been isolated.

The IR spectra of compounds No. 6, 8, and 9 indicate a trans position of the ^2D substituent with respect to the PR_2 bridging group. The $\nu(CO)$ bands between 1950 and 1975 cm^{-1} are assigned to the A mode. The band in the 1850 to 1900 cm^{-1} region belongs to the E mode, which is split by a slight deviation of the P–Fe–P' group from linearity. The remaining bands above 1980 cm^{-1} are assigned to the $C_5H_5(CO)_2Fe$ moiety. The IR spectra of compounds No. 11 to 14 also reveal a trans position of the P atoms in the $PR_2\text{-}Fe(CO)_3{}^2D$ unit and a coordination of the second ^2D ligand to the other Fe atom.

Compounds No. 3 and 6 to 10 do not react with PR_3 or $P(OR)_3$ in boiling benzene to give a type III compound. UV irradiation of the type III compounds displaces one ^2D ligand and leads to the type $C_5H_5(CO)Fe(\mu\text{-}CO)(\mu\text{-}PR_2)Fe(CO)_2{}^2D$ [3].

Explanations to Table 5: The ^{57}Fe Mössbauer spectra have been measured at 77 K. The isomer shifts refer to Fe at 295 K, and the first shift value always belongs to the C_5H_5 bonded Fe atom.

Table 5
$C_5H_5(CO)_2Fe(\mu\text{-}ER_2)Fe(CO)_4$ compounds (E=P and As) and donor derivatives.
Further information on compounds preceded by an asterisk is given at the end of the table. For abbreviations and dimensions see p. 170.

No.	ER$_2$ bridge Method of preparation (yield in %)	^2D ligands	Properties Explanations see above	Ref.
Compound type I:				
*1	P(CH$_3$)$_2$ II (77)	–	orange crystals; m.p. 43 to 45° ^1H NMR (C$_6$H$_6$): 1.62 (d, CH$_3$, J(P,H) = 8.5), 4.12 (d, C$_5$H$_5$, J(P,H) =1.7) IR (C$_6$H$_{12}$): (CO)$_4$Fe at 1916, 1928, 1964, 2065; C$_5$H$_5$(CO)$_2$Fe at 1984 and 2037	[7]

Table 5 [continued]

No.	ER$_2$ bridge Method of preparation (yield in %)	^{2}D ligands	Properties Explanations on p. 83	Ref.
*2	P(CF$_3$)$_2$ I (88)	−	cherry red; m.p. 118 to 121° ^{1}H NMR (CDCl$_3$): 5.20 (d, C$_5$H$_5$, J(P,H) =2.0) ^{19}F NMR (CDCl$_3$): 53.7 (d, J(P,F)=52) IR (C$_6$H$_{14}$): 1947, 1970, 1996, 2021, 2049, 2062	[10]
*3	P(C$_6$H$_5$)$_2$ I (40), II (40)	−	orange; m.p. 175 to 178° (dec.) ^{1}H NMR (CDCl$_3$): 5.04 (d, C$_5$H$_5$, J(P,H) =1.2) ^{57}Fe−γ: δ=0.09/−0.09, Δ=1.75/2.38 IR (C$_6$H$_{12}$): 1912, 1930, 1963, 1991, 2031, 2045; (CH$_2$Cl$_2$): 1910, 1918, 1923, 1955, 1987, 2012, 2028	[1 to 4, 9]
*4	As(CH$_3$)$_2$ II (5)	−	ruby red; m.p. 54 to 56° (dec.) ^{1}H NMR (C$_6$H$_6$): 1.53 (CH$_3$), 4.12 (C$_5$H$_5$) IR (C$_6$H$_{12}$): (CO)$_4$Fe at 1915, 1927, 1961, 2050; C$_5$H$_5$(CO)$_2$Fe at 1982 and 2030	[5]
*5	As(C$_6$H$_5$)$_2$ see further information	−	IR (−): 1927, 1948, 1988, 2025, 2040	[8]

Compound type II:

No.	ER$_2$ bridge Method of preparation (yield in %)	^{2}D ligands	Properties Explanations on p. 83	Ref.
6	P(C$_6$H$_5$)$_2$ III (<10), IV (≈40)	P(C$_2$H$_5$)$_3$	^{1}H NMR (CDCl$_3$): 1.05 (t, CH$_3$, J(H,H) =7), 1.32 (t, CH$_3$, J(H,H) =6), 5.05 (d, C$_5$H$_5$, J(P,H) =1.5) ^{57}Fe−γ: δ=0.09/−0.10, Δ=1.72/2.54 IR (C$_6$H$_{12}$): 1855, 1872, 1949, 1983, 2032	[3, 9]
7	P(C$_6$H$_5$)$_2$ IV (≈20)	P(OCH$_3$)$_3$	−	[3]
8	P(C$_6$H$_5$)$_2$ IV H≈20)	P(OC$_3$H$_7$-i)$_3$	^{1}H NMR (CDCl$_3$): 1.28 (d, CH$_3$, J(H,H) =6), 4.99 (d, C$_5$H$_5$, J(P,H) =1.5) IR (C$_6$H$_{12}$): 1865, 1879, 1963, 1980, 2030	[3]
9	P(C$_6$H$_5$)$_2$ IV (≈20)	P(OC$_6$H$_5$)$_3$	IR (C$_6$H$_{12}$): 1878, 1897, 1983, 2033	[3]
*10	P(C$_6$H$_5$)$_2$ see further information	(C$_6$H$_5$)$_2$PCH$_2$- CH$_2$P(C$_6$H$_5$)$_2$	IR (CH$_2$Cl$_2$): 1856, 1870, 1943, 1974, 2024	[3]

References on p. 86

Table 5 [continued]

No.	ER$_2$ bridge Method of preparation (yield in %)	^{2}D ligands	Properties Explanations on p. 83	Ref.

Compound type III:

*11	P(C$_6$H$_5$)$_2$ III (<10), IV (<10)	P(C$_2$H$_5$)$_3$	^{1}H NMR (C$_6$D$_6$): 0.8 (m, CH$_3$), 4.81 (t, C$_5$H$_5$, J(P,H) =1.9) ^{57}Fe-γ: δ=0.20/−0.09, Δ=1.72/2.56 IR (C$_6$H$_{12}$): 1839, 1859, 1930, 1951	[3, 9]
12	P(C$_6$H$_5$)$_2$ IV ($\approx$60)	P(OCH$_3$)$_3$	^{1}H NMR (CDCl$_3$): 3.42 (d, CH$_3$, J(P,H) =10.8), 3.72 (d, CH$_3$, J(P,H) =12.6), 4.88 (broad, C$_5$H$_5$) IR (CCl$_4$): 1867, 1878, 1953, 1972	[3]
13	P(C$_6$H$_5$)$_2$ IV ($\approx$50)	P(OC$_3$H$_7$-i)$_3$	^{1}H NMR (C$_6$D$_6$): 0.88 (d, CH$_3$, J(H,H) =6.2), 1.37 (d, CH$_3$, J(H,H) =6), 4.58 (broad, C$_5$H$_5$) IR(C$_6$H$_{12}$): 1857, 1869, 1944, 1963	[3]
14	P(C$_6$H$_5$)$_2$ IV ($\approx$50)	P(OC$_6$H$_5$)$_3$	^{1}H NMR (CDCl$_3$): 4.51 (broad, C$_5$H$_5$) ^{57}Fe-γ: δ=0.15/−0.11, Δ=1.77/2.65 IR (C$_6$H$_{12}$): 1873, 1892, 1960, 1980	[3, 9]

Compound type IV:

| *15 | P(C$_6$H$_5$)$_2$ IV ($\approx$60) | (C$_6$H$_5$)$_2$PCH$_2$- P(C$_6$H$_5$)$_2$ | ^{57}Fe-γ: δ=0.22/−0.10, Δ=1.78/2.15 IR (CH$_2$Cl$_2$): 1869, 1891, 1950, 1971 | [3, 9] |

* Further information:

C$_5$H$_5$(CO)$_2$Fe(μ-P(CH$_3$)$_2$)Fe(CO)$_4$ (Table 5, No. 1) decomposes slowly in air. Heating in benzene at reflux for 1 h gives a 58% yield of (CO)$_3$Fe(μ-P(CH$_3$)$_2$)$_2$Fe(CO)$_3$ and (C$_5$H$_5$Fe(CO)$_2$)$_2$. Irradiation in benzene for 2 h yields C$_5$H$_5$(CO)Fe(μ-CO)-(μ-P(CH$_3$)$_2$)Fe(CO)$_3$ [7].

C$_5$H$_5$(CO)$_2$Fe(μ-P(CF$_3$)$_2$)Fe(CO)$_4$ (Table 5, No. 2) shows the molecular ion in the mass spectrum. The solid is air-stable for short periods, but solutions are markedly sensitive to air [10].

C$_5$H$_5$(CO)$_2$Fe(μ-P(C$_6$H$_5$)$_2$)Fe(CO)$_4$ (Table 5, No. 3) has also been obtained in 80% yield from C$_5$H$_5$Fe(CO)$_2$Cl and (CO)$_4$FeP(C$_6$H$_5$)$_2$H in benzene/ether in the presence of (C$_2$H$_5$)$_2$NH at room temperature for 2 d. It is purified by chromatography on SiO$_2$ with hexane/methylene chloride (4:1) as eluent [2].

Bands at 1991 and 2031 cm^{-1} in the IR spectrum are assigned to the C$_5$H$_5$(CO)$_2$Fe part of the molecule [3]. The mass spectrum shows the molecular ion [M]$^+$ and the fragments [M−n CO]$^+$ with n=1 to 6 [1, 2, 3], the ion [M−6 CO]$^+$ being the most intense [2]. The compound is stable in air in the crystalline state. Solutions gradually change from orange to purple, even at room temperature under N$_2$, depositing C$_5$H$_5$(CO)-Fe(μ-CO) (μ-P(C$_6$H$_5$)$_2$)Fe(CO)$_3$ [2]. The same complex is almost quantitatively formed by irradiation of the title compound [2, 3].

References on p. 86

$C_5H_5(CO)_2Fe(\mu-As(CH_3)_2)Fe(CO)_4$ (Table 5, No. 4). In the preparation by method II the starting material $(CO)_4FeAs(CH_3)_2Cl$ is prepared in situ from $(CO)_4FeAs(CH_3)_2$-$N(CH_3)_2$ and HCl in ether.

A distortion of the local symmetry at the two Fe centres makes all CO vibrations IR active. However, the assignments of the $v(CO)$ bands to the two Fe centres is not certain. The solid compound is air-stable, but solutions are rather sensitive to air [5]. Disproportionation to $(CO)_3Fe(\mu-As(CH_3)_2)_2Fe(CO)_3$ and $(C_5H_5Fe(CO)_2)_2$ occurs when the complex is irradiated in cyclohexane for 1 h [6].

$C_5H_5(CO)_2Fe(\mu-As(C_6H_5)_2)Fe(CO)_4$ (Table 5, No. 5) has been obtained in good yield from a reaction of $Na_2[Fe(CO)_4]$ with $As(C_6H_5)_2Cl$ and $C_5H_5Fe(CO)_2Cl$ in tetrahydrofuran at room temperature. Details are not reported [8].

$C_5H_5(CO)_2Fe(\mu-P(C_6H_5)_2)Fe(CO)_3(C_6H_5)_2PCH_2CH_2P(C_6H_5)_2$ (Table 5, No. 10) forms as sole product from $C_5H_5(CO)Fe(\mu-CO)(\mu-P(C_6H_5)_2)Fe(CO)_3$ and an equimolar amount of $(C_6H_5)_2PCH_2CH_2P(C_6H_5)_2$ in benzene at room temperature. The compound decomposes in solution and thus could not be isolated pure [3].

$C_5H_5(CO)(P(C_2H_5)_3)Fe(\mu-P(C_6H_5)_2)Fe(CO)_3P(C_2H_5)_3$ (Table 5, No. 11). Preparations from $C_5H_5(CO)Fe(\mu-CO)(\mu-P(C_6H_5)_2)Fe(CO)_3$ and $P(C_2H_5)_5$ by either method III or method IV give only low yields. A good yield of 45% has been obtained by first forming $C_5H_5(CO)Fe(\mu-CO)(\mu-P(C_6H_5)_2)Fe(CO)_2P(C_2H_5)_3$ photochemically from the parent complex in benzene and then refluxing the solution with excess $P(C_2H_5)_3$ for 3 h. The monosubstituted derivative (type II) also present in the reaction mixture can be separated by fractional crystallization from benzene/petroleum ether [3].

$C_5H_5(CO)Fe(\mu-P(C_6H_5)_2)(\mu-(C_6H_5)_2PCH_2P(C_6H_5)_2)Fe(CO)_3$ (Table 5, No. 15). The irradiation of $C_5H_5(CO)Fe(\mu-CO)(\mu-P(C_6H_5)_2)Fe(CO)_3$ and the $^2D-^2D$ ligand molecule according to method III leads first to a product which probably is $C_5H_5(CO)Fe(\mu-CO)$-$(\mu-P(C_6H_5)_2)Fe(CO)_2{}^2D-{}^2D$ (the diphosphine as a monodentate ligand). This compound rearranges in solution giving the title complex [3].

References:

[1] R.J. Haines, C.R. Nolte, R. Greatrex, N.N. Greenwood (J. Organometal. Chem. 26 [1971] C45/C48). – [2] K. Yasufuku, H. Yamazaki (J. Organometal. Chem. 28 [1971] 415/21). – [3] R.G. Haines, C.R. Nolte (J. Organometal. Chem. 36 [1972] 163/75). – [4] R.J. Haines, A.L. du Preez, C.R. Nolte (J. Organometal. Chem. 55 [1973] 199/203). – [5] W. Ehrl, H. Vahrenkamp (Chem. Ber. 106 [1973] 2556/62).

[6] W. Ehrl, H. Vahrenkamp (Chem. Ber. 106 [1973] 2563/9). – [7] W. Ehrl, H. Vahrenkamp (J. Organometal. Chem. 63 [1973] 389/98). – [8] J.P. Collman, R.G. Komoto, W.O. Siegl (J. Am. Chem. Soc. 95 [1973] 2389/90). – [9] R. Greatrex, R.J. Haines (J. Organometal. Chem. 114 [1976] 199/211). – [10] R.C. Dobbie, P.R. Mason (J. Chem. Soc. Dalton 1976 189/91).

2.5.1.1.2 Compounds of the $C_5H_5(CO)Fe(\mu-CO)(\mu-ER_n)Fe(CO)_3$ Type (E=S and P) and Donor Substitution Products

Compounds with SR bridging groups (Formula I) are assumed to be intermediates in the photolysis of the $C_5H_5(CO)_2Fe(\mu-SR)Fe(CO)_4$ complex type, cf. 2.5.1.1.1. Only one compound has been isolated and characterized.

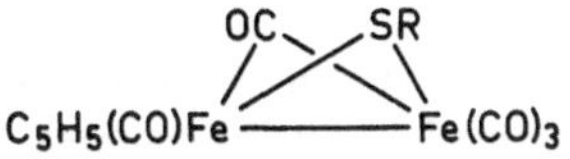

I

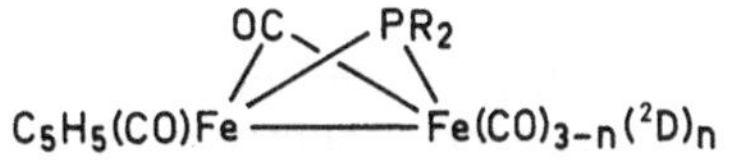

II

$C_5H_5(CO)Fe(\mu\text{-}CO)(\mu\text{-}SC_4H_9\text{-}t)Fe(CO)_3$ has been prepared in 20% yield by the reaction of $C_5H_5Fe(CO)_2SC_4H_9\text{-}t$ with $Fe_2(CO)_9$ in benzene at room temperature for 90 min. It is eluted from Al_2O_3 with petroleum ether. The green microcrystalline substance, which was not obtained completely pure, shows in its IR spectrum (CH_2Cl_2) $\nu(CO)$ bands at 1770 (bridging CO), 1967, and 2034 cm^{-1} (terminal CO). It is extremely reactive and decomposes rapidly in solution [1, 3].

The compounds collected in Table 6 contain PR_2 bridging groups (Formula II). They can be prepared by the following methods:

Method I:　Irradiation of $C_5H_5(CO)_2Fe(\mu\text{-}PR_2)Fe(CO)_4$ in benzene at 0 °C [2] or at room temperature [1, 3, 5] and purification by chromatography on Al_2O_3 [3] or SiO_2 [2] or recrystallization from benzene/petroleum ether [2, 3, 5].

Method II:　Irradiation of $C_5H_5(CO)Fe(\mu\text{-}CO)(\mu\text{-}PR_2)Fe(CO)_3$ in the presence of the donor molecule in benzene at room temperature and recrystallization from CH_3OH/CH_2Cl_2 [3].

Method III:　Heating of $C_5H_5(CO)Fe(\mu\text{-}CO)(\mu\text{-}PR_2)Fe(CO)_3$ with the donor molecule in benzene at reflux and recrystallization from benzene/petroleum ether, benzene/methanol, or methylene chloride/petroleum ether [3].

The rather unstable compounds No. 3, 8, and 9 have only been characterized by their IR spectra in solution [3]. The complexes are diamagnetic [3, 5]. Compounds No. 4 to 7 have no sharp melting points. The ^{1}H NMR and IR spectra of compounds No. 2 to 8 indicate the presence of isomers, possibly an equilibrium between III and IV, III being the major isomer. Thus, the ^{1}H NMR spectra have to C_5H_5 resonances and the intensity of the bridging CO band in the solution IR spectra is considerably weaker than that of the same band in the solid state spectrum (Nujol). The solid state spectra of compounds No. 4 to 7 have the highest energy band in the region 1960 to 1980 cm^{-1}, which is interpreted in terms of substitution of a CO group in the $Fe(CO)_3$ unit (see formula III). The presence of higher energy bands in the solution spectra (1990 to 2020 cm^{-1}) is consistent with structure IV for the second isomer [3].

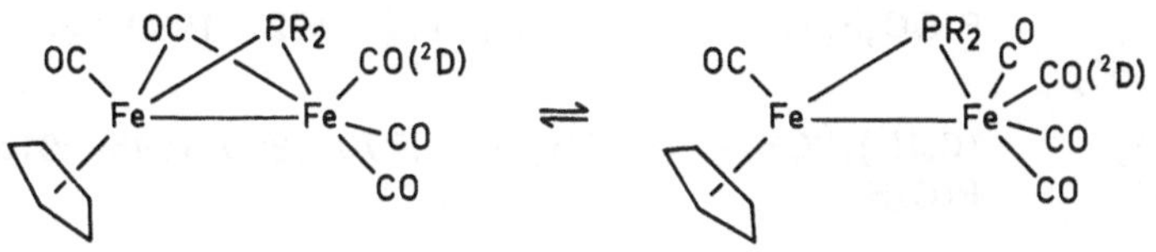

III　　　　　　　　　　　　　　　IV

Explanations to Table 6:　^{57}Fe Mössbauer spectra have been recorded at 77 K. The isomer shifts refer to Fe at 295 K. The assignment of the peaks to the two nonequivalent Fe atoms has been made by comparison with structurally similar compounds [7].

Table 6
$C_5H_5(CO)Fe(\mu\text{-}CO)\,(\mu\text{-}PR_2)Fe(CO)_3$ compounds and donor substitution products (Formula III). Further information on compounds preceded by an asterisk is given at the end of the table. For abbreviations and dimensions see p. 170.

No.	PR_2 bridge Method of preparation (yield in %)	^2D ligand	Properties Explanations on p. 87	Ref.
*1	$P(CH_3)_2$ I (60)	–	black, shiny, sword-like crystals; m.p. 105° ^1H NMR (C_6H_6): 1.48 (d, CH_3, J(P,H) = 12.1), 4.12 (d, C_5H_5, J(P,H) =0.9) IR (C_6H_{12}): 1806, 1946, 1965, 1972, 2040	[5]
*2	$P(C_6H_5)_2$ I (70, ≈100)	–	purple crystals; m.p. 120 to 125° (dec.) ^1H NMR $(CDCl_3)$: 4.93 and 5.81 (C_5H_5) $^{57}Fe\text{-}\gamma$: δ=0.19/0.05, Δ=1.80/1.39 IR (C_6H_{12}): 1797, 1956, 1970, 1980, 2035	[2, 3, 4, 7]
3	$P(C_6H_5)_2$ II (≈60)	$P(C_2H_5)_3$	IR (C_6H_{12}): 1775, 2893, 1930, 1989	[3]
4	$P(C_6H_5)_2$ III (≈60)	$P(C_6H_5)_3$	olive green ^1H NMR $(CDCl_3)$: 4.12 and 5.00 (C_5H_5) $^{57}Fe\text{-}\gamma$: δ=0.18/0.11, Δ=2.12/1.51 IR (C_6H_{12}): 1776, 1905, 1948, 1970, 2004	[3, 7]
5	$P(C_6H_5)_2$ III (≈80)	$P(OCH_3)_3$	green ^1H NMR $(CDCl_3)$: 3.83 (q, CH_3, J(P,H) =1 and 11.8), 4.67 and 5.06 (C_5H_5) $^{57}Fe\text{-}\gamma$: δ=0.18/0.07, Δ=1.98/1.51 IR (C_6H_{12}): 1782, 1917, 1958, 1980, 2014	[3, 7]
6	$P(C_6H_5)_2$ II (≈45)	$P(OC_2H_5)_3$	green ^1H NMR $(CDCl_3)$: 1.37 (t, CH_3, J(H,H) =7.3), 4.66 and 5.01 (C_5H_5) IR (C_6H_{12}): 1779, 1915, 1958, 1976, 2013	[3]
7	$P(C_6H_5)_2$ II (≈55)	$P(OC_3H_7\text{-}i)_3$	green ^1H NMR $(CDCl_3)$: 1.37 (d, CH_3, J(H,H) =6), 4.64 (C_5H_5) IR (C_6H_{12}): 1778, 1913, 1955, 1975, 2012	[3]
8	$P(C_6H_5)_2$ III (≈60)	$P(OC_6H_5)_3$	IR (C_6H_{12}): 1778, 1921, 1960, 1982, 2016	[3]
*9	$P(C_6H_5)_2$ II (–)	$(C_6H_5)_2PCH_2\text{-}$ $P(C_6H_5)_2$	IR $(CH_2Cl)_2$: 1770, 1900, 1945, 2015	[3]

* Further information:

$C_5H_5(CO)Fe(\mu\text{-}CO)\,(\mu\text{-}P(CH_3)_2)Fe(CO)_3$ (Table **6**, No. **1**) crystallizes in the monoclinic system with a=7.124±0.001, b=29.610±0.007, c=14.732±0.002 Å, and β= 109.38±0.01°; space group P2$_1$/c –C$^5_{2h}$, Z=8 gives D_c=1.70 g·cm^{-3}, D_m=1.71 g·cm^{-3}. The molecular structure is shown in **Fig. 22**. The $Fe_2PC(4)$ skeleton has approximately equal Fe–C and Fe–P bond lengths. The dihedral angles on the Fe–Fe bond and on the P–C(4) line are close to 130°. However, it appears that C(4)–O belongs somewhat more to Fe(1) than Fe(2) as indicated by the slightly shorter Fe(1)–C(4) bond and the larger Fe(1)–C(4)–O angle. The C atoms C(6) and C(7) are a little more distant from Fe(2) (2.14 Å mean) than the other ring atoms (2.09 Å), which may be explained by a steric hindrance of the CH₃ group C(11) [6].

Fig. 22

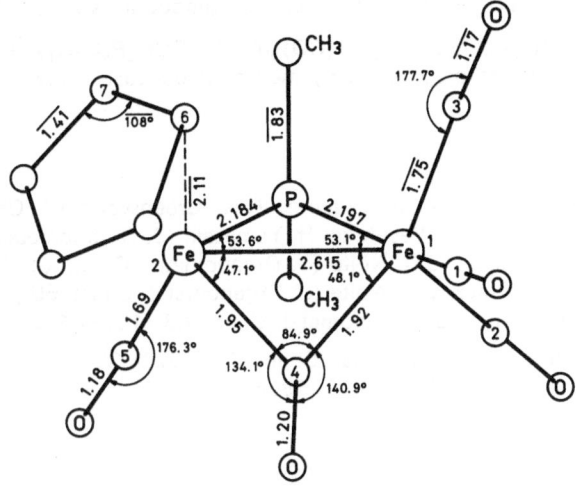

Molecular structure of $C_5H_5(CO)Fe(\mu\text{-}CO)(\mu\text{-}P(CH_3)_2)Fe(CO)_3$ [6].

Other selected bond angles (°). Cp indicates the middle of the C_5H_5 ring.

Fe(2)–Fe(1)–C(1)	111.1(4)	C(1)–Fe(1)–C(2)	101.8(5)
Fe(2)–Fe(1)–C(2)	129.4(4)	C(1)–Fe(1)–C(3)	88.1(5)
Fe(2)–Fe(1)–C(3)	111.7(4)	C(2)–Fe(1)–C(3)	106.7(4)
P–Fe(1)–C(1)	161.2(4)	Fe(1)–Fe(2)–C(5)	113.9(4)
P–Fe(1)–C(2)	96.8(4)	P–Fe(2)–C(5)	87.9(4)
P–Fe(1)–C(3)	89.3(4)	P–Fe(2)–Cp	132.4(1.2)
C(4)–Fe(1)–C(1)	85.8(5)	P–Fe(2)–C(4)	87.9(3)
C(4)–Fe(1)–C(2)	99.8(5)	C(4)–Fe(2)–C(5)	87.1(5)
C(4)–Fe(1)–C(3)	153.5(5)	C(4)–Fe(2)–Cp	125.7(1.2)
C(4)–Fe(1)–P	88.4(3)	C(5)–Fe(2)–Cp	122.3(1.2)
Fe(1)–P–Fe(2)	73.3(1)		

The complex is stable towards air and humidity. It is possibly an intermediate in the thermal decomposition of $C_5H_5(CO)_2Fe(\mu\text{-}P(CH_3)_2)Fe(CO)_4$ since it gives the same decomposition products $(C_5H_5Fe(CO)_2)_2$ and $(CO)_3Fe(\mu\text{-}P(CH_3)_2)_2Fe(CO)_3$, when refluxed in benzene for 1 h [5].

$C_5H_5(CO)Fe(\mu\text{-}CO)(\mu\text{-}P(C_6H_5)_2)Fe(CO)_3$ (Table 6, No. 2) forms slowly on storage of a solution of $C_5H_5(CO)_2Fe(\mu\text{-}P(C_6H_5)_2)Fe(CO)_4$ at room temperature. Only one ¹H NMR resonance of the C_5H_5 ligand at $\delta=4.87$ ppm (in $CDCl_3$) is reported in [2]. An illustration of the ⁵⁷Fe Mössbauer spectrum is given in [1]. The highest energy $\nu(CO)$ band at 2035 cm⁻¹ is assigned to carbonyl groups of the $Fe(CO)_3$ moiety, while the band at 1980 cm⁻¹ is assigned to the $C_5H_5(CO)Fe$ moiety [3].

The mass spectrum shows the molecular ion $[M]^+$ and the fragments $[M-5CO]^+$ [1, 3]. As the complex decomposes on heating to $C_5H_5(CO)Fe(\mu-P(C_6H_5)_2)Fe(CO)C_5H_5$ and $(CO)_3Fe(\mu-P(C_6H_5)_2)_2Fe(CO)_3$ the mass spectrum is contaminated with fragments of these molecules [2]. The reaction with I_2 gives $(CO)_3Fe(\mu-P(C_6H_5)_2)_2Fe(CO)_3$. The complex is unreactive towards carbon monoxide, pyridine, cyclooctα-1,5-diene, and bicyclo[2.2.1]heptadiene under the conditions of the preparative methods II and III. Treatment with $(C_6H_5)_2PCH_2CH_2P(C_6H_5)_2$ ($=^2D$) in benzene at room temperature gives $C_5H_5(CO)_2Fe(\mu-P(C_6H_5)_2)Fe(CO)_3{}^2D$ (Table 5, compound No. 10) [3].

$C_5H_5(CO)Fe(\mu-CO)(\mu-P(C_6H_5)_2)Fe(CO)_2(C_6H_5)_2PCH_2P(C_6H_5)_2$ (Table 6, No. 9) has not been isolated from the preparation by method II because it rearranges to compound No. 15 in Table 5 [3].

References:

[1] R.J. Haines, C.R. Nolte, R. Greatrex, N.N. Greenwood (J. Organometal. Chem. **26** [1971] C45/C48). – [2] K. Yasufuku, H. Yamazaki (J. Organometal. Chem. **28** [1971] 415/21). – [3] R. J. Haines, C.R. Nolte (J. Organometal. Chem. **36** [1972] 163/75). – [4] R.J. Haines, A.L. du Preez, C.R. Nolte (J. Organometal. Chem. **55** [1973] 199/203). – [5] W. Ehrl, H. Vahrenkamp (J. Organometal. Chem. **63** [1973] 389/98).

[6] H. Vahrenkamp (J. Organometal. Chem. **63** [1973] 399/406). – [7] R. Greatrex, R.J. Haines (J. Organometal. Chem. **114** [1976] 199/211).

2.5.1.1.3 Compounds with Nitrosyl Ligands

$C_5H_5(CO)_2Fe(\mu-P(CF_3)_2)Fe(CO)(NO)_2$ has been obtained in 74% yield from $Fe(CO)_2(NO)_2$ and $C_5H_5Fe(CO)_2P(CF_3)_2$ in $CFCl_3$ at room temperature (5 d). The deep red substance purified by sublimation melts at 110 to 112 °C. ^1H NMR spectrum (acetone): $\delta=5.60$ (d) ppm, $J(P,H)=2.0$ Hz. ^{19}F NMR spectrum (acetone): $\delta=52.7$ (d) ppm, $J(P,F)=55$ Hz. In the IR spectrum (CH_2Cl_2) accidental degeneracy produces only the two $\nu(CO)$ bands at 2015 and 2055 cm^{-1}, $\nu(NO)$ bands are at 1720 and 1770 cm^{-1}. The IR spectrum gives no indication of rotamers. The mass spectrum shows the molecular ion. The substance is stable in air for short periods but sensitive to air in solution.

Reference:

R.C. Dobbie, P.R. Mason (J. Chem. Soc. Dalton Trans. **1976** 189/91).

2.5.1.2 Carbonyl Compounds Containing One ^5L Bridging Ligand

In the compounds of the following section a $Fe_2(CO)_6$ group is bonded to five C atoms of an organic system as shown by Formulas I to IV.

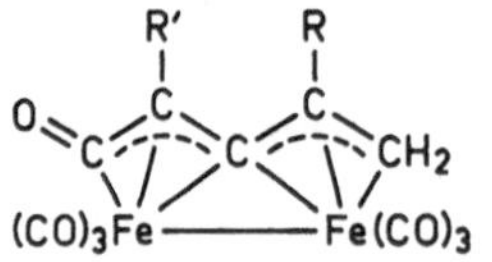

I

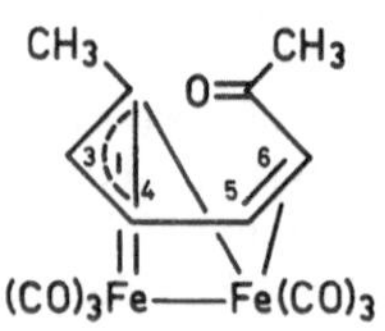

II

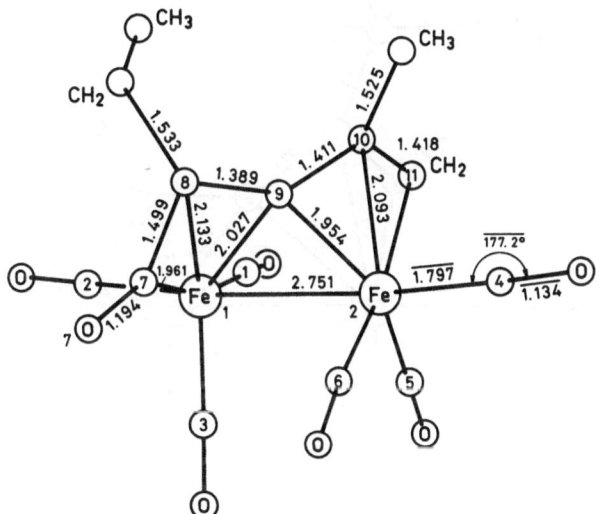

III   IV

$(CO)_3Fe(C_8H_{10}O)Fe(CO)_3$ (Formula I, $R=CH_3$, $R'=C_2H_5$) has been prepared from $Fe_2(CO)_9$ and $CH_2=C(CH_3)-C\equiv CC_2H_5$ (1:1 mole) in tetrahydrofuran at room temperature (3 h). In hexane the reaction requires 18 to 24 h. The separation by chromatography on SiO_2 with benzene and recrystallization from pentane at −40 °C yields dark red-orange crystals which have in their IR spectrum (hexane) $\nu(CO)$ bands at 1975, 1992, 2000, 2025, and 2060 cm^{-1} and a $\nu(C=O)$ band at 1785 cm^{-1}. The crystals belong to the monoclinic system with a = 9.354(3), b = 12.775(5), c = 13.840(5) Å, and β = 92.71(3)°; space group $P2_1/n-C_{2h}^5$. Z = 4 gives a D_c = 1.68 g·cm^{-3}. The molecular structure is

Fig. 23

Molecular structure of $(CO)_3Fe(C_8H_{10}O)Fe(CO)_3$ [6].

Selected bond angles (°):

C(7)-C(8)-C(9)	109.4(6)	Fe(2)-Fe(1)-C(2)	174.7(3)
C(8)-C(9)-C(10)	140.9(6)	Fe(2)-Fe(1)-C(3)	88.4(2)
C(9)-C(10)-C(11)	115.7(6)	C(7)-Fe(1)-C(1)	164.5(3)
Fe(1)-C(7)-O(7)	144.4(6)	C(7)-Fe(1)-C(2)	88.6(3)
Fe(1)-C(9)-C(8)	74.7(4)	C(7)-Fe(1)-C(3)	94.9(3)
Fe(1)-C(9)-C(10)	138.5(5)	Fe(1)-Fe(2)-C(4)	170.0(3)
Fe(1)-C(9)-Fe(2)	87.4(3)	Fe(1)-Fe(2)-C(5)	92.1(3)
Fe(2)-C(9)-C(8)	137.9(5)	Fe(1)-Fe(2)-C(6)	85.4(2)
Fe(2)-C(9)-C(10)	75.0(4)	C(11)-Fe(2)-C(4)	89.2(3)
Fe(2)-Fe(1)-C(9)	45.2(2)	C(11)-Fe(2)-C(5)	94.2(3)
Fe(1)-Fe(2)-C(9)	47.4(2)	C(11)-Fe(2)-C(6)	166.3(3)
Fe(2)-Fe(1)-C(1)	89.1(3)	C(8)-C(7)-O(7)	138.9(7)

References on p. 96

shown in **Fig. 23**, p. 91. The Fe atoms are bridged by a C_5 chain. The ligand part C(9)–C(10)–C(11) can be described as a π–allyl system bonded to Fe(2). However, there is considerable ambiguity how best to interpret the part C(7)–C(8)–C(9). Instead of a π–allyl system, the formulation as a σ,π–bonded C(7)–C(8)=C(9) unit appears to be a slightly better approximation. The mass spectrum shows no parent peak, but it shows the fragments [M–7CO] and [C_7H_{10}]$^+$ [6].

(CO)$_3$Fe(C$_{10}$H$_{14}$O)Fe(CO)$_3$ (Formula I, R=CH$_3$, R'=t-C$_4$H$_9$) has been isolated in small amounts from a reaction of Fe$_2$(CO)$_9$ with t-C$_4$H$_9$C≡CC$_4$H$_9$-t in hexane at 25 °C. The formation of the complex was due to a contamination of the alkyne with CH$_2$=C(CH$_3$)–C≡CC$_4$H$_9$-t [4, 6]. Thoroughly purified alkyne does not give the complex [6]. It is separated from the main product Fe$_2$(CO)$_6$(t-C$_4$H$_9$C$_2$C$_4$H$_9$-t) by chromatography on SiO$_2$. Recrystallization from pentane affords dark red–orange crystals [4, 6]. IR spectrum

Fig. 24

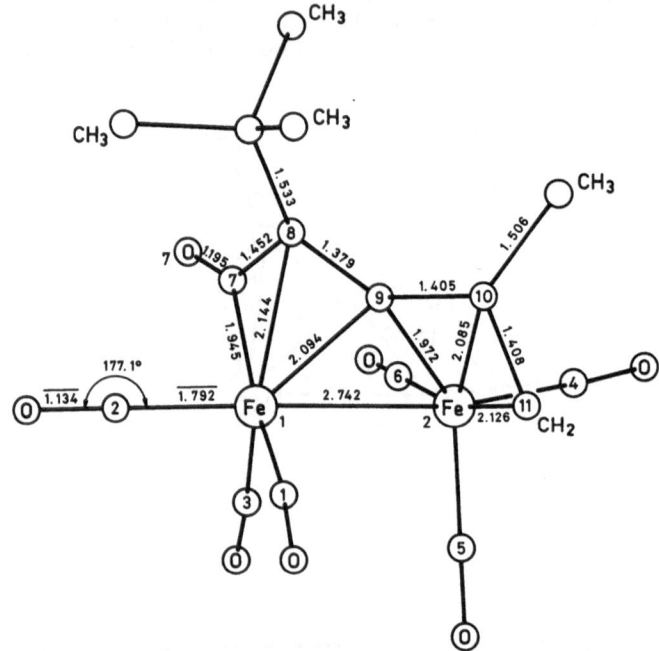

Molecular structure of (CO)$_3$Fe(C$_{10}$H$_{14}$O)Fe(CO)$_3$ [6].

Selected bond angles (°):

C(7)–C(8)–C(9)	109.1(3)	Fe(2)–Fe(1)–C(2)	174.7(3)
C(8)–C(9)–C(10)	143.1(4)	Fe(2)–Fe(1)–C(3)	82.7(1)
C(9)–C(10)–C(11)	115.9(4)	C(7)–Fe(1)–C(1)	159.7(2)
Fe(1)–C(7)–O(7)	144.3(3)	C(7)–Fe(1)–C(2)	87.4(2)
Fe(1)–C(9)–C(8)	74.6(2)	C(7)–Fe(1)–C(3)	99.5(2)
Fe(1)–C(9)–C(10)	136.6(3)	Fe(1)–Fe(2)–C(4)	169.1(1)
Fe(1)–C(9)–Fe(2)	86.0(1)	Fe(1)–Fe(2)–C(5)	93.7(1)
Fe(2)–C(9)–C(8)	137.7(3)	Fe(1)–Fe(2)–C(6)	85.4(2)
Fe(2)–C(9)–C(10)	74.2(2)	C(11)–Fe(2)–C(4)	92.3(2)
Fe(1)–Fe(2)–C(9)	48.2(1)	C(11)–Fe(2)–C(6)	166.3(2)
Fe(2)–Fe(1)–C(1)	93.7(1)	C(8)–C(7)–O(7)	138.3(4)

References on p. 96

(hexane): $\nu(CO)$ bands at 1780, 1975, 1990, 1985, 2025, and 2060 cm^{-1} [6]. The compound crystallizes in the monoclinic system with a = 10.025(3), b = 11.168(4), c = 16.482(7) Å, and β = 101.94(3)°; space group P2$_1$/n–C$_{2h}^5$. Z = 4 gives D$_c$ = 1.58 g·cm^{-3}. The molecular structure, shown in **Fig. 24**, and the bonding are the same as for the previous compound [6], see also [4].

The mass spectrum exhibits the same pattern of peaks as the previous complex [6].

(CO)$_3$Fe(C$_8$H$_{10}$O)Fe(CO)$_3$ (Formula II) has been found in about 0.7% yield among several products of the 3 h photolysis of 2,7–dimethyloxepine in the presence of Fe(CO)$_5$ in ether at –60 °C. The main products are o–xylene, 2,6–dimethylphenol, and C$_8$H$_{10}$O–Fe(CO)$_3$. The title complex is eluted from SiO$_2$ with benzene as a red oil, which crystallizes from pentane at –40 °C in red bunches, melting point 72 °C.

The assignment of the ^{1}H NMR spectrum (C$_6$H$_6$) refers to the numbering in **Fig. 25**: δ = 1.73 (COCH$_3$), 1.74 (H–12), 2.18 (CH$_3$ on C–8), 3.80 (H–11), 3.94 (H–9), and

Fig. 25

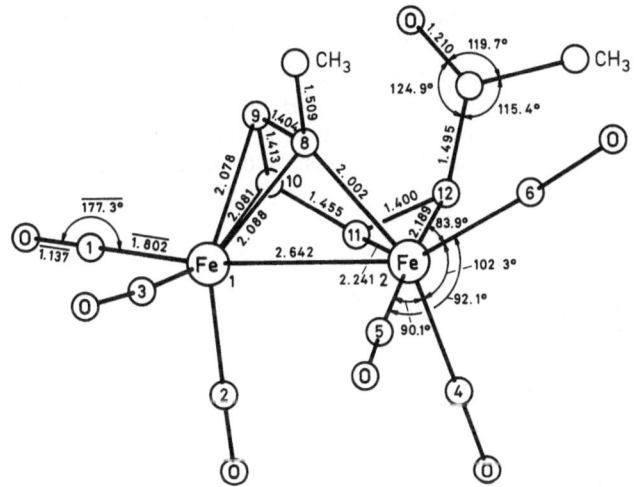

Molecular structure of (CO)$_3$Fe(C$_8$H$_{10}$O)Fe(CO)$_3$ [3].

Other selected bond angles (°):

C(1)–Fe(1)–C(2)	95.1(1)	C(4)–Fe(2)–Fe(1)	109.8(1)
C(1)–Fe(1)–C(3)	91.9(1)	C(5)–Fe(2)–Fe(1)	83.2(1)
C(2)–Fe(1)–C(3)	100.5(1)	C(6)–Fe(2)–Fe(1)	147.6(1)
C(1)–Fe(1)–Fe(2)	164.5(1)	C(8)–Fe(2)–Fe(1)	51.2(1)
C(2)–Fe(1)–Fe(2)	87.6(1)	C(11)–Fe(2)–Fe(1)	66.6(1)
C(3)–Fe(1)–Fe(2)	102.7(1)	C(12)–Fe(2)–Fe(1)	100.6(1)
C(1)–Fe(1)–C(8)	127.1(4)	C(11)–Fe(2)–C(12)	36.8(1)
C(1)–Fe(1)–C(9)	94.7(1)	Fe(2)–C(8)–C(9)	114.3(2)
C(1)–Fe(1)–C(10)	89.6(1)	C(8)–C(9)–C(10)	115.4(2)
C(8)–Fe(1)–Fe(2)	48.4(1)	C(9)–C(10)–C(11)	117.8(2)
C(9)–Fe(1)–Fe(2)	74.1(1)	C(10)–C(11)–Fe(2)	101.6(2)
C(10)–Fe(1)–Fe(2)	74.9(1)	C(10)–C(11)–C(12)	124.3(2)
Fe(1)–C(8)–Fe(2)	80.5(1)	C(11)–C(12)–Fe(2)	73.6(1)
Fe(1)–C(10)–C(11)	98.1(2)	C(12)–C(11)–Fe(2)	69.6(1)

References on p. 96

4.11 (H–10) ppm. Coupling constants J(H,H): J(9,10) =3.02, J(10,11) =3.25, J(11,12) = 8.34, and J(9,CH$_3$) =0.35 Hz. In the IR spectrum v(CO) bands occur at 1980, 1985, 1992, 1997, 2025, and 2065 cm^{-1}, v(C=O) at 1690 cm^{-1}.

The triclinic crystals have the parameters a=7.5135(5), b=8.5100(6), c= 12.8393(8) Å, α=72.132(6)°, β=85.877(6)°, and γ=83.514(8)°; space group P$\bar{1}$ –C$_i^1$. Z=2 gives D$_c$=1.72 g·cm^{-3}. Some bond lengths and angles are shown in Fig. 25. The π–allyl tail of the bridging ligand is symmetrically bonded to Fe(1), all Fe(1)–C–C angles being very close to the mean value of 70.2°, C(9)–Fe(1)–C(8,10) gives an average of 39.6°. C(8) also acts as a bridging carbon atom for the two Fe atoms. The double bond C(11)–C(12) is typically elongated by complexation to Fe(2).

The molecular ion [M]$^+$ occurs in the mass spectrum. Among the fragments [M –n CO]$^+$ with n=1 to 6, the [Fe$_2$(C$_8$H$_{10}$O]$^+$ ion is the most intense [3].

(CO)$_3$Fe(C$_7$H$_8$)Fe(CO)$_3$ (Formula III, R=H) is formed from compound VI and π– C$_6$H$_{10}$Fe(CO)$_4$ (C$_6$H$_{10}$=cyclohexene) in cyclohexene between –80 °C and room tempera- ture, 22% yield after purification on SiO$_2$ with petroleum ether [5, 7]. The reaction of Fe$_2$(CO)$_9$ with the hydrocarbon VII (R=H) in a mole ratio 2.5:1 in ether at 20 °C for 20 h gives the complex III in 4% yield together with the main products VIII (25%) and V (16%). Somewhat higher yields (12 to 14%) are obtained when the reaction is carried out in the presence of HCl/CH$_3$OH or CH$_3$ONa/CH$_3$OH. The formation of the title compound III is discussed in terms of a model reaction of a olefin metathesis going through the steps VIII and IX and possibly through an intermediate carbene com- plex X [5, 7].

(CO)$_3$Fe(C$_7$H$_7$D)Fe(CO)$_3$ (Formula III, R=D) has been prepared in an analogous way from Fe$_2$(CO)$_9$ and the 6–endo–D$_1$ form of VII (R=D) [7].

(CO)$_3$Fe(C$_7$H$_8$)Fe(CO)$_3$ forms red prisms from pentane at –70 °C. It decomposes slowly above 98 °C. The ^{1}H NMR spectrum (C$_6$H$_6$) of the compound and its monodeutero derivative are shown in [7] as diagrams: δ=1.05 (H–6, endo), 1.31 (H–6, exo), ≈2.2 (H–1,2), 2.70 (H–5), 3.10 (H–3), 3.89 (H–4), and 5.64 (H–7) ppm. Coupling constants were estimated on the basis of double resonance experiments: J(H,H) = –15 (endo, exo) 8 (1,7), 6 (4,5), 5 (3,4), 4 (2,3; 1, exo; 5, exo), 1.5 (3,5), and 1 (2,4) Hz. The

^{13}C NMR spectrum ($CDCl_3$) shows the presence of six CH and one CH_2 carbons: δ=30.0, 39.7 (C-6), 48.1, 64.7, 73.6, and 106.9; carbonyl peaks at 212.2 and 214.4 ppm. IR spectrum (hexane): ν(CO) bands at 1971, 1991 to 1992, 2016, and 2058 cm^{-1} [7].

The complex crystallizes in the orthorhombic system with a=19.273(3), b=13.858(1), and c=10.2347(7) Å; space group Pbca$-D_{2h}^{15}$, Z=8. Bond lengths and angles of the molecular structure are shown in **Fig. 26** [5].

Fig. 26

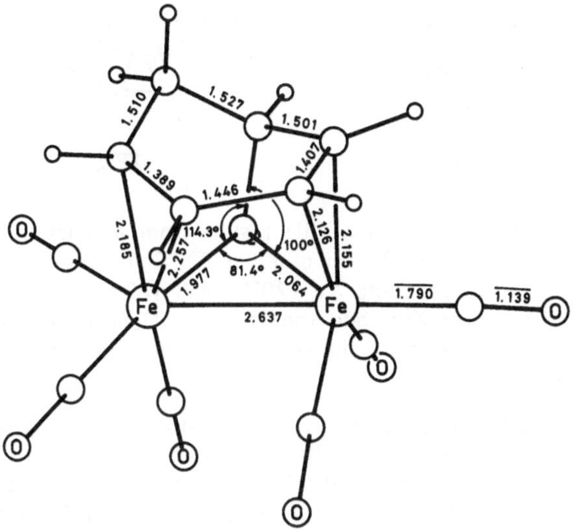

Molecular structure of $(CO)_3Fe(C_7H_8)Fe(CO)_3$ [5]. H atoms of the ^5L ligand are also shown.

The molecular ion $[M]^+$ and the fragments $[M-n\,CO]^+$ with n=1 to 6 occur in the mass spectrum. When the complex is dissolved in CF_3COOD at 25 °C, almost complete H/D exchange on the carbene carbon atom (7, Formula III) is observed within 24 h. Carbonylation leads to complex V, although in small yields. Carbonylation in benzene in the presence of methanol under 70 atm CO at 80 °C for 20 h gives the complex XI in 90% yield [7].

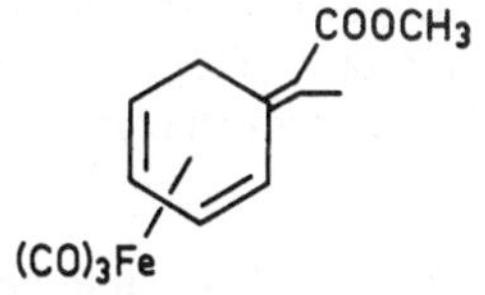

XI

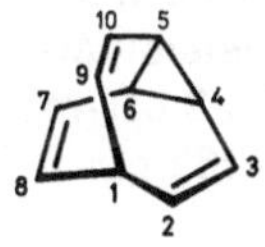

XII

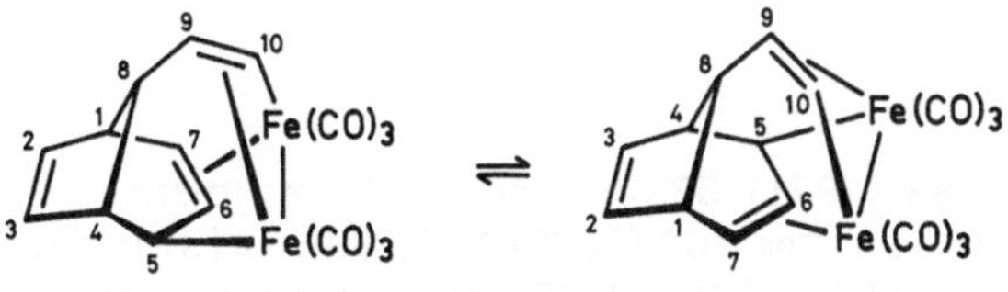

a b

XIII

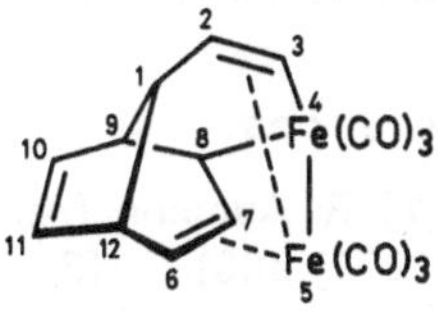

XIV

References on p. 96

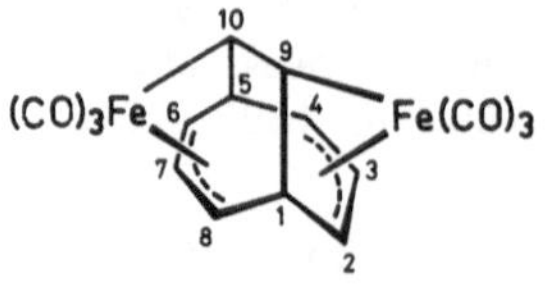

XV

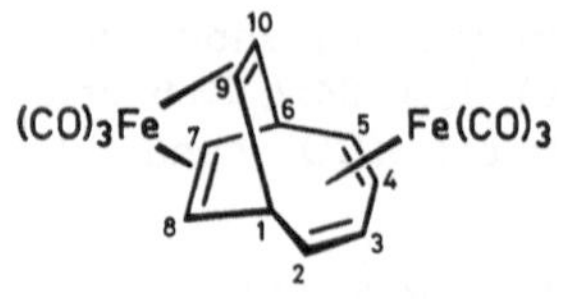

XVI

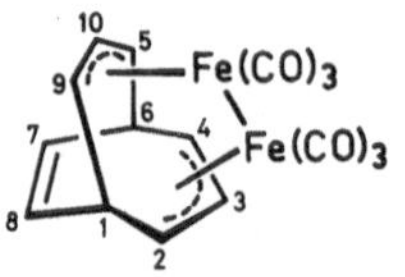

XVII

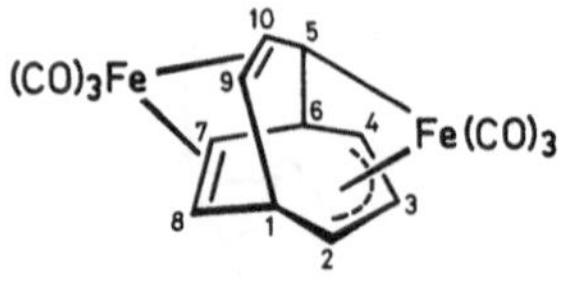

XVIII

$(CO)_3Fe(C_{10}H_{10})Fe(CO)_3$ (Formula XIII) results from a reaction of bullvalene (XII) with an fourfold excess of $Fe_2(CO)_9$ in ether at 30 °C for about 12 h, 3 to 10% yield [1, 2] along with the main products XV (20%), XVI (26%), and XVIII (39%) as well as small amounts of complex XVII and the ferrole $C_4H_4Fe_2(CO)_6$. The components of the reaction mixture are separated by chromatography on SiO_2 with petroleum ether/benzene (10:1) as eluents [2].

The title compound forms red octahedra from methanol, melting point 89 °C. Instead of XIII the valence bond formulation XIV has been preferred in [1] to account for the nonequivalence of the Fe atoms in the ^{57}Fe Mössbauer spectrum, $\delta=0.13$ and 0.24, $\Delta=0.90$ and 1.14 mm·s^{-1} [1]. Temperature dependent ^1H NMR spectra are shown as diagrams for the region from −6 to +86 °C. Due to a rapid interconversion of the enantiomeric forms XIIIa and XIIIb coalescence occurs for the proton pairs H-2,3, H-1,4, and H-7,5. The signals of the protons 6, 8, 9, and 10 remain unchanged because in the enantiomers they differ only in their chirality [1, 2]. ^1H NMR spectrum (C_6D_6): $\delta=1.18$ (H-1), 1.58 (H-8), 2.08 (H-5), 2.64 (H-6), 2.98 (H-9), 2.99 (H-7), 3.27 (H-4), 5.32 (H-3), 5.42 (H-2), and 7.30 (H-10); δ values measured in CS_2 are appreciably higher. Double resonance experiments gave the following coupling constants J(H,H): 9.5 (H-9,10), 7.5 (H-5,6), 7.0 (H-8,9), 6.5 (H-6,7), 5.9 (H-2,3), 4.5 (H-1,8 and H-4,8), and 2.5 (H-1,2, H-3,4, H-4,5, and H-1,7) Hz. Remote coupling of 1.3 to 1.5 Hz occurs between H-1,6, H-4,6, H-7,8, and H-5,8. Quantitative measurements on the proton pair H-1,4 between −12 and +13 °C gave for the valence isomerization XIIIa ⇌ XIIIb the parameters $\Delta G^*=16.0\pm0.2$ kcal·mol^{-1} [1, 2], $\Delta H^*=10.4\pm1$ kcal·mol^{-1}, and $\Delta S^*=-18.5\pm3$ cal·mol^{-1}·K^{-1} [1]. IR spectrum (hexane): $\nu(CO)$ at 1980, 1994, 1997, 2024, and 2062 cm^{-1} [2].

The mass spectrum shows the molecular ion $[M]^+$ and the fragments $[M-nCO]^+$ with n=1 to 6. The compound is soluble in hexane, benzene, and chloroform, and slightly soluble in cold methanol [2].

References:

[1] R. Aumann (Angew. Chem. **83** [1971] 583/4; Angew. Chem. Intern. Ed. Engl. **10** [1971] 560). − [2] R. Aumann (Chem. Ber. **108** [1975] 1974/88). − [3] R. Aumann, H. Averbeck, C. Krüger (Chem. Ber. **108** [1975] 3336/48). − [4] F.A. Cotton, J.D. Jamerson, B.R. Stults (J. Organometal. Chem. **94** [1975] C53/C55). − [5] R. Aumann,

H. Wörmann, C. Krüger (Angew. Chem. **88** [1976] 640/1; Angew. Chem. Intern. Ed. Engl. **15** [1976] 609).

[6] F.A. Cotton, J.D. Jamerson, B.R. Stults (Inorg. Chim. Acta **17** [1976] 235/42). — [7] R. Aumann, H. Wörmann, C. Krüger (Chem. Ber. **110** [1977] 1442/61).

2.5.2 Compounds with Two ^5L Ligands

2.5.2.1 Compounds without Carbonyl Groups

2.5.2.1.1 Compounds with Nitrosyl Bridging Groups

$C_5H_5Fe(\mu\text{-}NO)_2FeC_5H_5$ is prepared by introducing NO into a solution of $(C_5H_5Fe(CO)_2)_2$ in octane at 120 °C for 3 h. Black crystals are formed. They are extracted into benzene and, after concentration, precipitated with hexane, 74% yield. Toluene or xylene are less suitable as reaction media (maximum yields 55%) as the complex is more soluble in these solvents and further reacts with NO forming red–brown, insoluble by–products of unknown composition [1].

The compound forms dark green to black crystals which on heating decompose without melting. It sublimes in high vacuum with partial decomposition. The compound is diamagnetic. Unpaired electrons could not be detected by ESR, neither in a solution nor in a solid obtained from the gas phase by quenching at −196 °C. The ^1H NMR spectrum (C_6D_6) has one singlet at $\delta = 4.47$ ppm. The ^{57}Fe Mössbauer spectrum shows one quadrupole doublet. In the IR spectrum (KBr) one broad, intense $\nu(NO)$ band is split into peaks at 1490, 1498, 1506, and 1514 cm^{-1}. Bands of the C_5H_5 ligands are at 822, 839, 856, 1006, 1112, 1418, and 3096 cm^{-1}. UV spectrum (CH_2Cl_2): $\lambda_{max}(\varepsilon) =$ 249 (200000), 298 (35000), 352 (30000), 459 (5000), 580 (4500), and 1007 (750) nm [1].

The compound crystallizes in the monoclinic system with $a = 7.8257(9)$, $b = 5.9998(9)$, $c = 11.9875(13)$ Å, and $\beta = 105.548(9)°$; space group $P2_1/c - C_{2h}^5$. $Z = 2$ gives $D_c = 1.848$ g·cm^{-3}, $D_m = 1.85$ g·cm^{-3}. The molecular structure is shown in **Fig. 27**. The planar Fe_2N_2 unit stands perpendicular to the planes of the C_5H_5 rings. The rather short Fe–Fe distance [6] supports the earlier view of a double bond between the metal atoms [1].

Fig. 27

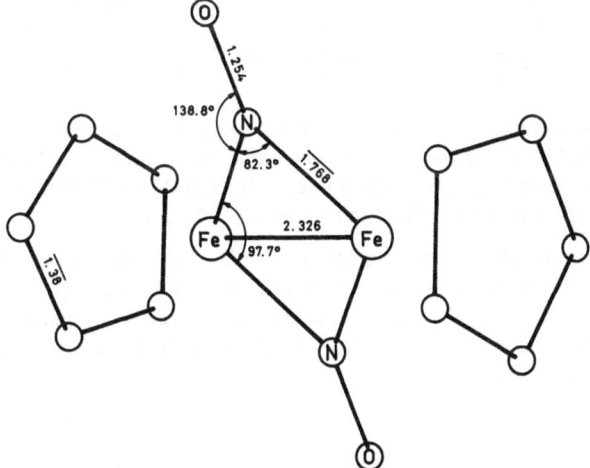

Molecular structure of $C_5H_5Fe(\mu\text{-}NO)_2FeC_5H_5$ [6].

The most intense peaks (relative intensity) in the mass spectrum are the molecular ion (100), $[(C_5H_5)_2Fe]^+$ (150), $[C_5H_5Fe_2O]^+$ (143), $[C_5H_5Fe]^+$ (98), and $[(C_5H_5)_2Fe_2NO]^+$ (79) [1], see also [3]. A metastable peak at m/e=137 has been attributed to the process $[(C_5H_5)_2Fe_2NO]^+ \rightarrow [C_5H_5Fe_2O]^+ + C_5H_5N$, in which the eliminated radicals probably combine to an azene and then rearrange to pyridine. The driving force of this unusual N$^\cdot$ elimination may be the positive charge on the Fe atoms, which tends to fix the more electronegative O atom of the NO ligand [3].

The complex is only slightly soluble in organic solvents. These solutions appear red in transmitted light. Chromatography in benzene is possible on oxygen-free SiO_2, but decomposition occurs immediately on admission of air. The solid discolors slowly to brown in contact with air. The thermal decomposition gives an iron mirror and small amounts of ferrocene [1]. $C_5H_5Fe(NO)I$ is formed in the reaction with an equimolar quantity of I_2 in CH_2Cl_2. The treatment with S in CH_2Cl_2 gives brown solutions which contain four compounds of unknown composition. Na/Hg in tetrahydrofuran reduces the complex to red-brown $Na_2[(C_5H_5FeNO)_2]$, which decomposed on attempted isolation. The reduction with $NaBH_4$ proceeds slowly giving metallic iron [2]. The treatment with $LiAlH_4/AlCl_3$ yields exclusively ferrocene [5]. A brown solution and a red-brown insoluble solid result from the further reaction with NO under the conditions of the preparation. These products do not contain terminal NO groups [1]. The treatment with CO at 30 atm in benzene or methylene chloride solutions at room temperature converts the complex to some extent into $(C_5H_5Fe(CO)_2)_2$. At 250 atm CO pressure and 140 °C the complex is completely reacted, giving a 60% yield of $(C_5H_5Fe(CO)_2)_2$. $P(C_6H_5)_3$ in boiling xylene replaces the C_5H_5 ligands and gives $Fe(NO)_2(P(C_6H_5)_3)_2$ in 78% yield [2]. Reactions with $Na_2S_2C_2(CN)_2/[P(C_6H_5)_4]Br$ and with $NaS_2CN(CH_3)_2$ in n-octane/methanol give $[P(C_6H_5)_4]_2[Fe(NO)(S_2C_2(CN)_2)_2]$ (67%) and $Fe(NO)(S_2CN(CH_3)_2)_2$ (60% yield), respectively [4].

$C_5H_5(CH_3)Fe(\mu\text{-NO})_2Fe(CH_3)C_5H_5$ has been prepared from the previous complex by reduction to $[C_5H_5FeNO]^{2-}$ with Na/Hg in tetrahydrofuran at room temperature and further reaction with CH_3I at 0 °C for a few minutes. The product can be purified by chromatography on SiO_2 in CH_2Cl_2, 37% yield.

The bright brown scale-like crystals decompose on heating without melting. They can be sublimed in high vacuum at 65 °C without decomposition. The ^1H NMR spectrum (C_6D_6) shows two singlets at $\delta = -0.04$ (CH_3) and 4.41 (C_5H_5) ppm. IR spectrum (medium not reported): $\nu(NO)$ bands at 1504 and 1572 cm^{-1}, other bands at 1000, 1114, 1414, 3067 (C_5H_5), 1157, and 2876 (CH_3) cm^{-1}.

Cryoscopic determinations of the molecular weight in benzene reveal the dimeric structure. The complex is diamagnetic in the solid state and in solution. However, the gas phase contains paramagnetic monomers $C_5H_5(CH_3)FeNO^\cdot$, which can be detected by ESR spectroscopy after quenching the vapor to -196 °C.

The mass spectrum shows no molecular ion but only the peak of the monomeric species (relative intensity 12). Other fragments are $[C_5H_5Fe]^+$ (100), $[C_5H_5Fe(NO)CH]^+$ (41.5), $[Fe]^+$ (41), $[C_5H_5FeCH]^+$ (32), $[C_5H_5FeNO]^+$ (23), $[C_3H_3Fe]^+$ (11), and $[C_2HFe]^+$ (6). Other metastable peaks are given. The complex is extremely air-sensitive [2].

References:

[1] H. Brunner (J. Organometal. Chem. **14** [1968] 173/8). – [2] H. Brunner, H. Wachsmann (J. Organometal. Chem. **15** [1968] 409/21). – [3] J. Müller (J. Organometal.

Chem. **23** [1970] C38/C40). – [4] J.A. McCleverty, T.A. James (J. Chem. Soc. A **1970** 3318/23). – [5] J. Müller, H. Dorner, G. Huttner, H. Lorenz (Angew. Chem. **85** [1973] 1117/8; Angew. Chem. Intern. Ed. Engl. **12** [1973] 1005/6).

[6] J.L. Calderon, S. Fontana, E. Frauendorfer, V.W. Day, S.D.A. Iske (J. Organometal. Chem. **64** [1974] C16/C18).

2.5.2.1.2 Compounds of the $[C_5H_5Fe(\mu\text{-}S_2)(\mu\text{-}SR)_2FeC_5H_5]^{n+}$ Type with $n=0$ to 2

The compounds in this section contain one S_2 and two SR bridging groups as shown by Formula I.

$$\left[C_5H_5Fe \overset{\displaystyle S-S}{\underset{\displaystyle S}{\overset{R}{\diagdown}} \overset{}{\underset{R}{\diagup}} } FeC_5H_5 \right]^{n+}$$

I

The neutral compounds are best prepared from $C_5H_5(CO)Fe(\mu\text{-}SR)_2Fe(CO)C_5H_5$ ($R=CH_3$, C_2H_5, and $CH_2C_6H_5$), and excess sulphur in methylcyclohexane at reflux for 3 h or overnight for $R=CH_2C_6H_5$. The crude products are purified by chromatography on Al_2O_3, about 10% yield [5]. This method was found to be preferable to the original synthesis (for $R=C_2H_5$) from $(C_5H_5Fe(CO)_2)_2$ and the polysulfide $(C_2H_5)_2S_x$ ($x=3$, 4) in boiling methylcyclohexane [1, 5], [6, p. 264].

The electrochemical oxidation gives the cations I with $n=1$ and 2. The half-wave potentials obtained by cyclic voltammetry in CH_3CN increase with bulkiness of the groups R ($E_{1/2}$ in V referred to the saturated calomel electrode at 25 °C):

R	$E^1_{1/2}$ reversible	$E^2_{1/2}$ quasi-reversible
CH_3	0.19	0.79
C_2H_5	0.21	0.90
$CH_2C_6H_5$	0.28	0.99

Salts of the monocation I ($n=1$, $R=C_2H_5$) could be isolated. The products of the second oxidation step are less stable and undergo chemical reaction soon after their production [5].

$C_5H_5Fe(\mu\text{-}S_2)(\mu\text{-}SCH_3)_2FeC_5H_5$ and $C_5H_5Fe(\mu\text{-}S_2)(\mu\text{-}SCH_2C_6H_5)_2FeC_5H_5$ have been mentioned in [5, 6]. Data other than their oxidation potentials are not available [5].

$C_5H_5Fe(\mu\text{-}S_2)(\mu\text{-}SC_2H_5)_2FeC_5H_5$ is regenerated from the reaction product of the second oxidation, $[C_5H_5(CH_3CN)Fe(\mu\text{-}SC_2H_5)_2Fe(NCCH_3)C_5H_5]^{2+}$, on treatment with aqueous Na_2S [6, p. 268].

The compound is dark green [1]. It is essentially diamagnetic at room temperature [1] and remains diamagnetic down to 4 K [4, 5]. A slight residual temperature independent paramagnetism was observed [5]. The ^1H NMR spectrum shows one sharp C_5H_5 resonance at $\delta=4.73$ ppm [1]. The substance crystallizes in the monoclinic system with $a=$ 17.374 ± 0.003, $b=8.125\pm0.001$, $c=12.782\pm0.002$ Å, and $\beta=108.37\pm0.03°$; space

Fig. 28

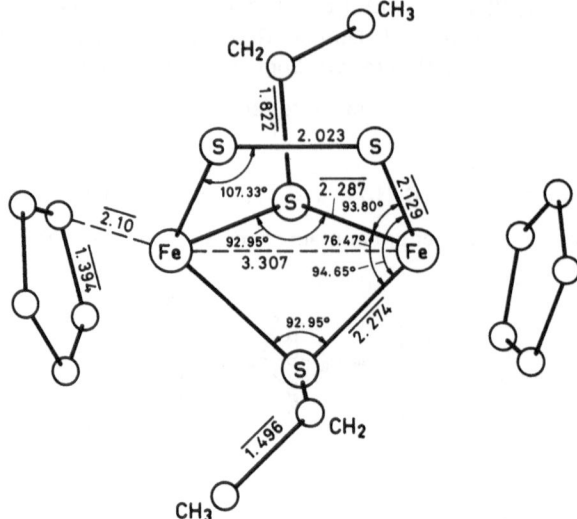

Molecular structure of $C_5H_5Fe(\mu-S_2)(\mu-SC_2H_5)_2FeC_5H_5$ [1, 2].

group $P2_1/c - C_{2h}^5$. $Z=4$ gives $D_c=1.661$ g·cm^{-3}, $D_m=1.63$ g·cm^{-3}. The molecular structure is shown in **Fig. 28**. There is a planar Fe-S-S-Fe system. The long Fe···Fe distance
does not permit bonding between the Fe atoms. The S-S distance (2.02 Å) is shorter
than expected for a cis disulfide (2.10 Å) but longer than in double bonded S_2^0 (1.889 Å).
This suggests a π bond order of about one-third. Other distances and angles are unexcep-
tional [1, 2]. A qualitative molecular orbital energy scheme is shown and discussed
in [4, 6]. It is consistent with an equivalent valence bond model in which the iron
atoms are formally Fe^{2+} and the S-S bridge is formulated as S_2^0. This picture involves
a sulfur to iron $p\pi-d\pi$ interaction that increases the Fe-S bond order [6]. The magnetic
measurements did not provide any evidence for the previous assumption [1] of a strong
magnetic coupling between iron centers and did not support the valence bond picture
of electron delocalization through several resonance forms [4].

The compound is stable towards air and X-rays [2]. The electrochemical reactions
observed by cyclic voltammetry in CH_3CN at 25 °C may be summarized as follows
(potentials in V referred to SCE):

$(C_5H_5Fe)_2(\mu-S_2)(\mu-SC_2H_5)_2$ $\xrightarrow{-1.2}$ reduction products (irreversible)

$\updownarrow +0.21$

$[(C_5H_5Fe)_2(\mu-S_2)(\mu-SC_2H_5)_2]^+$ $\xrightleftharpoons{+0.90}$ $[(C_5H_5Fe)_2(\mu-S_2)(\mu-SC_2H_5)_2]^{2+}$

$\downarrow \begin{matrix} + CH_3CN \\ - S_2 \end{matrix}$

oxidation products (irreversible) $\xleftarrow{+1.6}$ $[(C_5H_5(CH_3CN)Fe)_2(\mu-SC_2H_5)_2]^{2+}$

Voltammograms displaying the first two oxidation waves were also obtained in solutions
of CH_2Cl_2 and $HCON(CH_3)_2$ [5]. The acetonitrile solutions of the dication can be formed
by oxidation with $NOPF_6$ or air in CH_3CN [3, 4].

$[C_5H_5Fe(\mu-S_2)(\mu-SC_2H_5)_2FeC_5H_5]PF_6$ and $[C_5H_5Fe(\mu-S_2)(\mu-SC_2H_5)_2FeC_5H_5]SbF_6$ have been prepared from the parent neutral compound by electrolysis in CH_3CN or $HCON(CH_3)_2$ in the presence of NH_4PF_6 or NH_4SbF_6 (0.1 M) between 0.1 and 0.4 V and a maximum current of 200 mA. On concentration of the deep emerald-green solutions the salts crystallize giving microcrystalline black products (54% yield for the SbF_6^- salt). They can be recrystallized from dry acetonitrile/toluene [5].

Fig. 29

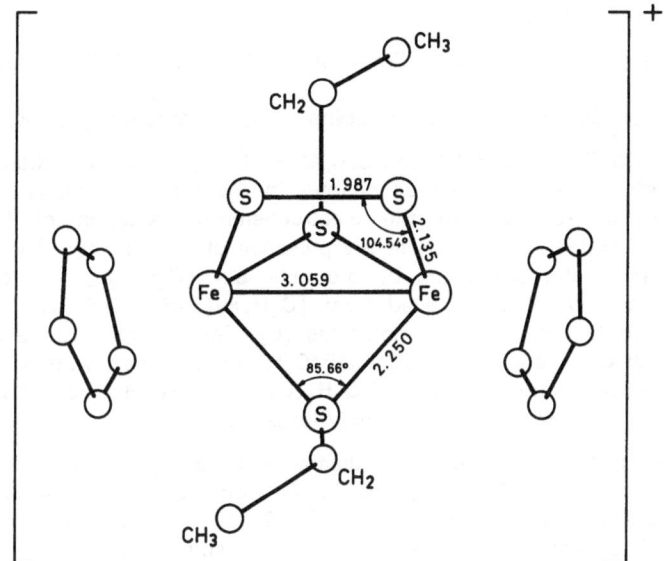

Molecular structure of $[C_5H_5Fe(\mu-S_2)(\mu-SC_2H_5)_2FeC_5H_5]^+$ in its SbF_6^- salt [4].

The salts are paramagnetic. Between 4 and 190 K the SbF_6^- salt obeys the Curie-Weiss law. The magnetic moment of $\mu=1.70+0.05$ B.M. corresponds to one unpaired electron. Frozen solutions in $HCON(CH_3)_2$ or CH_3CN at 103 K show an anisotropic ESR spectrum with $g_1=2.135$, $g_2=1.976$, and $g_3=1.934$. No hyperfine splitting was observable for a sample labeled with ^{33}S in the S_2 bridge. It had been assumed that the bridge was a possible site of a superexchange mechanism between the iron centers [5]. This indicates that the unpaired electron is in a molecular orbital which has essentially metal character [4, 5], probably a σ^* orbital [4].

Both salts crystallize in the monoclinic system, space group $P2_1/m-C_{2h}^2$.

PF_6^- salt, a=6.711(3), b=19.57(9), c=7.87(4) Å, and $\beta=91.2(4)°$, $D_c=1.75$ and $D_m=1.8$ g·cm^{-3}.

SbF_6^- salt, a=6.777(3), b=19.849(7), c=7.909(3) Å, and $\beta=91.56(3)°$. The structure of the cation is shown in **Fig. 29.**

The iron-sulfur core is similar to that in the neutral parent compound. The most salient difference is the decrease in the Fe-Fe distance to a value that is well within the range expected for a one-electron iron-iron interaction. The S-S and Fe-S bonding distances are practically unaffected by removal of an electron [4].

The solids are air-stable. In polar solvents they are rapidly oxidized by air. For example, in CH_3CN oxidation gives $[C_5H_5(CH_3CN)Fe(\mu-SC_2H_5)_2Fe(NCCH_3)C_5H_5]^{2+}$ [4, 5]. Solutions in organic solvents containing significant amounts of water slowly change to a different shade of green [5].

References on p. 102

References:

[1] G.J. Kubas, T.G. Spiro, A. Terzis (J. Am. Chem. Soc. **95** [1973] 273/4). –
[2] A. Terzis, R. Rivest (Inorg. Chem. **12** [1973] 2132/6). – [3] P.J. Vergamini, G.J.
Kubas, R.R. Ryan (Abstr. Papers 165th Natl. Meeting Am. Chem. Soc., Dallas, Tex., 1973,
INOR 85). – [4] P.J. Vergamini, R.R. Ryan, G.J. Kubas (J. Am. Chem. Soc. **98** [1976]
1980/2). – [5] G.J. Kubas, P.J. Vergamini, M.P. Eastman, K.B. Prater (J. Organometal.
Chem. **117** [1976] 71/9).

[6] P.J. Vergamini, G.J. Kubas (Progr. Inorg. Chem. **21** [1976] 261/82).

2.5.2.1.3 Other Compounds with Bridging Groups Containing S, N, and P Atoms

$[C_5H_5(CH_3CN)Fe(\mu-SC_2H_5)_2Fe(NCCH_3)C_5H_5][PF_6]_2$. The formation of this compound is mentioned in the previous chapter in the reaction scheme on p. 100. The
preparative method consists of the electrochemical oxidation of $[C_5H_5Fe(\mu-S_2)$-
$(\mu-SC_2H_5)_2FeC_5H_5]PF_6$ in CH_3CN in the presence of NH_4PF_6 at 1.2 V (vs. SCE), 85%
yield. Sulfur and sulfur–containing by–products, e.g. $(C_2H_5)_2S_4$, are formed from the
S_2 bridging group which is expelled from $[C_5H_5Fe(\mu-S_2)(\mu-SC_2H_5)_2FeC_5H_5]^{2+}$ and
replaced by two terminal CH_3CN molecules [6]. The complex can also be prepared
from the neutral $C_5H_5Fe(\mu-S_2)(\mu-SC_2H_5)FeC_5H_5$ by oxidation with $NOPF_6$ in CH_3CN
[3, 7]. Another way is the oxidation of $C_5H_5(CO)Fe(\mu-SC_2H_5)_2Fe(CO)C_5H_5$ with NO^+
or air in CH_3CN to a product, which slowly loses CO [7].

The black crystalline compound is diamagnetic [6, 7]. The molecular structure is
analogous to that given in **Fig. 30**. The Fe–Fe bond length is 2.649(7) Å; the angle
S–Fe–S is 105.2(3)° and the angle Fe–S–Fe is 73.9(3)° [7], see also [3]. The compound
is air-stable. The treatment with aqueous sodium sulfide regenerates $C_5H_5Fe(\mu-S_2)$-
$(\mu-SC_2H_5)FeC_5H_5$ [7]. Reactions with $X=CN^-$ and SCN^- give the neutral complex
type $C_5H_5(X)Fe(\mu-SC_2H_5)_2Fe(X)C_5H_5$ in yields higher than 50% [7].

Fig. 30

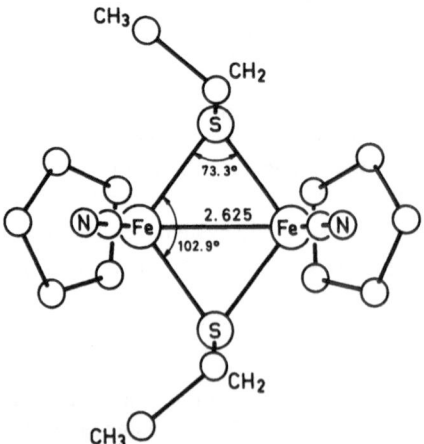

Molecular structure of $C_5H_5(CN)Fe(\mu-SC_2H_5)_2Fe(CN)C_5H_5$ [7].

$C_5H_5(CN)Fe(\mu-SC_2H_5)_2Fe(CN)C_5H_5$ and $C_5H_5(SCN)Fe(\mu-SC_2H_5)_2Fe(SCN)C_5H_5$.
The molecular structure of the CN derivative is given in [7], see Fig. 30. Other data
are not available.

I   II

[C₅H₅(²D-²D)Fe(μ-N₂)Fe(²D-²D)C₅H₅][BF₄]₂·2H₂O (²D-²D = (CH₃)₂PCH₂CH₂-P(CH₃)₂, proposed structure I) has been obtained by the reaction of C₅H₅Fe(²D-²D)I with TIBF₄ in acetone under N₂ at 0 °C or by introducing N₂ into a solution of [C₅H₅-Fe(²D-²D)CH₃COCH₃]BF₄ in acetone at 0 °C.

The orange crystalline substance is a 2:1 electrolyte in acetone. The intense Raman line of the solid at 2054 cm⁻¹ is assigned to the v(N=N) vibration. A rapid reaction with CO gives [C₅H₅Fe(²D-²D)CO]BF₄ in high yield. No NH₃ was detected in the products when the substance was converted to C₅H₅Fe(²D-²D)H with LiAlH₄ in tetrahydrofuran [1].

[C₅H₅(²D-²D)Fe(μ-N₂)Fe(²D-²D)C₅H₅][PF₆]₂ (²D-²D = (C₆H₅)₂PCH₂CH₂P(C₆H₅)₂) has been prepared by photolysis of [C₅H₅Fe(²D-²D)CO]PF₆ in acetone at −30 °C in a rapid stream of N₂, removal of the solvent, and further reaction of the intermediate [C₅H₅Fe(²D-²D)CH₃COCH₃]⁺ with N₂ in tetrahydrofuran, 75% yield.

The orange powder decomposes at 185 to 195 °C. ¹H NMR resonances are found (in acetone) at δ=2.32 and 2.53 (CH₂), 4.50 (t, C₅H₅), and 7.70 (m, C₆H₅) ppm. A Raman line of the solid in KBr at 2040 cm⁻¹ indicates the centrosymmetric structure I. The compound is air-stable at room temperature. With NaBH₄ in tetrahydrofuran it gives C₅H₅Fe(²D-²D)H. The complex is only soluble in acetone, but above −20 °C it reacts to give [C₆H₅Fe(²D-²D)CH₃COCH₃]⁺. With HCl, HBr, Zn/HCl, and Na₂S₂O₄ decomposition products are formed which do not absorb CO [5, 8].

[C₅H₅(P(C₆H₅)₂H)Fe(μ-(C₆H₅)₂PP(C₆H₅)₂)₂Fe(P(C₆H₅)₂H)C₅H₅][PF₆]₂ (proposed structure II) has been obtained by UV irradiation of [C₅H₅Fe(CO)₂P(C₆H₅)₂H]PF₆ in tetrahydrofuran for 4 h, chromatography on Al₂O₃ with acetone, repeated precipitation from ethanol with water in the presence of NH₄PF₆, and slow crystallization from methanol/ether at −20 °C, 19% yield.

The diamagnetic yellow crystals melt at 183 to 184 °C. ¹H NMR spectrum (CD₃COCD₃): δ=4.93 (q, C₅H₅, J(P,H)=1.7 Hz), 6.56 (d, PH, J(P,H)=369 Hz), and 7.65 (m, C₆H₅) ppm. IR spectrum (Nujol/Fluorolube): v(PH) band at 2320 cm⁻¹, other bands between 690 and 1588 cm⁻¹ reported without assignment. The anion shows a strong band at 835 cm⁻¹.

C₅H₅Fe(μ-PF₂)(μ-PF₂N(CH₃)PF₂)(μ-N(CH₃)=PF₂)FeC₅H₅ (Formula III) is formed together with complex IV by UV irradiation of (C₅H₅Fe(CO)₂)₂ and CH₃N(PF₂)₂ in tetrahydrofuran or pentane. The two products were separated on SiO₂ with mixtures of dichloromethane and hexane as eluents. The total yield is less than 10%. Compound V may be an intermediate in this photochemical reaction.

III   IV   V

The red crystals (from dichloromethane/hexane) melt at 162 to 164 °C. ^{1}H NMR spectrum (probably CH_2Cl_2): $\delta = 2.43$ (dd, CH_3, $J_1(P,H) = 18$ Hz, $J_2(P,H) = 4$ Hz), 3.05 (t, CH_3, $J(P,H) = 7$ Hz), 4.50, and 4.83 (C_5H_5) ppm. ^{13}C NMR (probably CH_2Cl_2): $\delta = 27.4$, 37.0 (CH_3, weak broad signals), 79.2, and 81.6 (C_5H_5) ppm. The complex crystallizes in the monoclinic system with a = 9.903(2), b = 13.789(3), c = 18.167(2) Å, and $\beta = 128.72(1)°$; space group $P2_1/c - C_{2h}^5$, Z = 4. The molecular structure is shown in **Fig. 31**. The iron–iron distance (3.646 Å) is clearly nonbonding. The P–N distance in the bridging $N(CH_3)PF_2$ ligand suggests a phosphorous–nitrogen double bond [9].

Fig. 31

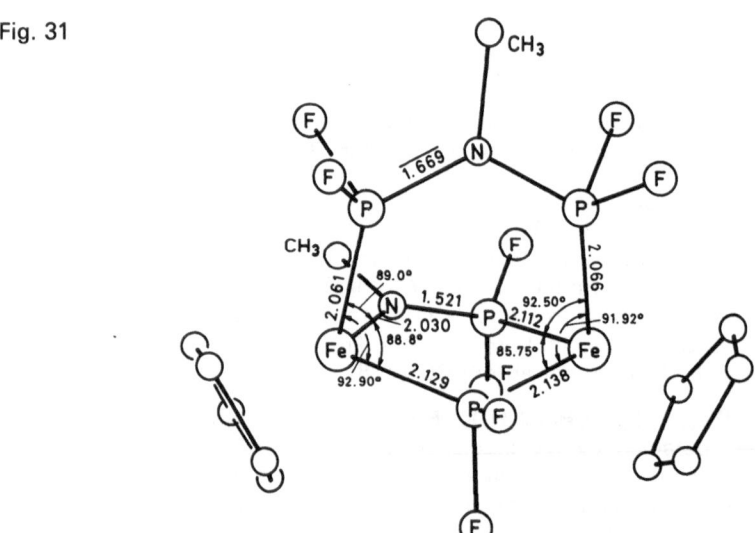

Molecular structure of
$C_5H_5Fe(\mu\text{-}PF_2)(\mu\text{-}PF_2N(CH_3)PF_2)(\mu\text{-}N(CH_3)=PF_2)FeC_5H_5$ [9].

References:

[1] W.E. Silverthorn (J. Chem. Soc. Chem. Commun. **1971** 1310/1). – [2] P.M. Treichel, W.K. Dean, W.M. Douglas (J. Organometal. Chem. **42** [1972] 145/58). – [3] P.J. Vergamini, G.J. Kubas, R.R. Ryan (Abstr. Papers 165th Natl. Meeting Am. Chem. Soc., Dallas, Tex., 1973, INOR 85). – [4] E. Abel (Ann. N.Y. Acad. Sci. **239** [1974] 306/14). – [5] D. Sellmann, E. Kleinschmidt (Angew. Chem. **87** [1975] 595/6; Angew. Chem. Intern. Ed. Engl. **14** [1975] 571).

[6] G.J. Kubas, P.J. Vergamini, M.P. Eastman, K.B. Prater (J. Organometal. Chem. **117** [1976] 71/9). − [7] P.J. Vergamini, G.J. Kubas (Progr. Inorg. Chem. **21** [1976] 261/82). − [8] D. Sellmann, E. Kleinschmidt (J. Organometal. Chem. **140** [1977] 211/9). − [9] M.G. Newton, R.B. King, M. Chang, J. Gimeno (J. Am. Chem. Soc. **100** [1978] 1632/4).

2.5.2.2 Carbonyl Compounds with Single Bridging Groups

The majority of compounds described in this section contains two $C_5H_5Fe(CO)_2$ units linked by various types of bridging groups A; thus in general terms, they are of the $C_5H_5Fe(CO)_2$-A-$(CO)_2FeC_5H_5$ type with σ-bonds between the Fe atoms and many elements of the Main Groups III to VII (except carbon) and with Fe-M bonds where M represents Mg and transition metals. A subdivision is made between neutral compounds (2.5.2.2.1 to 2.5.2.2.7) and cationic derivatives (2.5.2.2.8 to 2.5.2.2.10) the latter including the cation $[C_5H_5Fe(CO)_2(\mu\text{-H})(CO)_2FeC_5H_5]^+$.

Further compounds of various types are compiled in Chapters 2.5.2.2.11 to 2.5.2.2.13. The rare examples of compounds containing ^5L ligands other than C_5H_5 are included in Chapters 2.5.2.2.3.2, 2.5.2.2.5.2, and 2.5.2.2.9. The CO-substituted ^2D derivatives are compiled together with their parent compounds in 2.5.2.2.3.3, 2.5.2.2.8, and 2.5.2.2.12.

2.5.2.2.1 Bridging Groups with Fe-S Bonds

The following compounds can be divided into three kinds by the method of preparation and the bonding character. For the preparation of the second, third, and fourth compounds neutral SO_2 is inserted into a Fe-Fe or Fe-C bond, formally donating one electron to the Fe-S σ-bond. A strong Fe-S π bonding is discussed for these compounds. In one case this has been confirmed by an X-ray study [6, 7]. The multiple bond character is also discussed in terms of the CO and SO bands in the IR spectrum and the ^1H NMR shifts of the C_5H_5 ligands, but this argumentation is less clear-cut in the literature [2, 8]. The next two compounds are prepared from dithiolate derivatives, the sulfur ligand thus having anionic character [3, 4]. The strong electron-withdrawing property of the sulfur groups is indicated by chemical shifts to low field of the C_5H_5 protons [2, 4].

The last complex contains sulfur atoms as three-electron donors within a $(CO)_4$-$Mn(\mu\text{-S})_2Mn(CO)_4$ system.

$(C_5H_5Fe(CO)_2)_2S_3$. A complex with this formula is only briefly mentioned. Reactions of $(C_5H_5Fe(CO)_2)_2$ with cyclohexene sulfide have been found to give several sulfur-containing substances, including the title compound [1].

$C_5H_5Fe(CO)_2$-SO_2-$(CO)_2FeC_5H_5$ is prepared by treating one mole of $Na[C_5H_5$-$Fe(CO)_2]$ in tetrahydrofuran with 1.5 moles of SO_2 at -78 °C and warming up to room temperature over 45 to 60 min. Chromatography on SiO_2 gives with chloroform/acetone small amounts of $C_5H_5Fe(CO)_2(\mu\text{-CO})(\mu\text{-SO}_2)(CO)_2FeC_5H_5$ and then with acetone the title compound in 20 to 30% yield [6, 10]. The report of a 30-fold excess SO_2 for this preparation [6] is incorrect [8]. The complex can also be obtained by photolytic expulsion of SO_2 from the following compound in tetrahydrofuran (26% yield), though the reaction is accompanied by formation of some $(C_5H_5Fe(CO)_2)_2$ and extensive decomposition [7].

The red crystals decompose at 142 °C [6, 10]. The ^1H NMR spectrum (CDCl$_3$ or CD$_3$COCD$_3$) shows a C_5H_5 singlet at $\delta = 4.94$ ppm [10]. IR spectrum (Nujol): ν(CO) bands at 1953, 1965, 2015, and 2027 cm^{-1}; ν(SO$_2$) bands at 993 and 1135 cm^{-1}

Fig. 32

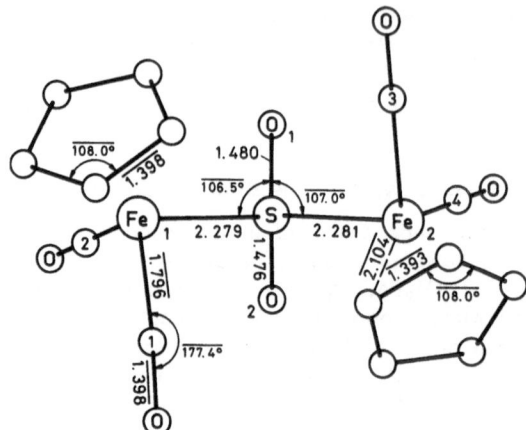

Molecular structure of $C_5H_5Fe(CO)_2-SO_2-(CO)_2FeC_5H_5$ [6, 7].

Selected bond angles (°):

Fe(1)-S-Fe(2)	118.00(02)	O(1)-S-O(2)	112.91(08)
S-Fe(1)-C(1)	92.11(06)	S-Fe(2)-C(3)	90.95(07)
S-Fe(1)-C(2)	85.99(07)	S-Fe(2)-C(4)	85.87(08)
C(1)-Fe(1)-C(2)	93.08(09)	C(3)-Fe(2)-C(4)	95.24(11)

[6, 10]. The complex crystallizes in the monoclinic system with a=7.7028(17), b= 17.1559(45), c=12.8488(27) Å, and β=117.48(1)°; space group P2$_1$/c−C$_{2h}^5$. Z=4 corresponds to D$_c$=1.844 g·cm^{-1}. The molecular structure is shown in **Fig. 32**. The central S atom is in a distorted tetrahedral environment. The mean Fe–S distance (2.280 Å) is shorter than that expected for an unit bond order (2.38 Å). This, in conjunction with rather low ν(SO) frequencies, provides evidence for multiple bond character in the Fe–S linkages. Other distances and angles are as expected for a $C_5H_5Fe(CO)_2$ derivative [6, 7].

The solid is reasonably stable. It dissolves in C_6H_6, $CHCl_3$, and CH_3COCH_3. The solutions decompose quite rapidly when exposed to air. Attempted sublimation at 110 °C/ 0.15 Torr for 17 h or heating in tetrahydrofuran at reflux for 2 h resulted in the formation of only $(C_5H_5Fe(CO)_2)_2$. The yields were 66% and 60%, respectively. The mass spectrum is essentially that of $(C_5H_5Fe(CO)_2)_2$. Photolysis in tetrahydrofuran gives the same complex. Only one Fe–S bond is broken upon reaction with neat CH_3I or in tetrahydrofuran solution, affording $C_5H_5Fe(CO)_2I$ (54 to 56%) and $C_5H_5Fe(CO)_2SO_2CH_3$ (21 to 26%) [6, 10]. For the reactions with SO_2 see the next compound.

$C_5H_5Fe(CO)_2-SO_2-SO_2-(CO)_2FeC_5H_5$ has been prepared by the procedure for the preceeding complex, but with an 11–fold excess of SO_2. The substance is removed from SiO_2 with chloroform/acetone (5:1) in 17% yield before the previous compound (7% yield) [8, 9]. The formation from the previous compound by insertion of SO_2 proceeds either in neat SO_2 at reflux or in tetrahydrofuran solution [8]. Small amounts (3%) have also been obtained from $K[C_5H_5Fe(CO)_2SO_2]·0.5\ SO_2$ and CH_3OSO_2F along with the main product $C_5H_5Fe(CO)_2-SO-OCH_3$ [9].

The orange crystalline substance melts at 140 to 146 °C with decomposition. The single C_5H_5 resonance in the ^1H NMR spectrum (CDCl$_3$) at δ=5.36 ppm does not broaden down to −40 °C in CD_3COCD_3. ^{13}C NMR spectrum (CDCl$_3$): δ=86.176 (C_5H_5)

and 208.381 (CO) ppm. The former resonance remains sharp down to −70 °C in CD_3COCD_3. The IR spectrum ($CDCl_3$) has $\nu(CO)$ bands at 2024, 2059, and 2070 cm^{-1} and $\nu(SO)$ bands (Nujol) at 1040 and 1223 cm^{-1} [8].

Photolysis in tetrahydrofuran leads to SO_2 expulsion. Cleavage with CH_3I in tetrahydrofuran at reflux yields $C_5H_5Fe(CO)_2SO_2CH_3$ and $C_5H_5Fe(CO)_2I$ in a 1: 1.8 mole ratio for a total yield of 62%. No evidence was found for the formation of $C_5H_5Fe(CO)_2SO$-OCH_3 [8].

$C_5H_5Fe(CO)_2$-SO_2-$CH_2CH_2CH_2$-SO_2-$(CO)_2FeC_5H_5$ has been obtained from $C_5H_5Fe(CO)_2$-$CH_2CH_2CH_2$-$(CO)_2FeC_5H_5$ by treatment with liquid SO_2 at −40 °C for 12 h, warming the product up to room temperature, and chromatography on Al_2O_3 with $CHCl_3$ as eluent. Upon addition of pentane to the concentrated eluate, the compound precipitates, 61.5% yield. Longer reaction times increase the yield.

The yellow crystals melt sharply at 185 °C. ^1H NMR spectrum ($CDCl_3$): $\delta = 2.49$ to 2.69 (m, CCH_2C), 3.52 (t, CH_2SO_2), and 5.40 (s, C_5H_5) ppm. IR spectrum ($CHCl_3$): $\nu(CO)$ bands at 2016 and 2060 cm^{-1}; $\nu(SO)$ bands (in Nujol) at 1049, 1169, and 1185 cm^{-1}; other bands at 733, 855, and 1128 cm^{-1}.

The solid tends to become gummy in moist air. It dissolves less readily than the mononuclear sulfinate complexes, especially in benzene and chloroform [2].

$C_5H_5Fe(CO)_2$-SCH=CHS-$(CO)_2FeC_5H_5$ forms immediately and precipitates when $C_5H_5Fe(CO)_2Cl$ in CH_3OH/H_2O is treated with $Na_2[SCH$=$CHS]$ (2:1 mole), 64% yield of crude product. The compound is difficult to recrystallize because it decomposes readily in solution. But a pure sample was obtained by crystallization from benzene/hexane.

The black crystals melt at 106 to 108 °C. ^1H NMR spectrum ($CDCl_3$): $\delta = 4.45$ (C_5H_5) and 7.21 (CH=CH) ppm. IR spectrum (KBr): $\nu(CO)$ bands at 1952, 1966, and 2018 cm^{-1}. Three bands with similar wave numbers have also been observed for benzene solutions. The spectrum is completely reported from 570 to 1485 cm^{-1}.

The solid decomposes slowly even under N_2 at 0 °C over a period of 4 to 6 weeks. It could not be sublimed. The solubility is low in aliphatic hydrocarbons and ethanol. Solutions in CH_2Cl_2 or CH_3COCH_3 decompose rapidly [4].

$C_5H_5Fe(CO)_2$-$SC(CN)$=$(CN)CS$-$(CO)_2FeC_5H_5$ has been prepared from C_5H_5Fe-$(CO)_2Cl$ and $Na_2[SC(CN)$=$(CN)CS]$ (about 1:1 mole) by shaking the materials in water for 12 h. The solid formed is reprecipitated from acetone with water and recrystallized from acetone/ethanol, 28% yield.

The red crystals melt at 136 to 138 °C with decomposition. The ^1H NMR spectrum (CH_3COCH_3) shows a single sharp C_5H_5 resonance at $\delta = 5.36$ ppm. The IR spectrum is completely given for KBr disk's and solutions in $CHCl_3$. In $CHCl_3$, $\nu(CO)$ bands are at 1960, 2010, and 2050 cm^{-1}, and $\nu(CN)$ is at 2200 cm^{-1}. The presence of a band at 1454 cm^{-1} assigned to the $\nu(C$=$C)$ of the bridging group suggests a cis configuration.

The compound reacts rapidly with CCl_4, $CHCl_3$, and CH_2Cl_2. Treatment with excess $Na_2[SC(CN)$=$(CN)CS]$ cleaves the C_5H_5 ligand and presumably forms $[Fe(SC(CN)$=$(CN)CS)_3]^{2+}$ [3].

$(C_5H_5Fe(CO)_2$-$SMn(CO)_4)_2$ has been prepared from $C_5H_5Fe(CO)_2Cl$ and $(CO)_4Mn(\mu$-$SSn(CH_3)_3)_2Mn(CO)_4$ in benzene at room temperature (2 d). The $(CH_3)_3SnCl$ formed is removed by shaking the solution with N_2 saturated H_2O. The product precipitates in 24% yield on addition of hexane.

References on p. 108

The deep red-brown powder decomposes above 250 °C without melting. ^{1}H NMR spectrum (C_6H_6): $\delta = 4.15$ ppm. IR spectrum (C_6H_6): ν(CO) bands at 1932, 1992, 2000, 2024, and 2051 cm^{-1}. The compound is soluble in all solvents except alkanes [5].

References:

[1] P.M. Treichel, G.R. Wilkes, M. Brauner (Proc. 2nd Intern. Conf. Organometal. Chem., Madison, Wisc., 1965, p. 120). — [2] J.P. Bibler, A. Wojcicki (J. Am. Chem. Soc. **88** [1966] 4862/70). — [3] J. Locke, J.A. McCleverty (Inorg. Chem. **5** [1966] 1157/61). — [4] R.B. King, C.A. Eggers (Inorg. Chem. **7** [1968] 340/5). — [5] H. Vahren-kamp (Chem. Ber. **103** [1970] 3580/90).

[6] M.R. Churchill, B.G. DeBoer, K.L. Kalra, P. Reich-Rohrwig, A. Wojcicki (J. Chem. Soc. Chem. Commun. **1972** 981/2). — [7] M.R. Churchill, B.G. DeBoer, K.L. Kalra (Inorg. Chem. **12** [1973] 1646/50). — [8] N.H. Tennent, S.R. Su, C.A. Poffenberger, A. Wojcicki (J. Organometal. Chem. **102** [1975] C46/C48). — [9] C.R. Jablonski (J. Organometal. Chem. **142** [1977] C25/C30). — [10] P. Reich-Rohrwig, A.C. Clark, R.L. Downs, A. Wojcicki (J. Organometal. Chem. **145** [1978] 57/68).

2.5.2.2.2 Bridging Groups with Fe-P, Fe-As, and Fe-Sb Bonds

Compound types I to III are listed in Table 7:

I $C_5H_5Fe(CO)_2-As(CH_3)_2 \rightarrow M \leftarrow As(CH_3)_2-(CO)_2FeC_5H_5$,

II $C_5H_5Fe(CO)_2-Sb(X,R)-(CO)_2FeC_5H_5$, and

III $C_5H_5Fe(CO)_2-SbX-(CO)_2FeC_5H_5$; M represents a transition metal carbonyl group.

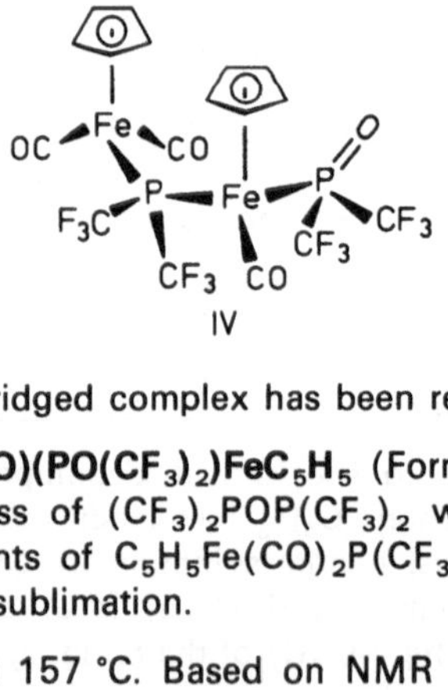

IV

Only one example of a P bridged complex has been reported:

$C_5H_5Fe(CO)_2-P(CF_3)_2 \rightarrow (CO)(PO(CF_3)_2)FeC_5H_5$ (Formula IV) has been obtained in 63% yield by shaking an excess of $(CF_3)_2POP(CF_3)_2$ with $(C_5H_5Fe(CO)_2)_2$ in CFCl$_3$ at 40 °C for 24 h. Small amounts of $C_5H_5Fe(CO)_2P(CF_3)_2$ and $C_5H_5Fe(CO)_2PO(CF_3)_2$ can be separated by fractional sublimation.

The orange crystals melt at 157 °C. Based on NMR and IR evidence, the structure IV is proposed. The ^{1}H NMR spectrum (in CFCl$_3$ or CCl$_4$) shows two slightly broadened singlets of equal intensity at $\delta = 5.06$ and 5.42 ppm. ^{19}F NMR spectrum (CCl$_4$ or CFCl$_3$, shifts referred to CFCl$_3$): $\delta = 48.3$, 49.3 (P(CF$_3$)$_2$, J(P,F) = 43 and 42 Hz), 67.7, 69.3 (PO(CF$_3$)$_2$, J(P,F) = 63 and 62 Hz) ppm. The spectrum is temperature dependent which can be explained by hindered rotation about the bridging P atom. Spectra over the range 233 to 301 K are shown in a figure. IR spectrum (CCl$_4$): ν(CO) bands at 1976, 2022, and 2071 cm^{-1}. An absorption at 1207 cm^{-1} is tentatively assigned to ν(P=O).

The mass spectrum does not show the parent ion but [M−CO]$^+$ [7].

Compound type I with M=Cr(CO)$_4$, Mo(CO)$_4$, or W(CO)$_4$ is prepared by stirring a solution of π-C$_7$H$_8$M(CO)$_4$ (C$_7$H$_8$=bicyclo[2.2.1]heptadiene) and C$_5$H$_5$Fe(CO)$_2$-As(CH$_3$)$_2$ (1:2 mole) in cyclohexane at room temperature. The products separate from the solutions. Another way is the reaction of cis-M(CO)$_4$(As(CH$_3$)$_2$Cl)$_2$ with [C$_5$H$_5$Fe(CO)$_2$]$^-$ in tetrahydrofuran, but it gives larger amounts of decomposition products and pure samples are difficult to isolate [6].

In the IR spectra the ν(CO) bands of the M(CO)$_4$ units clearly reveal the cis geometry around the transition metal atom. Under protecting atmosphere the solids are stable for a long period of time; however, solutions are unstable, especially for M=Cr(CO)$_4$. In air, the solids slowly decompose, while the solutions decompose rapidly. The compounds are soluble in polar solvents. The solubility depends only slightly on temperature [6].

Compounds No. 4, 6, and 7 are obtained by reacting Na[C$_5$H$_5$Fe(CO)$_2$] with C$_5$H$_5$Fe(CO)$_2$-SbBr$_2$ (No. 4) or C$_5$H$_5$Fe(CO)$_2$SbX$_2$-Cr(CO)$_5$ (Nos. 6 and 7) in benzene [4] or cyclohexane [5] at 25 °C for 48 h. After filtration and evaporation of the solvent the residues are recrystallized from boiling cyclohexane or benzene/cyclohexane [5]. Other preparations are given in the further information to compounds No. 5 and 8.

It has been reported that the analogous complexes (C$_5$H$_5$Fe(CO)$_2$)$_2$SbCl and (C$_5$H$_5$Fe(CO)$_2$)$_2$SbI can be prepared by the method described for No. 4 [1]. But details are not available.

For cationic derivatives with Fe–As and Fe–Sb bonds see 2.5.2.2.10, p. 157.

Table 7
Compounds containing bridging groups with Fe–As and Fe–Sb bonds.
Further information on compounds preceded by an asterisk is given at the end of the table. For abbreviations and dimensions see p. 170.

No.	X or R	Group M (yield in %)	Properties	Ref.
Type I. C$_5$H$_5$Fe(CO)$_2$-As(CH$_3$)$_2$→M←As(CH$_3$)$_2$-(CO)$_2$FeC$_5$H$_5$:				
1	–	Cr(CO)$_4$ (46)	dark red; m.p. 111° (dec.) ^{1}H NMR (C$_6$H$_6$): 1.61 (CH$_3$), 4.36 (C$_5$H$_5$) IR (–): 1970, 1986 (FeCO); 1862, 1879, 1898, 2015, 2018 (CrCO)	[6]
2	–	Mo(CO)$_4$ (38)	yellow brown; m.p. 120° ^{1}H NMR (C$_6$H$_6$): 1.61 (CH$_3$), 4.30 (C$_5$H$_5$) IR (–): 1970, 2002 (FeCO); 1864, 1893, 1909, 2017, 2020 (MoCO)	[6]
3	–	W(CO)$_4$ (49)	brown; m.p. 127° (dec.) ^{1}H NMR (C$_6$H$_6$): 1.74 (CH$_3$), 4.40 (C$_5$H$_5$) IR (–): 1970, 1998 (FeCO); 1856, 1879, 1899, 2018, 2020 (WCO)	[6]
Type II. C$_5$H$_5$Fe(CO)$_2$-Sb(X,R)-(CO)$_2$FeC$_5$H$_5$:				
*4	Br	– (57.2)	violet-brown needles; m.p. 150 to 152° (dec.) ^{1}H NMR (C$_6$D$_6$): 4.35 (s, C$_5$H$_5$) IR (CH$_2$Cl$_2$): 1959, 1988, 2018	[1, 5]

References on p. 111

Table 7 [continued]

No.	X or R	Group M (yield in %)	Properties	Ref.
*5	CH₃	— (39.3)	dark brown crystals; m.p. 80 to 82° (dec.) ^{1}H NMR (C_6D_6): 1.92 (s, CH_3), 4.28 (s, C_5H_5) IR (C_6H_{12}): 1942, 1950, 1981, 2005	[3]

Type III. $C_5H_5Fe(CO)_2$-SbX-$(CO)_2FeC_5H_5$:

$$\downarrow$$
$$M$$

No.	X or R	Group M (yield in %)	Properties	Ref.
*6	F	$Cr(CO)_5$ (−)	orange–red needles; m.p. 171 to 173° (dec.) ^{1}H NMR (C_6D_6): 4.50 (s, C_5H_5) ^{19}F NMR (THF): 233 IR (THF): 1926, 1935, 1965, 1981, 2007, 2021, 2054	[4]
*7	Br	$Cr(CO)_5$ (61)	black needles; m.p. 176 to 178° (dec.) ^{1}H NMR (C_6D_6): 4.51 (s, C_5H_5) IR (THF): 1966, 1984, 2008, 2022 (FeCO); 1926, 1935, 1973, 2054 (CrCO)	[5]
*8	Br	$Mn(CO)_2C_5H_5$ (51.5)	deep black crystals; m.p. 193 to 195° (dec.) ^{1}H NMR (C_6D_6): 4.59 (s, FeC_5H_5), 4.63 (s, MnC_5H_5) IR (THF): 1958, 1973, 2002, 2019 (FeCO); 1862, 1914 (MnCO)	[5]

* Further information:

$(C_5H_5Fe(CO)_2)_2SbBr$ (Table 7, No. 4) is formed in addition to $C_5H_5Fe(CO)_2SbBr_2$ in the reaction of $SbBr_3 \cdot C_4H_8O$ (C_4H_8O = tetrahydrofuran) with $(C_5H_5Fe(CO)_2)_2$ (2:1 mole ratio) in benzene or tetrahydrofuran as solvent. In this context, the corresponding substances with SbCl and SbI bridges have been mentioned [1, 2].

The compound is further metallated by reaction with $Na[C_5H_5Fe(CO)_2]$ in cyclohexane at room temperature to give $(C_5H_5Fe(CO)_2)_3Sb$ [4]; see also compound No. 8.

$(C_5H_5Fe(CO)_2)_2SbCH_3$ (Table 7, No. 5). For its preparation a suspension of $Na[C_5H_5Fe(CO)_2]$ and CH_3SbBr_2 (2.4:1 mole ratio) in cyclohexane is warmed at 50 °C for 33 h. After evaporation of the solvent in vacuum, the compound is extracted from the residue with pentane and crystallized at −78 °C, 39.3% yield. It is also formed from the heteronuclear $C_5H_5Fe(CO)_2$-Sb(CH_3)-$(CO)_3MoC_5H_5$ via symmetrization (the second product is $(C_5H_5Mo(CO)_3)_2SbCH_3$ or via metallate exchange with $Na[C_5H_5-Fe(CO)_2]$, yields of about 40%.

The compound is much more soluble in tetrahydrofuran, benzene, cyclohexane, and chloroform than the corresponding Cr, Mo, and W analogues. It crystallizes even from pentane only at low temperature. It is stable at room temperature under an inert gas atmosphere [3].

$(C_5H_5Fe(CO)_2)_2SbX \rightarrow Cr(CO)_5$ (X = F and Br, Table **7**, Nos. **6** and **7**). Further metallation of the fluoro compound No. 6 by $Na[C_5H_5Fe(CO)_2]$ yields the trinuclear complex $(C_5H_5Fe-(CO)_2)_3Sb \rightarrow Cr(CO)_5$. However, similar treatment of No. 7 at 60 °C for 3 d gives mainly $(C_5H_5Fe(CO)_2)_2$ [5].

$(C_5H_5Fe(CO)_2)_2SbBr \rightarrow Mn(CO)_2C_5H_5$ (Table **7**, No. **8**) is obtained from compound No. 4 and $C_5H_5Mn(CO)_2 \cdot C_4H_8O$ (C_4H_8O = tetrahydrofuran) at room temperature for 2 to 3 d. It is recrystallized from hot cyclohexane [5].

The IR spectrum of the CS_2 solution shows practically identical $\nu(CO)$ bands to those given in the table. An approximate value for a CO stretching force constant has been calculated. The substance is air-stable for some time. It dissolves in tetrahydrofuran to give a green solution [5].

References:

[1] W.R. Cullen, D.J. Patmore, J.R. Sams, L.K. Thompson (J. Chem. Soc. Chem. Commun. **1971** 952/3). − [2] W.R. Cullen, D.J. Patmore, J.R. Sams (Inorg. Chem. **12** [1973] 867/72). − [3] W. Malisch, P. Panster (Chem. Ber. **108** [1975] 700/15). − [4] W. Malisch, P. Panster (Angew. Chem. **88** [1976] 680/1; Angew. Chem. Intern. Ed. Engl. **15** [1976] 618). − [5] P. Panster, W. Malisch (Chem. Ber. **109** [1976] 692/704).

[6] M. Börner, H. Vahrenkamp (J. Chem. Res. M **1977** 0801/17; J. Chem. Res. S **1977** 74/5). − [7] R.C. Dobbie, M.J. Hopkinson (J. Chem. Soc. Dalton Trans. **1974** 1290/3).

2.5.2.2.3   Bridging Groups with Fe–Si, Fe–Ge, Fe–Sn, and Fe–Pb Bonds

General References:

M.J. Newlands, Organotin Compounds with Tin − Other Metal Bonds, in: A.K. Sawyer, Organotin Compounds, Vol. 3, Dekker, New York 1972, p. 881/930.
J.J. Zuckerman, Application of ^{119m}Sn Mössbauer Spectroscopy to the Study of Organotin Compounds, Advan. Organometal. Chem. **9** [1970] 21/134.
N.E. Kolobova, A.B. Antonova, K.N. Anisimov, Derivatives of Metal Carbonyls Containing a Bond between Atoms of the Transition Metals and Group IVB Elements, Uspekhi Khim. **38** [1969] 1802/19; Russ. Chem. Rev. **38** [1969] 822/39.
F.G.A. Stone, Transition Metal Derivatives of Silicon, Germanium, Tin, and Lead, in: E.A.V. Ebsworth, A.G. Maddock, A. Sharpe, New Pathways in Inorganic Chemistry, Univ. Press, Cambridge 1968, p. 283/302.
N.S. Vyazankin, G.A. Razuvaev, O.A. Kruglaya, Organometallic Compounds with Metal–Metal Bonds between Different Metals, Organometal. Chem. Rev. A **3** [1968] 323/423.

General Remarks

The compounds described in the following two sections are collected in Table 8 (for E = Si and Ge) and in Table 9 (for E = Sn and Pb). They are crystalline if not otherwise specified. They are rather stable at ambient temperature in the solid state but decompose comparatively readily in solution in the presence of air [1]. The presence of Si–Si bonds in the polysilanyl derivatives (Table 8, Nos. 7 to 9) does not introduce additional instability [10].

The complexes are soluble in solvents like aliphatic and aromatic hydrocarbons, halohydrocarbons, alcohols, and ethers but are insoluble in water [1].

The nature of the Fe-E and E-(X,R) bonds is extensively discussed in the literature, especially for the complexes with Sn as bridging atom. It is reported that the ^{57}Fe Mössbauer spectra are quite similar for various compounds of this type. The isomer shifts δ and quadrupole splittings Δ are primarily determined by the structure of the $C_5H_5Fe(CO)_2$ group, and less by the Fe-E bond. δ is in the region 0.30 to 0.37 mm·s^{-1} at 80 K and in the region 0.25 to 0.33 mm·s^{-1} at 300 K. Δ varies from 1.60 to 1.83 mm·s^{-1} at 80 K and from 1.55 to 1.75 mm·s^{-1} at 300 K. The slight temperature dependence of Δ is characteristic for half-filled $3d^5$ or filled $3d^{10}$ orbitals. Δ increases slightly with the number of electronegative groups on the Sn atom. This is explained by the drawing of electrons from the iron to the tin atom, which makes the electric field gradient positive at the Fe nucleus [5, 12].

More information on the electronic structure of the molecules is provided by the ^{119}Sn Mössbauer spectra. In compounds of the $(C_5H_5Fe(CO)_2)_nSnCl_{4-n}$ type, Δ is close to 0.35 mm·s^{-1} in all cases. This value is less than that in the corresponding $(C_6H_5)_nSnCl_{4-n}$ compounds. The authors [5] arrive at the conclusion, that the metal carbonyl group is a better electron acceptor than the phenyl group. This is in contrast to the usual assumption that the $C_5H_5Fe(CO)_2$ group is a donor, which has less π-acceptor character than the C_6H_5 group [15, 16].

In compounds of the $(C_5H_5Fe(CO)_2)_2SnR_2$ type a quadrupole splitting of the ^{119}Sn signal is absent when the substituents R have no lone electron pair, i.e., R=alkyl, phenyl. In these cases only σ bonds are formed between Sn and the ligand and no deviation from the sp^3 hybridization of Sn occurs [5]. An increase in δ with an increasing number of metal atoms attached to Sn in the series $(C_5H_5Fe(CO)_2)_nSn(X,R)_{4-n}$ is explained by an increase in the ionic character of the Fe-Sn bonds in [5] and by an enhanced s character of the Fe-Sn bond in [7].

For the series of tin compounds with X=Cl, Br, I, CH$_3$COO, NCS, and HCOO (Table 9 Nos. 2, 3, 4, 13, 16, 17), a very small variation in δ and Δ in the ^{119}Sn resonance is observed with different substituents. This shows that the bonding electrons of Sn are mainly located in the Fe-Sn bond. The large Δ values in these compounds indicate the presence of an electric field gradient. The origin of the field gradient has been discussed [12].

The Mössbauer data show that the Fe-Sn bonds are single bonds without involvement of the Sn 5 d electrons in π bonding [7, 13]. But it is also stressed that the method is rather insensitive to the participation of 5 d electrons [13].

The IR range of ν (CO) bands in solution (1800 to 2200 cm^{-1}) has been intensively investigated. The structure of the ν(CO) bands can change significantly with the nature of the substituents on the atom E. From the pattern of the CO bands the molecular symmetry C_s is proposed for the compounds No. 17, 18, 20, 22 (Table 8) and No. 18, 19, 20, 22, 23 (Table 9) and the symmetry C_2 is proposed for Nos. 8, 9, 10 (Table 8) and Nos. 2, 3, 4, 13, 14, 16 (Table 9) [3, 8, 11]. The different configurations arise from rotation of the $C_5H_5Fe(CO)_2$ group around the Fe-E bond. Due to this rotation the geometric configuration in solution can be expected to differ somewhat from the rigid crystal structure [3]. See [3, 8] for a discussion of possible multiple bonds between E and the ligands X and a d_π-d_π interaction in the Fe-E bond. The CO bands shift to lower wave numbers in the series $Sn(C_2H_5)_2 < SnBr_2 \leqq SnCl_2$. The influence of the solvent polarity on ν(CO) is not significant [14].

X-ray diffraction patterns of the Ge and Sn compounds reveal that the group IV metal has an approximately tetrahedral coordination with different degrees of distortion.

The coordination around Fe is of the "piano-stool" type characteristic for the half-sandwich compounds. Electronic effects of substituents on Sn are transferred via the Fe-Sn bond. As the electronegativity of the substituent increases in the series $CH_3 < ONO < Cl$, an increase in C-O bond order (cf. [14]), a shortening of the Fe-Sn bonds, and an increase in the Fe-Sn-Fe angles are observed [4, 6, 9]. The effect is explained by a greater s character in the Sn hybrid orbitals participating in the Sn-Fe bond (Bent's rule) [6, 7]. Another explanation is based on the hypothesis that the d_π-d_π interaction between the filled 3 d orbitals of Fe and the vacant 5 d orbitals of Sn is intensified, when the π-acceptor ability of the SnX_2 group increases as a result of an increase in the electronegativity of X [9]. However, this latter argument has been questioned [7].

The compounds with E=Ge, Sn, and Pb can be used as plasticisers and stabilizers for synthetic resins like polyvinyl chloride. They are potential antiknock and antiwear agents in motor fuels and lubricants [1]. By thermal decomposition of the vapor alloys can be plated on heated substrates such as steel [2].

References:

[1] R.D. Gorsich, Ethyl Corp. (U.S. 3069449 [1962]; C.A. **58** [1963] 10241). – [2] T.P. Whaley, V. Norman, Ethyl Corp. (U.S. 3071493 [1963]; C.A. **58** [1963] 4235). – [3] N. Flitcroft, D.A. Harbourne, J. Paul, P.M. Tucker, F.G.A. Stone (J. Chem. Soc. A **1966** 1130/3). – [4] B.P. Bir'yukov, K.N. Anisimov, Yu.T. Struchkov, N.E. Kolobova, V.V. Skripkin (Zh. Strukt. Khim. **8** [1967] 556/7; J. Struct. Chem. [USSR] **8** [1967] 498/9).

[5] V.I. Gol'danskii, B.V. Borshagovskii, E.F. Makarov, R.A. Stukan, K.N. Anisimov, N.E. Kolobova, V.V. Skripkin (Teor. Eksperim. Khim. **3** [1967] 478/82; Theor. Exptl. Chem. [USSR] **3** [1967] 275/7). – [6] J.E. O'Connor, E.R. Corey (Inorg. Chem. **6** [1967] 968/71). – [7] D.E. Fenton, J.J. Zuckerman (J. Am. Chem. Soc. **90** [1968] 6226/8). – [8] K.N. Anisimov, B.V. Lokshin, N.E. Kolobova, V.V. Skripkin (Izv. Akad. Nauk SSSR Ser. Khim. **1968** 1024/30; Bull. Acad. Sci. USSR Div. Chem. Sci. **1968** 978/81). – [9] B.P. Bir'yukov, Yu.T. Struchkov (Zh. Strukt. Khim. **9** [1968] 488/502; J. Struct. Chem. [USSR] **9** [1968] 412/25). – [10] R.B. King, K.H. Pannell, C.R. Bennett, M. Ishaq (J. Organometal. Chem. **19** [1969] 327/37).

[11] A.N. Nesmeyanov, K.N. Anisimov, B.V. Lokshin, N.E. Kolobova, F.S. Denisov (Izv. Akad. Nauk SSSR Ser. Khim. **1969** 758/63; Bull. Acad. Sci. USSR Div. Chem. Sci. **1969** 690/3). – [12] S.R.A. Bird, J.D. Donaldson, A.F. Le C. Holding, B.J. Senior, M.J. Tricker (J. Chem. Soc. A **1971** 1616/21). – [13] B.A. Goodman, R. Greatrex, N.N. Greenwood (J. Chem. Soc. A **1971** 1868/72). – [14] S. Cenini, B. Ratcliff, A. Fusi, A. Pasini (Gazz. Chim. Ital. **102** [1972] 141/63). – [15] Yu.V. Kolodyazhnyi, V.V. Skripkin, N.E. Kolobova, A.D. Garnovskii, B.V. Lokshin, O.A. Osipov, K.N. Anisimov, M.G. Gruntfest (Zh. Strukt. Khim. **13** [1972] 160/2; J. Struct. Chem. [USSR] **13** [1972] 148/50).

[16] H.E. Sasse, M.L. Ziegler (Z. Naturforsch. **30b** [1975] 30/2).

2.5.2.2.3.1 Compounds with Fe-Si and Fe-Ge Bonds

Compounds with Si or Ge bridging atoms are listed in Table 8. They are prepared by the methods summarized below. Other procedures applied for some of the compounds are indicated with the term "special" in the table and are described under further information.

Method I: A tetrahydrofuran solution containing $Na[C_5H_5Fe(CO)_2]$ and the corresponding $X-(SiR_2)_n-X$ derivative $(X=Cl$ and $I)$ is stirred in a mole ratio 1:1 to 1:2 at room temperature for 5 to 16 h. The solvent is removed under reduced pressure and the residue chromatographed on Al_2O_3 and eluted with hexane/ether [11].

Method II: $Na[C_5H_5Fe(CO)_2]$, prepared from $(C_5H_5Fe(CO)_2)_2$ and Na/Hg in tetrahydrofuran, and the corresponding GeX_4 (2:1 mole) in tetrahydrofuran are heated at reflux for 20 h. The solvent is removed under reduced pressure and the residue is recrystallized from benzene/petroleum ether [3]. The dimethyl derivative, No. 19, has been prepared with $(CH_3)_2GeCl_2$ in ether at $-10\,°C$ for 5 min. After evaporation of the solvent, the product is taken up in CH_2Cl_2 and filtered through SiO_2 [21].

Method III: A mixture of $(C_5H_5Fe(CO)_2)_2$ and the dioxane complex of $GeCl_2$ (about 1:1 mole) in dioxane/ether is refluxed for 2 h. Filtration and solvent removal in vacuum gives a residue which is recrystallized from ether/tetrahydrofuran [8]. The reaction with GeI_2 has been carried out in refluxing benzene for 24 h [3].

Method IV: $(C_5H_5Fe(CO)_2)_2GeCl_2$ is converted into compounds with other GeX_2 groups by treatment with CH_3ONa in methanol/acetone, CH_3COOK in glacial acetic acid, KSCN in acetone, C_2H_5SNa in ethanethiol/acetone, C_5H_5Na in tetrahydrofuran, or RMgBr in ether. The mixture is filtered and the solvent removed in vacuum [8].

Method V: $(C_5H_5Fe(CO)_2)_2Ge(OOCCH_3)_2$ is treated with HF in ether or dry HBr or HI in CCl_4/tetrahydrofuran at room temperature. The crystals are filtered off and washed with petroleum ether [8].

Explanations to Table 8: ^1H NMR chemical shifts referred to $(CH_3)_3SiOSi(CH)_3$ are indicated by an asterisk, ^1H NMR*. ^{29}Si NMR shifts are referred to $Si(CH_3)_4$.

Table 8
Compounds of the $C_5H_5Fe(CO)_2-(E(X,R)_2)_n-(CO)_2FeC_5H_5$ Type with $E=Si$ and Ge. Further information on compounds preceded by an asterisk is given at the end of the table. For abbreviations and dimensions see p. 170.

No.	Bridging group $E(X,R)_2$ Method of preparation (yield in %)	Properties and further remarks Explanations see above	Ref.
	With $E=Si$:		
*1	SiH_2 I (11.4)	dark yellow; m.p. 119° (in vacuum) ^1H NMR $(C_6D_6?)$: 4.31 (s, C_5H_5), 5.01 (s, SiH, $J(Si,H)=167$) IR (Nujol): 1919, 1925, 1972, 1990; $\delta(FeCO)$ at 598, 603, 648; $\nu(SiH)$ at 2030, 2037; $\delta(SiH_2)$ at 935	[23]

Table 8 [continued]

No.	Bridging group $E(X,R)_2$ Method of preparation (yield in %)	Properties and further remarks Explanations on p. 114	Ref.
2	Si(Cl)H see compound No. 1	^1H NMR (CDCl$_3$): 4.98 (C$_5$H$_5$), 6.54 (SiH) identified only as an intermediate by ^1H NMR, see No. 1	[23]

*3	SiCl$_2$
	see compound No. 1
		bright yellow; m.p. 187° (in vacuum)
		^1H NMR (CDCl$_3$): 4.98 (C$_5$H$_5$)
		IR (Nujol): 1927, 1962, 2003, 2016;
		  δ(FeCO) at 638; δ(FeC) at 467, 510;
		  δ(SiCl) at 415; ν(CC) at 1358, 1416,
		  1435; ν(CH) at 3112 and δ(CH) at 835, 844
			[23]

*4	Si(CH$_3$)H
	special
		orange yellow; m.p. 89 to 90°
		^1H NMR (C$_6$H$_6$): 1.14 (d, CH$_3$, ^3J(H,H) =3.6),
		  4.37 (s, C$_5$H$_5$), 5.46 (q, SiH)
		^{13}C NMR (C$_6$D$_6$): 12.0 (CH$_3$), 84.3 (C$_5$H$_5$),
		  216.52, 216.65 (CO)
		^{29}Si NMR (C$_6$D$_6$): 62.8
		IR (C$_6$H$_{12}$): 1935, 1943, 1952, 1990;
		  ν(SiH) at 2025
			[24]

*5	Si(CH$_3$)Cl	see further information	[24]
*6	Si(C$_6$H$_5$)$_2$	see further information	[15]
*7	(–Si(CH$_3$)$_2$–)$_2$ I (40)	yellow; m.p. 150° ^1H NMR (CHCl$_3$): 0.45 (s, CH$_3$), 4.79 (s, C$_5$H$_5$) IR (KBr): 1929, 1977; (C$_6$H$_{12}$): 1949, 1997	[10, 11]
*8	(–Si(CH$_3$)$_2$–)$_3$ I (33)	yellow; m.p. 172 to 173° ^1H NMR (CDCl$_3$): 0.30 (s, CH$_3$SiSi), 0.52 (s, CH$_3$SiFe), 4.78 (s, C$_5$H$_5$) IR (KBr): 1913, 1977; (C$_6$H$_{12}$): 1939, 1990	[10, 11, 12]
*9	(–Si(CH$_3$)$_2$CH$_2$–)$_2$ I (30)	yellow; m.p. 155 to 156° ^1H NMR (CDCl$_3$): 0.32 (s, CH$_3$), 0.86 (s, CH$_2$), 4.50 (s, C$_5$H$_5$) IR (KBr): 1925, 1983; (C$_6$H$_{12}$): 1948, 2005	[11]

With E = Ge:

*10	GeH$_2$ special	yellow; m.p. 110° (dec.) ^1H NMR (CDCl$_3$): 3.87 (GeH), 4.80 (C$_5$H$_5$) IR (C$_6$H$_{12}$): 1940, 1952, 1986, 2003, 2012 (ν(CO) and ν(GeH) region)	[3]
11	GeF$_2$ V (75)	yellow; m.p. 196 to 198° (dec.) ^1H NMR* (THF): 5.12 (C$_5$H$_5$) IR (C$_6$H$_{12}$): 1955, 1972, 2001, 2027	[8, 9]

References on p. 121

Table 8 [continued]

No.	Bridging group E(X,R)$_2$ Method of preparation (yield in %)	Properties and further remarks Explanations on p. 114	Ref.
*12	GeCl$_2$ II (32), III (85)	yellow orange; m.p. 192 to 195° (dec.), 194 to 197 (in a sealed tube) ^{1}H NMR (THF): 5.15 (C$_5$H$_5$) ^{57}Fe$-\gamma$ (77 K): $\delta=0.36$, $\Delta=1.66$ IR (Nujol): 1942, 1978, 2015, 2030; (C$_6$H$_{12}$): 1962, 1985, 2010, 2036; (THF): 1970, 2003, 2009, 2025	[3, 4, 7, 8, 9, 13, 16]
*13	GeBr$_2$ V (92.3)	orange red, rust colored; m.p. 198 to 201°, dec. at 160° without melting ^{1}H NMR* (THF): 5.17 (C$_5$H$_5$) IR (C$_6$H$_{12}$): 1960, 1987, 2008, 2033	[8, 9, 25]
*14	GeI$_2$ II (60), III (76), V (71.4)	red; m.p. 231 to 233° (dec.) ^{1}H NMR* (THF): 5.17 (C$_5$H$_5$) IR (C$_6$H$_{12}$ or CS$_2$): 1963, 1985, 2008, 2033	[3, 8]
*15	Ge(OCH$_3$)$_2$ IV at $-35°$ (98.5)	yellow; m.p. 122 to 123° ^{1}H NMR* (CCl$_4$): 3.62 (CH$_3$), 5.96 (C$_5$H$_5$) IR (C$_6$H$_{12}$): 1942, 1954, 1960, 1988, 2002, 2013, 2021	[8, 9]
16	Ge(OOCCH$_3$)$_2$ IV (83)	yellow plates; m.p. 160 to 162° ^{1}H NMR* (CCl$_4$): 2.31 (s, CH$_3$), 4.98 (s, C$_5$H$_5$) IR (C$_6$H$_{12}$): 1965, 1974, 2007, 2029; (KBr): ν(C−O) at 1270, 1290; ν(C=O) at 1670; δ(CH$_3$) at 1365 crystallization from THF/petroleum ether slowly turns green in light	[8, 9]
17	Ge(NCS)$_2$ IV at 25° (90.5)	yellow; m.p. 204 to 207° (dec.) ^{1}H NMR* (THF): 5.28 (C$_5$H$_5$) IR (C$_6$H$_{12}$): 1965, 1985, 2007, 2033; (KBr): ν(NCS) at about 2050 and 2070 crystallization from THF/petroleum ether	[8, 9]
18	Ge(SC$_2$H$_5$)$_2$ IV at $-40°$ (84)	orange; m.p. 139 to 141° ^{1}H NMR* (CCl$_4$): 1.33 (t, CH$_3$), 2.79 (d, CH$_2$, J(CH$_2$,CH$_3$)=7.2), 4.99 (s, C$_5$H$_5$) IR (KBr): C$_2$H$_5$ bands at 1250, 1380, 1460, 2860, 2930, 2980 crystallization from benzene/ether	[8]
*19	Ge(CH$_3$)$_2$ II (22)	yellow; m.p. 129 to 130° ^{1}H NMR (CS$_2$): 0.89 (CH$_3$), 4.77 (C$_5$H$_5$) (C$_6$D$_6$): 0.93 (CH$_3$), 4.20 (C$_5$H$_5$) IR (C$_6$H$_{12}$): 1941, 1952, 1987, 1999, 2008	[3, 21]

References on p. 121

Table 8 [continued]

No.	Bridging group E(X,R)$_2$ Method of preparation (yield in %)	Properties and further remarks Explanations on p. 114	Ref.
20	Ge(C$_2$H$_5$)$_2$ IV (76)	yellow; m.p. 122 to 124° ^{1}H NMR (CCl$_4$): 1.42 (m, C$_2$H$_5$), 4.92 (C$_5$H$_5$) IR (C$_6$H$_{12}$): 1930, 1940, 1978, 1988; (KBr): C$_2$H$_5$ bands at 1380, 1460, 2870, 2905, 2925, 2965	[8, 9]
21	Ge(CH$_2$CH=CH$_2$)$_2$ IV (26.6)	orange red; m.p. 98 to 99° ^{1}H NMR* (THF): 5.05 (C$_5$H$_5$)	[8]
22	Ge(C$_4$H$_9$)$_2$ IV (80.5)	yellow; m.p. 117 to 118° ^{1}H NMR* (THF): 5.01 (C$_5$H$_5$) IR (C$_6$H$_{12}$): 1931, 1941, 1978, 1989	[8, 9]
*23	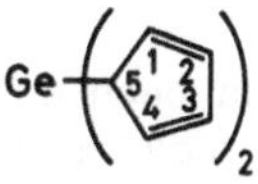 Ge(C$_5$H$_5$-cyclopentadienyl ring)$_2$ IV at −40° (32.6)	reddish brown; m.p. 139 to 141° (dec.) IR (KBr): v(C$_5$H$_5$-σ) at 750, 900	[8]
*24	Ge(C$_6$H$_5$)$_2$ II (−), IV (30.2)	yellow; m.p. 136 to 137° ^{1}H NMR* (THF): 4.71 (C$_5$H$_5$) IR (C$_6$H$_{12}$): 1938, 1953, 1964, 1987, 1997, 2006, 2016	[8, 14, 15]
*25	(−Ge(CH$_3$)$_2$−)$_2$ special	yellow; m.p. 148 to 149.5° ^{1}H NMR (CS$_2$): 0.59 (s, CH$_3$), 4.6 (s, C$_5$H$_5$) IR (KBr): 1920, 1960	[18]
*26	(−Ge(C$_2$H$_5$)$_2$−)$_2$ special	yellow; m.p. 98 to 101° ^{1}H NMR (CCl$_4$): 1.23 (m, C$_2$H$_5$), 4.81 (s, C$_5$H$_5$) IR (KBr): 1925, 1980	[18]
*27	(Ge(CH$_3$)$_2$−)$_2$O special	off−white ^{1}H NMR (C$_6$H$_6$): 0.89 (CH$_3$), 4.25 (C$_5$H$_5$) (CD$_3$COCD$_3$): 0.67, 0.73 (CH$_3$, ratio 1:2), 4.90, 4.97 (C$_5$H$_5$, ratio 2:1)	[22]

* Further information:

(C$_5$H$_5$Fe(CO)$_2$)$_2$SiH$_2$ (Table 8, No. 1) is recrystallized from benzene/isopentane (1:1). All reactions are carried out under vacuum.

The IR spectrum of the cyclohexane solution shows v(CO) bands at 1953, 1991, and 2005 cm^{-1}. A band at 329 cm^{-1} is assigned to the asymmetric v(FeSiFe). The SiH$_2$ wagging vibration is observed at 799 cm^{-1}.

The mass spectrum contains the molecular ion [M]$^+$, [M−n CO]$^+$ with n=1 to 4, [Fe$_2$(C$_5$H$_5$)$_2$]$^+$, [SiFe$_2$]$^+$, and other fragments with one Fe atom. If a solution in CDCl$_3$ is briefly exposed in a sealed tube to sunlight, the original ^{1}H NMR peaks diminish after 10 min and new resonances appear. These are assigned to (C$_5$H$_5$Fe(CO)$_2$)$_2$Si(Cl)H and (C$_5$H$_5$Fe(CO)$_2$)$_2$SiCl$_2$. After 6 h without further exposure to light, the latter can be isolated in 100% yield [23].

(C₅H₅Fe(CO)₂)₂SiCl₂ ($C_5H_5Fe(CO)_2)_2SiCl_2$ (Table **8**, No. **3**). For its quantitative formation see the previous compound. The mass spectrum shows [M]⁺, [M −n CO]⁺ with n=1 to 4, [M −Cl]⁺, [M −C₅H₅Fe(CO)₂]⁺, and [C₅H₅Si]⁺ [23].

$(C_5H_5Fe(CO)_2)_2Si(CH_3)H$ (Table **8**, No. **4**) has been obtained by the reaction of $C_5H_5Fe(CO)_2Si(CH_3)(Cl)H$ with $Na[C_5H_5Fe(CO)_2]$ (about 1:1 mole) in methylcyclo-hexane at 25 °C for 6 d in the dark. The residue from the evaporation of the solvent is extracted with pentane. The complex crystallizes from pentane at −78 °C in 59% yield.

The solubility of the compound increases in the series pentane < cyclohexane < ben-zene < toluene. In daylight, the solid as well as its solutions loose CO to form the CO bridged compound $C_5H_5Fe(CO)(\mu-CO)(\mu-Si(CH_3)H)(CO)FeC_5H_5$. The electronega-tive H atom in the $Si(CH_3)H$ group is readily exchanged by treatment with CCl_4 or $[(C_6H_5)_3C]^+BF_4^-$ in benzene at room temperature [24].

$(C_5H_5Fe(CO)_2)_2Si(CH_3)Cl$ (Table **8**, No. **5**) has been obtained from compound No. 4 by halogenation with CCl_4 in benzene at 25 °C. UV irradiation converts it in benzene to $C_5H_5Fe(CO)(\mu-CO)(\mu-Si(CH_3)Cl)(CO)FeC_5H_5$ [24].

$(C_5H_5Fe(CO)_2)_2Si(C_6H_5)_2$ (Table **8**, No. **6**). The preparation has not been described [25]. Irradiation in benzene at 25 °C gives $(C_5H_5Fe(CO)_2)_2$ and a violet substance, which is believed to be $C_5H_5Fe(CO)(\mu-CO)(\mu-Si(C_6H_5)_2)(CO)FeC_5H_5$ [15].

$(C_5H_5Fe(CO)_2)_2(Si(CH_3)_2)_n$ (n=2 and 3, Table **8**, Nos. **7** and **8**). The complete IR spectra are reported in [11]. The two complexes are at least as stable as $C_5H_5Fe(CO)_2^-$ $Si(CH_3)_3$ to both air oxidation and thermal decomposition [11].

$(C_5H_5Fe(CO)_2)_2(Si(CH_3)CH_2)_2$ (Table **8**, No. **9**) is crystallized from ether/pentane. The complete IR spectrum is reported in [11]. UV irradiation in benzene gives some $(C_5H_5Fe(CO)_2)_2$ [11].

$(C_5H_5Fe(CO)_2)_2GeH_2$ (Table **8**, No. **10**) is prepared by reducing $(C_5H_5Fe(CO)_2)_2GeCl_2$ with $NaBH_4$ in tetrahydrofuran/methanol at 0 °C. It is purified by sublimation at 120 °C/ 10^{-4} Torr, 55% yield. The air-sensitive solid reacts with $CHCl_3$ to give the dichloro compound No. 12, probably via $(C_5H_5Fe(CO)_2)_2Ge(Cl)H$ [3].

$(C_5H_5Fe(CO)_2)_2GeCl_2$ (Table **8**, No. **12**) has also been prepared in 85% yield by heating $(C_5H_5Fe(CO)_2)_2$ with a 50% excess of $[(CH_3)_3NH][GeCl_3]$ in tetrahydrofuran for 18 h. Slightly reduced yields have been obtained with $CsGeCl_3$ [13]. It is formed in low yields together with $C_5H_5Fe(CO)_2GeCl_3$ in the reaction of $HGeCl_3$ with $(C_5H_5Fe(CO)_2)_2$ [4].

The ⁵⁷Fe Mössbauer data agree within the experimental error with those of the corresponding Sn compound [7, 16], see Table 9, compound No. 2. The IR spectrum in cyclohexane shows six $\nu(CO)$ bands at 1959, 1981, 1985, 2006, 2021, and 2033 cm⁻¹ suggesting that at least two conformers exist in solution [13]. For other IR bands (in KBr) see also [4].

The compound crystallizes in the monoclinic system with a=14.79±0.03, b= 7.63±0.02, c=15.04±0.03 Å, and β=96°5′±15′; space group $C2/c - C_{2h}^6$. Z=4 gives $D_c=D_m=1.96$ g·cm⁻³. The molecular structure is shown in **Fig. 33**. The point group symmetry of 2 (C_2) agrees with the interpretation of the IR spectrum in solution [9]. The C₅H₅ rings are regular pentagonal and planar. The Ge–Cl bond length is significantly longer than that reported for $GeCl_4$ (2.09 Å), and the Fe–Ge bond is shorter than expected from the covalent radii of Ge and Fe [5, 6]. A drawing of the crystal packing is given in [6].

References on p. 121

Fig. 33

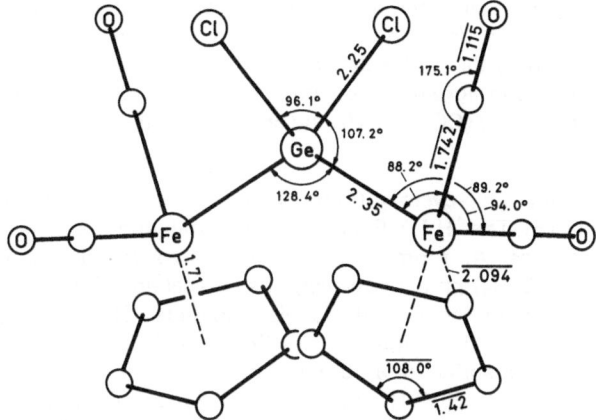

Molecular structure of $(C_5H_5Fe(CO)_2)_2GeCl_2$ [5, 6].

The substance is stable in air. It is very slightly soluble in petroleum ether, somewhat more soluble in ether, and soluble in acetone, tetrahydrofuran and other strongly polar solvents. On heating, it begins to darken at 140 °C [4]. The Cl atoms can be exchanged with other X groups [8], see method of preparation IV. It can be reduced with $NaBH_4$ to $(C_5H_5Fe(CO)_2)_2GeH_2$ [3], see compound No. 10.

$(C_5H_5Fe(CO)_2)_2GeBr_2$ (Table 8, No. 13) has recently been prepared in quantitative yield by reacting $(C_5H_5Fe(CO)_2)_2$ with $GeBr_2$ in tetrahydrofuran at reflux for 12 h. The hot solution gives the product on cooling. It is recrystallized from tetrahydrofuran [25] or tetrahydrofuran/petroleum ether [8]. The IR spectrum (KBr) shows $\nu(CO)$ bands at 1957, 1971, and 2005 cm^{-1} [25].

$(C_5H_5Fe(CO)_2)_2GeI_2$ (Table 8, No. 14). The preparation by method III was mentioned for the first time in [1]. The crystals obtained directly from the reaction mixture were red brown and decomposed at 150 °C [3]. A higher decomposition point is reported for a sample recrystallized from tetrahydrofuran/petroleum ether (1:3) [8]. The reaction between $(C_5H_5Fe(CO)_2)_2Hg$ and GeI_2 in boiling benzene for 6 h gives the complex in 21% yield [19].

The IR spectrum in CS_2 is similar to that given in the table, but in C_6H_{12} the third band disappears. This may be due to the presence of rotational isomers and their mutual transitions [9].

The crystals are air-stable, but solutions decompose in contact with air [3]. The compound can be alkylated with CH_3Li, see compound No. 19.

$(C_5H_5Fe(CO)_2)_2Ge(OCH_3)_2$ (Table 8, No. 15). Characteristic IR bands (KBr) of the OCH_3 group are reported in [8]. Seven bands in the $\nu(CO)$ region appear also in heptane, CS_2, and CCl_4, but only three bands appear in $CHCl_3$. The great number of bands may be associated with the existence of rotational isomers in solution [9].

$(C_5H_5Fe(CO)_2)_2Ge(CH_3)_2$ (Table 8, No. 19) is also prepared by treating a solution of $(C_5H_5Fe(CO)_2)_2GeI_2$ in tetrahydrofuran with CH_3Li in ether at −80 °C for 3 h. The product is chromatographed on Al_2O_3 and is eluted with petroleum ether/ether, 74% yield [3].

The compound can be sublimed at 100 °C/0.1 Torr [3]. It is appreciably air–sensitive [3] and thermally unstable decomposing rapidly in the dark to give $(C_5H_5Fe(CO)_2)_2$. The mass spectrum shows the molecular ion $[M]^+$ and the fragments $[M-n\ CO]^+$ with $n=1$ to 4. UV photolysis in ligroin gives within 20 min the CO bridged complex C_5H_5Fe-$(CO)(\mu-CO)(\mu-Ge(CH_3)_2)(CO)FeC_5H_5$ and small amounts of $(C_5H_5Fe(CO)_2)_2$ [21], see also [20].

$(C_5H_5Fe(CO)_2)_2Ge(C_5H_5-cyclo)_2$ (Table **8**, No. **23**) is recrystallized from petroleum ether/benzene (3:1) [8]. The ^1H NMR spectrum (in tetrahydrofuran) has been investigated at different temperatures. At room temperature $\delta=4.78$ (singlet, $\pi-C_5H_5$) and about 6.3 (broad signal, $\sigma-C_5H_5$) ppm. However, the δ scaling of Fig. 1 in [17] must be incorrect and $\delta=4.18$ ppm for $\pi-C_5H_5$ is also possible. The $\pi-C_5H_5$ peak does not depend on temperature in the range from -94 to $+75$ °C. The other signal narrows on heating at 75 °C and shifts to $\delta\approx6$ ppm. At -50 °C it splits into three multiplets of a AA'BB'X system with $\delta=3.38$ (H-5), 6.49 (H-2,3), and 6.76 (H-1,4) ppm (intensity ratio 1:2:2). Thus the compound undergoes a degenerate metallotropic rearrangement. Further changes of the AA'BB' part down to -94 °C indicate a relatively high barrier of rotation about the Ge–C and Ge–Fe axes [17].

$(C_5H_5Fe(CO)_2)_2Ge(C_6H_5)_2$ (Table **8**, No. **24**). The preparation by method II has been carried out with $(C_6H_5)_2GeCl_2$ and $(C_6H_5)_2GeBr_2$ at room temperature, but details are not available. The presence of seven $v(CO)$ bands in the IR spectrum indicates a restricted rotation about the Ge–Fe bonds as it is also proposed for compound No. 12 [15]. Four bands are reported in [9].

Photolysis in a degassed benzene solution at 25 °C for 45 min gives cis– and trans–$C_5H_5Fe(CO)(\mu-CO)(\mu-Ge(C_6H_5)_2)(CO)FeC_5H_5$, which may be formed from different rotamers of the starting material [15].

$(C_5H_5Fe(CO)_2)_2(Ge(CH_3)_2)_2$ (Table **8**, No. **25**) is formed in 24.9% yield along with $C_5H_5Fe(CO)_2Ge(CH_3)_3$ (47%) when $C_5H_5Fe(CO)_2GeCl_3$ reacts with CH_3MgI in ether at 20 °C. The two components are separated by chromatography on Al_2O_3 with petroleum ether and recrystallization from hexane [18].

The IR spectrum (KBr) is completely given from 468 to 3119 cm^{-1}. In solutions of cyclohexane two pairs of $v(CO)$ bands are observed close to 1945 and 1995 cm^{-1}. The compound is slightly soluble in nonpolar solvents and readily soluble in polar organic solvents. It slowly decomposes on long standing in light and is unstable in solution [18].

$(C_5H_5Fe(CO)_2)_2(Ge(C_2H_5)_2)_2$ (Table **8**, No. **26**) is obtained in 14.5% yield together with 65.5% $C_5H_5Fe(CO)_2Ge(C_2H_5)_3$ by the method described for the compound No. 25 but with C_2H_5MgBr.

The IR spectrum (KBr) is completely reported from 465 to 3110 cm^{-1}. In cyclohexane two pairs of $v(CO)$ bands are close to 1940 and 1985 cm^{-1}. Their intensity ratio differs somewhat from that of compound No. 25, which can be associated with different conformations of the two compounds in solution [18]. The solubility and stability are similar to those of No. 25.

$(C_5H_5Fe(CO)_2)_2(Ge(CH_3)_2-)_2O$ (Table **8**, No. **27**) is prepared in about 30% yield by oxidation of $C_5H_5Fe(CO)(\mu-CO)(\mu-Ge(CH_3)_2)(CO)FeC_5H_5$ in toluene. Air is blown over the surface. The residue from evaporation of the solvent is chromatographed on Al_2O_3, eluted with CH_2Cl_2/acetone, and recrystallized from pentane at -78 °C [22].

The ^{1}H NMR spectrum in CD_3COCD_3 suggests the presence of two species in a ratio of about 2:1. A ratio 3:1 is obtained in CF_2Cl_2. On cooling the CF_2Cl_2 solution the ratio increases to 10:1 at $-5\,°C$ and to an extremely large value at $-33\,°C$. Warming to 25 °C restores the original equilibrium ratio. The two species may be rotational isomers with a barrier imposed by $d\pi-p\pi$ interaction in the Ge–O bonds or by solvation by polar solvents [22].

Fig. 34

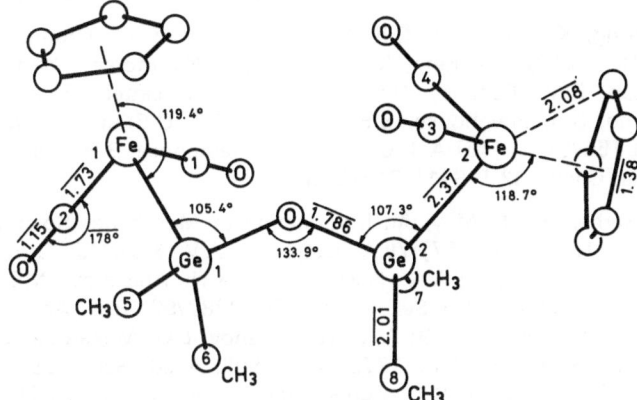

Molecular structure of $(C_5H_5Fe(CO)_2)_2(Ge(CH_3)_2-)_2O$ [22].

Other bond angles (°):

Ge(1)–Fe(1)–C(1)	85.7(4)	C(5)–Ge(1)–C(6)	107.4(6)
Ge(1)–Fe(1)–C(2)	86.6(4)	C(5)–Ge(1)–O	106.7(5)
C(1)–Fe(1)–C(2)	85.6(6)	C(6)–Ge(1)–O	109.7(5)
Ge(2)–Fe(2)–C(3)	86.2(5)	Fe(2)–Ge(2)–C(7)	110.0(4)
Ge(2)–Fe(2)–C(4)	85.4(5)	Fe(2)–Ge(2)–C(8)	112.7(4)
C(3)–Fe(2)–C(4)	92.8(7)	C(7)–Ge(2)–C(8)	116.3(6)
Fe(1)–Ge(1)–C(6)	113.3(4)	C(7)–Ge(2)–O	106.4(5)
Fe(1)–Ge(1)–C(5)	114.1(5)	C(8)–Ge(2)–O	103.4(6)

The compound crystallizes in the monoclinic system with $a=8.056(2)$, $b=12.506(2)$, $c=22.631(3)$ Å, and $\beta=98.01(1)°$; space group $P2_1/n-C_{2h}^5$. $Z=4$ gives $D_c=1.692\ g\cdot cm^{-3}$. Two rotational isomers are present in the unit cell in a disordered fashion. Approximately 88% of the molecules in the crystal have the conformation shown in **Fig. 34**. The remaining molecules differ from this by a rotation about the Ge(1)–O bond and probably also by a rotation about the Ge(2)–Fe(2) bond.

The complete mass spectrum is listed in a table in [22]. The molecular ion is not observed.

References:

[1] D. Harbourne, F.G.A. Stone (unpublished results according to [2]). – [2] F.G.A. Stone (Pure Appl. Chem. **10** [1965] 37/51, 44). – [3] N. Flitcroft, D.A. Harbourne, J. Paul, P.M. Tucker, F.G.A. Stone (J. Chem. Soc. A **1966** 1130/3). – [4] A.N. Nesmeyanov, K.N. Anisimov, N.E. Kolobova, F.S. Denisov (Izv. Akad. Nauk SSSR Ser. Khim. **1966** 2246; Bull. Acad. Sci. USSR Div. Chem. Sci. **1966** 2185). – [5] M.A. Bush, P. Woodward (Chem. Commun. **1967** 166/7).

[6] M.A. Bush, P. Woodward (J. Chem. Soc. A **1967** 1833/8). – [7] R.H. Herber, Y. Goscinny (Inorg. Chem. **7** [1968] 1293/8). – [8] A.N. Nesmeyanov, K.N. Anisimov, N.E. Kolobova, F.S. Denisov (Izv. Akad. Nauk SSSR Ser. Khim. **1968** 142/5; Bull. Acad. Sci. USSR Div. Chem. Sci. **1968** 133/6). – [9] A.N. Nesmeyanov, K.N. Anisimov, B.V. Lokshchin, N.E. Kolobova, F.S. Denisov (Izv. Akad. Nauk SSSR Ser. Khim. **1969** 758/63; Bull. Acad. Sci. USSR Div. Chem. Sci. **1969** 690/3). – [10] R.B. King, K.H. Pannell, M. Ishaq, C.R. Bennett (4th Intern. Conf. Organometal. Chem., Bristol 1969, A3).

[11] R.B. King, K.H. Pannell, C.R. Bennett, M. Ishaq (J. Organometal. Chem. **19** [1969] 327/37). – [12] R.B. King, K.H. Pannell (Z. Naturforsch. **24b** [1969] 262). – [13] J.D. Cotton, R.M. Peachey (Inorg. Nucl. Chem. Letters **6** [1970] 727/31). – [14] A.J. Cleland, S.A. Fieldhouse, B.H. Freeland, R.J. O'Brien (J. Chem. Soc. Chem. Commun. **1971** 155/6). – [15] A.J. Cleland, S.A. Fieldhouse, B.H. Freeland, R.J. O'Brien (J. Organometal. Chem. **32** [1971] C15/C18).

[16] R.H. Herber (in: L. May, An Introduction to Mössbauer Spectroscopy, London 1971, p. 138/54, 150/2). – [17] Yu.A. Ustynyuk, A.V. Kisin (J. Organometal. Chem. **33** [1971] C61/C63). – [18] A.N. Nesmeyanov, K.N. Anisimov, N.E. Kolobova, F.S. Denisov (Izv. Akad. Nauk SSSR Ser. Khim. **1971** 2287/90; Bull. Acad. Sci. USSR Div. Chem. Sci. **1971** 2158/60). – [19] A.N. Nesmeyanov, L.G. Makarova, V.N. Vinogradova (Izv. Akad. Nauk SSSR Ser. Khim. **1972** 1449; Bull. Acad. Sci. USSR Div. Chem. Sci. **1972** 1406). – [20] R.C. Job, M.D. Curtis (Inorg. Chem. **12** [1973] 2510/3).

[21] R.C. Job, M.D. Curtis (Inorg. Chem. **12** [1973] 2514/9). – [22] R.D. Adams, F.A. Cotton, B.A. Frenz (J. Organometal. Chem. **73** [1974] 93/101). – [23] B.J. Aylett, H.M. Colquhoun (J. Chem. Res. S **1977** 148; J. Chem. Res. M **1977** 1677/93). – [24] W. Malisch, W. Ries (Angew. Chem. **90** [1978] 140/1; Angew. Chem. Intern. Ed. Engl. **17** [1978] 120). – [25] J.V. Scibelli, M.D. Curtis (Syn. Reactiv. Inorg. Metal–org. Chem. **8** [1978] 399/405).

2.5.2.2.3.2 Compounds with Fe–Sn and Fe–Pb Bonds

The compounds are listed in Table 9. In the case of E=Sn different ^{5}L ligands are bonded to the Fe atoms, e.g. C_5H_5 (Nos. 1 to 32), $C_5H_4CH_3$ (Nos. 33 to 36), and C_9H_7 (indenyl, Nos. 37 to 40). Compounds No. 41 to 47 are examples with a different ^{5}L ligand on each Fe atom. Only one example for E=Pb has been described (No. 48). Attempts to prepare $(C_5H_5Fe(CO)_2)_2PbCl_2$ and $(C_5H_5Fe(CO)_2)_2Pb(OOCCH_3)_2$ by method II were unsuccessful [3, 40].

General methods of preparation are given below. Other methods, indicated by the term "special", are described under further information.

Method I: A solution of $(^5LFe(CO)_2)_2$ and $SnX_2 \cdot 2\ H_2O$ (or anhydrous SnX_2, equimolar or excess) in alcohol, tetrahydrofuran, or benzene is heated at reflux temperature for several hours. The product separates upon cooling or by addition of a hydrocarbon solvent, see [3, 12, 52] for X=Cl, Br, I and [41] for X=NCS, HCOO, CH_3COO. Under rigorous exclusion of air and moisture the direct insertion reaction predominates for all halides, and it is markedly enhanced by diffuse light [59].

Method II: The $(^5LFe(CO)_2)_2$ (in excess) is reacted with SnX_4 (X=Cl, Br, I) in refluxing xylene or n–butanol. With X=Cl further reaction to $(^5LFe(CO)_2)_3SnX$ takes place [51].

Method III: A solution of R_2EX_2 (E=Sn and Pb) in tetrahydrofuran is added over a 1 to 2 h period to $Na[C_5H_5Fe(CO)_2]$ in tetrahydrofuran. The reaction usually occurs at room temperature. The solvent is evaporated in a vacuum, and the residue is extracted with a hydrocarbon solvent [1]. With SnI_4 the mixture is refluxed for 24 h [9].

Method IV: Anhydrous SnX_2 (X=Cl, Br) in acetone is added to $(C_5H_5Fe(CO)_2)_2Hg$ in acetone. After refluxing for 5 to 6 h and evaporation to dryness, the residue is extracted with CH_2Cl_2, and the solvent removed in a vacuum. The solid product is washed with methanol [15].

Method V: $(C_5H_5Fe(CO)_2)_2SnCl_2$ and a salt KX or NaX are refluxed in methanol or acetone for 1 to 2 h. After removal of the solvent under reduced pressure, the residue is suspended in benzene and filtered. The product precipitates from the concentrated benzene solution upon addition of ligroin [41]. Or the KCl is filtered off, and water is added to separate the crude product [34].

Method VI: $(C_5H_5Fe(CO)_2)_2SnCl_2$ is reacted in benzene with thiols in the presence of $N(C_2H_5)_3$. After 2 h at room temperature and removal of $[(C_2H_5)_3NH]Cl$ the solvent is evaporated [34].

Method VII: For the preparation of compounds containing two different ^5L ligands, $C_5H_5Fe(CO)_2SnX_3$ and the $(^5LFe(CO)_2)_2$ are refluxed in petroleum ether (100 to 120 °C) or in xylene [51].

Mechanistic aspects of the formation in methods I and II are discussed in [38, 51, 52, 59].

Explanations to Table 9: Dipole moments, μ_D, have been measured in benzene at 25 °C. The atomic polarization has not been taken into account [53].

Table 9
Compounds of the $^5LFe(CO)_2-E(X,R)_2-(CO)_2Fe^5L'$ Type with E=Sn and Pb.
Further information on compounds preceded by an asterisk is given at the end of the table. For abbreviations and dimensions see p.170.

No.	Bridging group $E(X,R)_2$ Method of preparation (yield in %)	Properties and further remarks Explanations see above	Ref.
With E=Sn, $^5L=^5L'=C_5H_5$:			
1	SnF₂ I (80)	m.p. 175 to 177° (in a sealed tube) IR (CS_2): 1955, 1974, 1999, 2025 preparation in $n-C_4H_9OH$ at reflux for 12 h	[51, 52]
*2	SnCl₂ I (90), II (−), IV (25)	yellow needles; m.p. 168 to 169° ^1H NMR $(CDCl_3)$: 5.02 (s, C_5H_5) $^{57}Fe-\gamma$ (77 K): $\delta=0.36$, $\Delta=1.68$ $^{119}Sn-\gamma$ (77 K): $\delta=1.95$ (SnO_2), $\Delta=2.38$ IR (CS_2): 1951, 1971, 1999, 2023 (Nujol): 1972, 2027	[3, 8, 15, 22, 23, 32, 41, 49, 51, 52]

References on p. 137

Table 9 [continued]

No.	Bridging group $E(X,R)_2$ Method of preparation (yield in %)	Properties and further remarks Explanations on p. 123	Ref.
*3	$SnBr_2$ I (70), II (−), IV (not pure)	orange needles; m.p. 183 to 185°, dec. at 182° ^1H NMR $(CDCl_3)$: 5.02 (s, C_5H_5) $^{119}Sn-\gamma$ (80 K): $\delta=0.11$ (α-Sn), $\Delta=2.42$ IR (C_6H_{12}): 1955, 1974, 1999, 2025	[12, 15, 22, 23, 41, 49, 51, 53]
*4	SnI_2 I (48), II (−), III (86)	dark orange to brown; m.p. 238 to 240° (sealed tube) ^1H NMR (THF): 5.13 (s, C_5H_5) $^{119}Sn-\gamma$ (80 K): $\delta=0.10$ (α-Sn), $\Delta=2.25$ IR (C_6H_{12}): 1955, 1972, 1998, 2025	[9, 12, 22, 23, 41, 44, 51, 53]
5	SnS −	yellow; m.p. 268° preparation apparently like method V, but no details reported	[11]
*6	Sn⟨ S–S / S–S ⟩ VI (60)	orange microcrystals; m.p. 133 to 135° ^1H NMR $(CDCl_3, 28°)$: 4.85 (s, C_5H_5) IR (KBr): 1943, 1990, 2050; bands of C_5H_5 at 860, 1005, 1020, 1420, 1440, 3150	[55]

7	$Sn(SC_2H_5)_2$
	 (−)
		yellow; m.p. 129 to 131°
		 ^1H NMR (THF): 5.10 (s, C_5H_5)
		 IR (C_6H_{12}): 1950, 1959, 1987, 2007;
		    (THF): 1945, 1952, 1988, 2004
		 preparation like method V, no details
		    given
			[11, 22]

*8	$Sn(SC_6H_5)_2$ VI (84)	ochre crystals from CH_3COCH_3/H_2O; m.p. 132 to 134° IR $(CHCl_3)$: 1961, 1994, 2017	[34]

9	Sn⟨ S / S ⟩
	 VI (60)
		orange-yellow microcrystals from aqueous
		    CH_3OH; m.p. 160 to 165°
		 IR $(CHCl_3)$: 1965, 1990, 2011;
		    (C_6H_{12}): 1951, 1963, 1991, 1997, 2015
		 stability like No. 8
			[34]

10	Sn⟨ S / S ⟩ C_6H_3-CH_3
	 VI (72)
		deep orange plates from $CHCl_3/CH_3OH$;
		    m.p. 172 to 173°
		 IR $(CHCl_3)$: 1950, 1962, 1993, 2018;
		    (C_6H_{12}): 1950, 1968, 1990, 2020
		 stability like No. 8
			[34]

*11	$Sn(OS(O)C_6H_5)_2$ V (79)	orange yellow; m.p. 153 to 155° ^1H NMR $(CDCl_3)$: 5.09 (s, C_5H_5), 7.6 (m, C_6H_5) $^{119}Sn-\gamma$ (80 K): $\delta=1.91$ (SnO_2), $\Delta=2.54$ IR (KBr): 1926, 1953, 1966, 1968, 2008	[29, 44, 47]

References on p. 137

Table 9 [continued]

No.	Bridging group $E(X,R)_2$ Method of preparation (yield in %)	Properties and further remarks Explanations on p.123	Ref.
12	$Sn(OH)_2$ (−)	orange; m.p. 155° no details of the preparation available	[11]
*13	$Sn(OOCCH_3)_2$ I (−), V (100)	yellow; m.p. 215.5 to 216.5° ^{1}H NMR (THF): 5.05 (s, C_5H_5) ^{119}Sn-γ (80 K): $\delta = -0.47$ (α-Sn), $\Delta = 2.60$ IR (C_6H_{12}): 1955, 1971, 2000, 2025	[11, 22, 41]
*14	$Sn(NO_2)_2$ −	yellow to orange, plate-like crystals; m.p. 166° ^{1}H NMR (THF): 4.93 (s, C_5H_5) IR (THF): 1964, 1984, 2004, 2027 $\mu_D = 5.45$ D	[11, 14, 22, 27, 53]
15	$Sn(NO_3)_2$ presumably by V (−)	yellow; m.p. 191° $\mu_D = 4.82$ D	[11, 53]
*16	$Sn(NCS)_2$ I (−), V (77)	orange needles; m.p. 197° (dec.) ^{1}H NMR (THF): 5.33 (s, C_5H_5) ^{57}Fe-γ (77 K): $\delta = 0.12$ (Fe), $\Delta = 1.69$ ^{119}Sn-γ (77 K): $\delta = 1.87$ ($BaSnO_3$), $\Delta = 2.57$ IR (C_6H_{12}): 1980, 2005, 2027, 2050 (CHCl$_3$): 1970, 1983, 2007, 2022 $\mu_D = 6.26$ D	[11, 22, 34, 41, 45, 53]
17	$Sn(OOCH)_2$ I (−), V (−)	^{119}Sn-γ (80 K): $\delta = -0.49$ (α-Sn), $\Delta = 2.19$ IR (CHCl$_3$): 1967, 1974, 2005, 2028	[41]
*18	$Sn(CH_3)_2$ II (78)	orange to yellow; m.p. 106 to 108° ^{1}H NMR (CS_2): 0.59 (CH_3, J(Sn,CH_3) = 36), 4.76 (s, C_5H_5) ^{57}Fe-γ (78 K): $\delta = 0.380$, $\Delta = 1.715$ ^{119}Sn-γ (78 K): $\delta = 1.68$ (SnO_2), $\Delta = 0$ IR (C_6H_{12}): 1930, 1938, 1974, 1985	[1, 9, 18, 19, 22, 39, 54, 60]
*19	$Sn(C_2H_5)_2$ II (32)	orange to yellow; m.p. 87 to 89°, 76° (dec.) ^{1}H NMR (CCl$_4$): 1.31 (C_2H_5), 4.81 (C_5H_5) (THF): 4.99 (C_5H_5) ^{57}Fe-γ (78 K): $\delta = 0.373$, $\Delta = 1.88$ ^{119}Sn-γ (78 K): $\delta = 1.74$ (SnO_2), $\Delta = 0$ IR (C_6H_{12}): 1928, 1938, 1974, 1983; (CH$_3$CN): 1918, 1965, ≈1982, similar in CH_2Cl_2 $\mu_D = 3.32$ D	[3, 9, 11, 18, 19, 22, 48, 49, 53]

References on p. 137

Table 9 [continued]

No.	Bridging group E(X,R)$_2$ Method of preparation (yield in %)	Properties and further remarks Explanations on p. 123	Ref.
*20	Sn(C$_4$H$_9$-n)$_2$ —	m.p. 90 to 92° IR (C$_6$H$_{12}$): 1928, 1936, 1973, 1983, similar in THF μ_D=3.42 D	[22, 48, 53]
*21	Sn(C$_4$H$_9$-i)$_2$ —	μ_D=3.63 D	[48, 53]
22	Sn(CH$_2$CH=CH$_2$)$_2$ probably like No. 19 (−)	m.p. 77 to 80° IR (C$_6$H$_{12}$): 1932, 1942, 1977, 1988; (THF): 1927, 1935, 1973, 1985	[22]
*23	Sn(C$_5$H$_5$-σ)$_2$ —	orange needles; m.p. 149 to 150° ^{1}H NMR (THF): 5.00 (C$_5$H$_5$Fe), (CDCl$_3$): 6.31 (C$_5$H$_5$Sn, J(Sn,C$_5$H$_5$) =13.8) IR (C$_6$H$_{12}$): 1941, 1952, 1983, 1993 (THF): 1946, 1954, 1980, 1986	[11, 22], 39]
*24	Sn(C$_5$H$_4$CH$_3$-σ)$_2$ —	^{1}H NMR (CCl$_4$): 2.17 (CH$_3$), 5.60, 6.28 (C$_5$H$_4$CH$_3$, positions α and β to CH$_3$)	[39]
*25	Sn(C$_6$H$_5$)$_2$ II (−)	orange; m.p. 148 to 150° ^{1}H NMR (THF): 4.82 (s, C$_5$H$_5$) ^{119}Sn-γ (−): δ= −0.36 (α-Sn), Δ=0 IR (C$_6$H$_{12}$): 1933, 1947, 1980, 1998; (THF): 1926, 1941, 1975, 1994 μ_D=3.39 D	[1, 11, 16, 22, 43, 48, 53]
*26	Sn(C$_5$H$_5$Mo(CO)$_3$)$_2$ special	red orange to red; dec. at 140 to 150° IR (CH$_2$Cl$_2$): 1883.3, 1916.6, 1950, 1988.8, 2011, 2016.7	[5]
27	Sn(Cl)CH$_2$CH=CH$_2$	small quantities isolated in a reaction of No. 2 with CH$_2$=CHCH$_2$MgI; not characterized	[54]
*28	Sn(Cl)C$_6$H$_5$ special	yellow orange; 128 to 130° IR (CH$_3$COCH$_3$): 1948, 1956, 1989, 2015; (CS$_2$): 1943, 1959, 1987, 2011	[58]
*29	Sn(Cl)C$_5$H$_5$Mo(CO)$_3$ special	purple; dec. at 128 to 136° IR (CH$_2$Cl$_2$): 1886, 1921, 1949, 1988, 2008, 2029, 2057	[21]
30	Sn(Cl)Mn(CO)$_5$	brown; m.p. 141 to 143° ^{119}Sn-γ (−): δ=0 (α-Sn), Δ=2.02 preparation not described	[11, 16, 30, 47]
31	Sn(Cl)Re(CO)$_5$	brown; m.p. 143 to 144° preparation not described	[11]

References on p. 137

Table 9 [continued]

No.	Bridging group E(X,R)$_2$ Method of preparation (yield in %)	Properties and further remarks Explanations on p.123	Ref.
*32	Sn(Br)CH$_3$ II at −78° (−)	red; m.p. 127 to 129° ^{1}H NMR (C$_6$D$_6$): 1.38 (CH$_3$), 4.23 (C$_5$H$_5$) IR (−): 1940, 1952, 1988, 2000	[60]

With E = Sn, ^{5}L = ^{5}L′ = C$_5$H$_4$CH$_3$:

No.			
33	SnF$_2$ I (−)	prepared by method I in refluxing C$_4$H$_9$OH for 15 h, purification not possible	[51]
34	SnCl$_2$ I (80), II (−)	m.p. 102 to 103° (sealed tube) IR (CS$_2$): 1948, 1971, 1993, 2020	[51]
35	SnBr$_2$ I (27), II (−)	m.p. 119 to 120° (sealed tube) IR (CS$_2$): 1950, 1970, 1993, 2020	[51]
36	SnI$_2$ I (36 to 60), II (−)	m.p. 139 to 142° (sealed tube) IR (CS$_2$): 1949, 1970, 1992, 2020 reaction in benzene gives the best yield for method I	[51]

With E = Sn, ^{5}L = ^{5}L′ = C$_9$H$_7$ (indenyl):

No.			
37	SnF$_2$ I (−)	formed along with (C$_9$H$_7$Fe(CO)$_2$)$_3$SnF by method I in refluxing C$_4$H$_9$OH/3 h; not isolated, but identified by IR spectrum	[51]
38	SnCl$_2$ I (75), II (−)	m.p. 218 to 220° (sealed tube) IR (CS$_2$): 1950, 1973, 1995, 2021	[51]
39	SnBr$_2$ I (44), II (−)	dec. at 220° (sealed tube) IR (CS$_2$): 1951, 1973, 1996, 2121	[51]
40	SnI$_2$ I (50 to 70), II (−)	dec. at 215° (sealed tube) IR (CS$_2$): 1950, 1972, 1993, 2020 the preparation in benzene by method I gives the best yield	[51]

With E = Sn, ^{5}L = C$_5$H$_5$, ^{5}L′ = C$_5$H$_4$CH$_3$:

No.			
41	SnCl$_2$ VII (55)	m.p. 109 to 111° (sealed tube) IR (CS$_2$): 1952, 1972, 1996, 2023	[51]
42	SnBr$_2$ VII (−)	m.p. 122 to 125° (sealed tube) IR (CS$_2$): 1953, 1972, 2000, 2022	[51, 52]
43	SnI$_2$ VII (21)	m.p. 148 to 149° (dec.) IR (CS$_2$): 1953, 1974, 1996, 2022	[51]
44	Sn(Cl) (CO)$_3$CoP(C$_4$H$_9$)$_3$ VII (−)	m.p. 106 to 108° IR (CS$_2$): 1928, 1956, 1951, 1980, 1991, 2016	[52]

References on p. 137

Table 9 [continued]

No.	Bridging group $E(X,R)_2$ Method of preparation (yield in %)	Properties and further remarks Explanations on p.123	Ref.

With E = Sn, $^5L = C_5H_5$, $^5L' = C_9H_7$ (indenyl):

45	$SnCl_2$ VII (22)	dec. at 160° IR (CS_2): 1951, 1972, 1997, 2021	[51]
46	$SnBr_2$ VII (−)	dec. at 130° IR (CS_2): 1953, 1974, 1996, 2023	[51]
47	SnI_2 VII (8)	dec. at 140° IR (CS_2): 1955, 1973, 1995, 2021	[51]

With E = Pb, $^5L = ^5L' = C_5H_5$:

| *48 | $Pb(CH_3)_2$
III (−) | black | [1, 24] |

* Further information:

$(C_5H_5Fe(CO)_2)_2SnCl_2$ (Table **9**, No. **2**) has also been obtained from the reaction of $(C_5H_5Fe(CO)_2)_2Sn(OS(O)C_6H_5)_2$ with HCl gas in C_6H_6 for 1 h, 79% yield [44]. The formation from $(C_5H_5Fe(CO)_2)_2Sn(C_6H_5)_2$ (No. 25) or $(C_5H_5Fe(CO)_2)_3SnC_6H_5$ and HCl in CCl_4 is mentioned without details in [11]. Also briefly reported is the formation from $(C_5H_5Fe(CO)_2)_2$ and $(C_6H_5)_2SnCl_2$ (14% yield) together with ferrocene, C_5H_5Fe-$(CO)_2Sn(C_6H_5)Cl_2$, and an unidentified tin compound [8]. In boiling tetrahydrofuran, anhydrous $SnCl_2$ slowly inserts into the Fe–C bond of $C_5H_5Fe(CO)_2C_2H_5$, giving the title compound and $C_5H_5Fe(CO)_2SnCl_3$ as major products (50% conversion after 104 h). $(C_5H_5Fe(CO)_2)_2SnCl_2$ and $(CH_3)_2SnCl_2$ are products of the thermal decomposition of $C_5H_5Fe(CO)_2Sn(CH_3)Cl_2$ during its sublimation [56]. The compound has also been found in a mixture of products obtained from $C_5H_5Fe(CO)(\mu\text{-}CO)_2(P(OC_6H_5)_3)FeC_5H_5$ and $SnCl_2$ [31].

$(C_5H_5Fe(CO)_2)_2SnCl_2$ can be crystallized from CH_3OH as yellow flat needles several millimeters long [50]. The diamagnetic compound [4] is nonconducting in acetone [3]. The dipole moment is $\mu = 4.67$ D (in C_6H_6 at 25 °C, P_A not taken into account). Dipole moments calculated by means of the Fe–Sn bond moment of $C_5H_5Fe(CO)_2SnCl_3$ are much higher than the experimental value even for the least polar of the possible conformations. Thus, the Fe–Sn bond moment of the complex should be considerably lower than in $C_5H_5Fe(CO)_2SnCl_3$ and the $C_5H_5Fe(CO)_2$ group must be a stronger donor than the Cl atom [53], see also [48].

The ^{57}Fe Mössbauer data remain essentially unchanged when the spectra are recorded in methyltetrahydrofuran or polymethyl methacrylate matrices. Therefore, the anomalous bond lengths and angles do not arise from intermolecular forces in the crystalline state. However, the ^{119}Sn Mössbauer spectrum changes in poly(methyl methacrylate) at 77 K. Then it consists of two quadrupole doublets with $\delta = 1.31$ and 1.96, $\Delta = 2.29$ and 2.25 mm·s^{-1}. This is explained by the presence of two rotational isomers [32], see also [36, 46]. For discussions of the Mössbauer spectra see also [17 to 19]. Other numerical values of $\delta(^{57}Fe)$ referred to Fe [45] and $\delta(^{119}Sn)$ referred to α-Sn [16, 41] are reported. The ^{119}Sn parameters are slightly smaller at room temperature [57].

References on p. 137

The quadrupole coupling constant, e^2qQ, was found to be positive for ^{57}Fe and ^{119}Sn [45]. The asymmetry parameter of the electric field gradient tensor is $\eta = 0.65 \pm 0.05$ as determined in a 50 kG magnetic field at 4.2 K. The use of an oriented matrix of single crystals established the direction of V_{zz} to be at 48° to the crystal bc plane [50]. For a calculation of Δ values from the point-charge model see [47].

The IR spectrum in Nujol from 838 to 3115 cm^{-1} is given in [3]. In a poly(methyl methacrylate) matrix there is an additional band at 2043 cm^{-1} [32]. The $v(CO)$ region has been studied in CH_2Cl_2, CH_3CN, CS_2, C_6H_{12}, and $CH_2Cl\text{-}CHCl_2$. The band positions are nearly independent of the solvent and are at 1951 to 1958, 1965 to 1975, 1998 to 2000, and 2020 to 2027 cm^{-1}. No significant change in the $v(CO)$ pattern and intensity occurs between 20 and 60 °C in $CH_2Cl\text{-}CHCl_2$ solution, which seems to exclude the presence of conformers [49]. Similar $v(CO)$ bands are reported for solutions in $CHCl_3$ [4, 41], CS_2 [51, 52, 58], acetone [58], and tetrahydrofuran [22]. A C_1 or C_2 molecular symmetry is consistent with the intensity pattern of the CO bands in cyclohexane [9, 22]. Vibrations below 630 cm^{-1} observed in the IR [23, 41] or Raman spectrum [23] are listed in Table 10.

Fig. 35

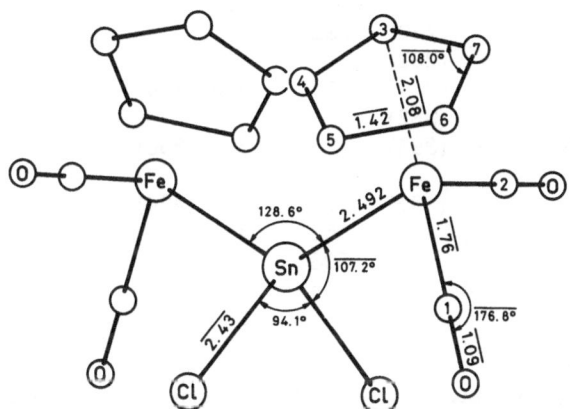

Molecular structure of $(C_5H_5Fe(CO)_2)_2SnCl_2$ [20].

Selected angles (°):

Sn–Fe–C(1)	89.0(1.2)	Sn–Fe–C(4)	86.2(1.0)
Sn–Fe–C(2)	89.0(1.2)	Sn–Fe–C(5)	103.2(1.0)
C(1)–Fe–C(2)	98.4(1.7)	Sn–Fe–C(6)	145.2(1.1)
Sn–Fe–C(3)	105.8(0.9)	Sn–Fe–C(7)	145.2(1.1)

The compound crystallizes in the monoclinic system with $a = 14.98 \pm 0.03$, $b = 7.63 \pm 0.02$, $c = 15.18 \pm 0.03$ Å, and $\beta = 94°27' \pm 20'$; space group $C2/c\text{–}C_{2h}^6$. $Z = 4$ gives $D_c = 2.08$ g·cm^{-3}, $D_m = 2.02$ g·cm^{-3}. The most unusual features of the molecule are the bond lengths and angles associated with the coordination environment of the Sn atom, see **Fig. 35**. The Fe–Sn distance is shorter than any previously reported tin to transition metal bond. The Sn–Cl distances is greater than the usual Sn–Cl bond length in 4-coordinate Sn compounds. The Fe–Sn–Fe and Cl–Sn–Cl angles show the largest deviation from ideal tetrahedral coordination yet observed for a Sn atom. These data are consistent with an increased s character in the tetrahedral hybrid orbitals used in Fe–Sn bond formation and increased p character in the orbitals used in Sn–Cl bond formation [20].

References on p. 137

Table 10
Vibrational frequencies (cm^{-1}) of $(C_5H_5Fe(CO)_2)_2SnX_2$ compounds, X=Cl, Br, and I. Solid state (Nujol) with assignment [23]; in parentheses, in $CHCl_3$ without assignment [41].

X=Cl		Br		I		
IR	Raman	IR	Raman	IR	Raman	Assignment
	106		109			
114						possibly δ (CFeC) and δ (FeSnFe)
123		(122)			121	
				143		
			149			
155 (154)	155	(154)		(164)		
	196	201	197	196 (196)	198	ν(Sn–Fe)sym.
233 (230)	234	235 (230)	234	232 (232)	233	ν(Sn–Fe)asym.
				167	167	ν(Sn–X)sym.
286 (280)	287	(280)	182		177	ν(Sn–X)asym.
(292)		(292)				
370 (374)	371	372	374	370 (376)	371	ν(Fe–C$_{ring}$)
(380)		(382)				
393	393	387 (394)		390 (390)		
	(422)					
	(432)	(436)		(434)		
456	454	452	453	454	452	ν(Fe–CO)
504	506	504	508			
	584	574	574			
596	599			596		δ(FeCO)
627	626	624		626		

The mass spectrum shows a very weak molecular peak, ions resulting from fission of one Sn–Fe bond, but mainly ions which still contain the three metal atoms. Loss of CO groups and C_5H_5 transfer from Fe to Sn predominate [10].

The compound is insoluble in petroleum ether and water, sparingly soluble in CCl_4, CS_2, ether, ethyl acetate, and methanol, but fairly soluble in benzene and acetone. It is air–stable and decomposes thermally only well above its melting point, yielding ferrocene and CO in nearly quantitative yield. It is recovered unchanged from boiling pyridine after 2 h and is unaffected by boiling 3 M hydrochloric acid [3].

The Cl atoms can readily be replaced by other groups to make, for example, compounds No. 6, 8, 9, 10, 11, 16, 18, 19, and 26. With $LiAlH_4$ in tetrahydrofuran only dark pyrophoric products have been obtained. Treatment with $SnCl_4$ in toluene at 100 to 110 °C for 1 h gave $C_5H_5Fe(CO)_2SnCl_3$ [3]. No reaction occured with SO_2 between room temperature and 50 °C [44]. A product from the reaction with $Na[Co(CO)_4]$ contained Fe, Co, and Sn, but no pure compound could be isolated [5, 6]. An attempt to bridge the Fe–Sn bonds with PR_2 groups by reacting the compound with $(C_6H_5)_2PLi$ in tetrahydrofuran failed and gave only small amounts of $(C_5H_5Fe(CO)_2)_3SnCl$ [58]. For the reaction with $(C_5H_5Fe(CO)_2)_2Sn(C_6H_5)_2$ see compound No. 28.

$(C_5H_5Fe(CO)_2)_2SnBr_2$ (Table 9, No. 3). The yields obtained by method I depend on the reaction condition. For more details and for the kinetics and the mechanism

References on p. 137

of the $SnBr_2$ insertion, see [38, 51, 52, 59]. The compound can be recrystallized from benzene/n-hexane [15], toluene/n-pentane, and boiling methanol [12].

The ^1H NMR resonance in tetrahydrofuran has been found at $\delta = 5.20$ ppm [22]. The IR spectrum ($CHCl_3$) from 580 to 3108 cm^{-1} is reported in [12]. The $\nu(CO)$ bands vary only slightly in solvents like CH_2Cl_2, CS_2, CH_3CN, and C_6H_{12}. They are in the same regions as those of the previous dichloro complex [49]. On the basis of intensity pattern a C_1 or C_2 molecular symmetry has been proposed. Additional data are given for Nujol [15], C_6H_{12} [22], tetrahydrofuran [22], CS_2 [51, 52], and $CHCl_3$ [12, 41]. Vibrations in the region below 630 cm^{-1} from IR [23, 41] and Raman spectra [23] are listed in Table 10.

$(C_5H_5Fe(CO)_2)_2Fe\,I_2$ (Table 9, No. 4) is formed in high selectivity by method II when the reaction is carried out in refluxing benzene [51]. For discussions of the mechanism of formation see [51, 52, 59]. The compound is recrystallized from CH_2Cl_2/pentane [12].

The IR spectrum ($CHCl_3$) from 735 to 3040 cm^{-1} is given in [12]. The $\nu(CO)$ bands remain nearly unchanged in CCl_4, $CHCl_3$, CS_2, and C_6H_6 [22]. Further data in the $\nu(CO)$ region is given for C_6H_{12} [9, 22], CS_2 [9, 51], $CHCl_3$ [41], and tetrahydrofuran [22]. In strongly polar solvents like CH_3CN and $HCON(CH_3)_2$, the bands broaden. A change in the intensity ratio of the two high-frequency bands is interpreted as a change in the conformation of the molecule [22]. The low-frequency regions of the IR [23, 41] and Raman spectra [23] are given in Table 10.

$(C_5H_5Fe(CO)_2)_2SnS_4$ (Table 9, No. 6) has been prepared by method V with Na_2S_4 (or K_2S_5) in benzene/methanol at room temperature. It is recrystallized from chloroform or benzene [55]. The Fe-Fe bond depicted in the Fig. in [55] is obviously wrong. The complex appears to be air-stable in the crystalline state but decomposes in solution [55].

$(C_5H_5Fe(CO)_2)_2Sn(SC_6H_5)_2$ (Table 9, No. 8) is air-stable as a solid and in organic solvents. It is somewhat light-sensitive [34].

$(C_5H_5Fe(CO)_2)_2Sn(OS(O)C_6H_5)_2$ (Table 9, No. 11) has been prepared in 70% yield by passing SO_2 for 48 h through a solution of $(C_5H_5Fe(CO)_2)_2Sn(C_6H_5)_2$ in benzene at room temperature. After removal of the solvent, the residual oil crystallized from ether/petroleum ether, see also [29, 40]. The reaction can also be carried out in liquid SO_2 in an evacuated sealed tube at 22 °C for 24 h, 60% yield. The compound is recrystallized from ethanol or methylene chloride/petroleum ether [44].

The IR spectrum (KBr) is completely reported from 235 to 3112 cm^{-1}. The pattern at 800 to 900 cm^{-1} indicates an O-sulfinate group [44]. Probable assignments are given for $\nu(S-O)$ at 1088 and 1103 cm^{-1} and for $\nu(S=O)$ at 853 and 869 cm^{-1} [29].

The structure proposed from ^{119}Sn Mössbauer [47] and IR spectra [29, 44] has been confirmed by an X-ray analysis. The compound crystallizes in the orthorhombic system with $a = 17.08$, $b = 12.24$, and $c = 13.38$ Å (all ± 0.02 Å); space group $P2_12_12_1$ $-D_2^4$ and $Z = 4$. A figure of the molecular structure has not been published. The two Sn-Fe distances are much shorter than those in the compound No. 14. The geometry at the S atom is approximately tetrahedral with the lone electron pair presumably occupying the fourth arm of the tetrahedron. The following angles are given: Fe-Sn-Fe 119°, O-Sn-O 81°, and average values for O-S=O 115°, O-S-C 110°, and O=S-C 98° (each with $\pm 3°$) [28].

References on p. 137

The thermal decomposition of the complex at 180 °C for 48 h gives $C_6H_5SSC_6H_5$ (31% yield) but no SO_2. Nor is SO_2 detected when HCl gas is passed through a solution of the complex in benzene for 1 h, but $C_6H_5SO_2SC_6H_5$ (80%) and $(C_5H_5Fe(CO)_2)_2SnCl_2$ (79%) are formed [44].

$(C_5H_5Fe(CO)_2)_2Sn(OOCCH_3)_2$ (Table 9, No. 13). The preparation by method I is carried out with $Sn(OOCCH_3)_2$ in acetic anhydride as solvent [41]. In the IR spectrum $v(CO)$ bands are similar to the values in Table 9 for solutions of tetrahydrofuran [22] or $CHCl_3$ [41]. Low frequency vibrations ($CHCl_3$) from 96 to 434 cm^{-1} are given with the assignments $v(Fe-Sn)$ at 174 and $v(Sn-O)$ at 240 (asym.) and 278 (sym.) cm^{-1} [41].

$(C_5H_5Fe(CO)_2)_2Sn(NO_2)_2$ (Table 9, No. 14) has been prepared in quantitative yield [11], presumably by method V. The complex crystallizes in the monoclinic system with a $=20.26\pm0.04$, b $=7.30\pm0.02$, c $=14.65\pm0.04$ Å, and $\beta=123.4\pm0.5°$; space group C2/c $-C_{2h}^6$. Z$=4$ gives $D_c=2.10$ and $D_m=2.08$ g·cm^{-3} [13, 14, 27]. The molecular structure is shown in **Fig. 36**. The Fe–Sn bond length is considerably less than the sum of the octahedral radius of Fe and the tetrahedral radius of Sn. The NO_2 ligand is bonded as the O–N=O group, the N–O distances corresponding to the bond length in the free nitrite ion. The CO groups are bent towards the Sn atom [27].

Fig. 36

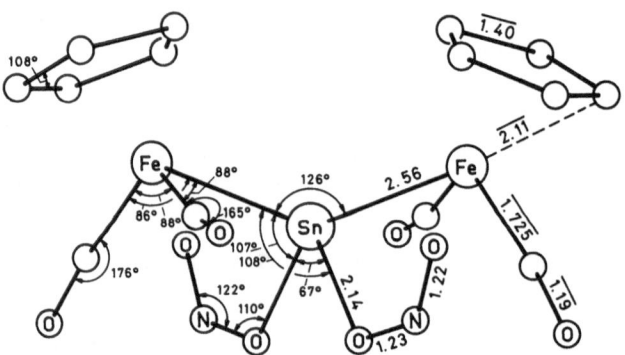

Molecular structure of $(C_5H_5Fe(CO)_2)_2Sn(NO_2)_2$ [14, 27].

The complex is insoluble in cyclohexane [22].

$(C_5H_5Fe(CO)_2)_2Sn(NCS)_2$ (Table 9, No. 16). The preparation by method V is carried out with KSCN in refluxing acetone for 2 h. Recrystallization is from $CHCl_3/CH_3OH$ [34].

The ^{57}Fe and ^{119}Sn Mössbauer data are similar at 77 and 4.2 K. The quadrupole coupling constants are positive for both nuclei [45]. For ^{119}Sn Mössbauer data see also [41]. The $v(CO)$ region of the IR spectrum is also given for solutions in tetrahydrofuran [22]. An Sn–NCS bonding is proposed since no absorptions are found between 690 and 72 cm^{-1}, the region where S bonded thiocyanates absorb [34]. Low-frequency vibrations from 142 to 432 cm^{-1} are reported. Some of them are assigned to $v(Fe-Sn)$ (168 or 204 cm^{-1}), and to $v(Sn-NCS)$ (248 (asym.) and 274 (sym.) cm^{-1}). A shoulder at 2045 cm^{-1} may be the $v(C=N)$ [41].

The substance is sensitive to light [34].

References on p. 137

(C₅H₅Fe(CO)₂)₂Sn(CH₃)₂ (Table **9**, No. **18**) is also prepared by treating (C₅H₅Fe-(CO)₂)₂SnCl₂ with excess CH₃Li in tetrahydrofuran at −80 °C. The product is isolated by chromatography on Al₂O₃ with ether/petroleum ether and sublimation at 100 °C/ 0.1 Torr, 86% yield [9]. The reaction of (C₅H₅Fe(CO)₂)₂SnCl₂ with CH₃MgI in ether gives the complex in 43% yield [54]. The compound is formed together with Sn(CH₃)₄ by disproportionation of C₅H₅Fe(CO)₂Sn(CH₃)₃ at 115 °C [42].

The ¹H NMR shift of the CH₃ groups is reported to be $\delta=0.64$ ppm (no solvent given) with J(^{117}Sn, CH₃) =35.8 and J(^{119}Sn, CH₃) =37.3 Hz [60]. The plate-like crystals belong to the monoclinic system with a =9.69 ±0.01, b =15.59 ±0.03, c =11.66 ±0.02 Å, and $\beta=107.1 \pm 1°$; space group P2₁/n−C$_{2h}^{5}$. Z=4 gives D_c=1.85 and D_m=1.79 g·cm⁻³ [13, 26, 27]. The molecular structure is shown in **Fig. 37**. The Fe-Sn bonds are longer than those in the compounds No. 2 and 14 but still shorter than the sum of the covalent radii (2.66 Å). The C₅H₅ rings are planar. The short nonbonding intramolecular distances and the distribution of the molecules in the crystal are given [27].

Fig. 37

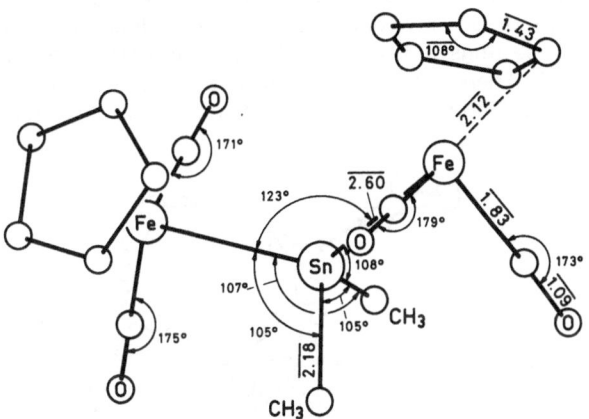

Molecular structure of (C₅H₅Fe(CO)₂)₂Sn(CH₃)₂ [26, 27].

(C₅H₅Fe(CO)₂)₂Sn(CH₃)₂ is sensitive to air and moisture [9]. Photolysis in cyclohexane for 52 min proceeds with loss of one CO ligand and ring closure to C₅H₅Fe(CO)-(μ-CO)(μ-Sn(CH₃)₂)(CO)FeC₅H₅ [60]. The polarographic reduction in CH₃OCH₂-CH₂OCH₃ occurs in a two-electron step at E$_{1/2}$= −2.7 V (referred to 10⁻³ M Ag⁺/Ag). The products are [C₅H₅Fe(CO)₂]⁻ and Sn(CH₃)₂ [7]. With SO₂ in ether, C₅H₅Fe(CO)₂-SO₂CH₃ is formed. It may be a decomposition product of (C₅H₅Fe(CO)₂)₂Sn(O-S(O)CH₃)₂ [44]. The Fe-Sn bonds are cleaved by I₂ in (CD₃)₂SO; with an excess of I₂, the end-products are C₅H₅Fe(CO)₂I and (CH₃)₂SnI₂. With HgCl₂ in CD₃COCD₃ almost quantitative yields of C₅H₅Fe(CO)₂HgCl and C₅H₅Fe(CO)₂Sn(CH₃)₂Cl are obtained. A similar reaction occurs with CH₃HgCl [54].

(C₅H₅Fe(CO)₂)₂Sn(C₂H₅)₂ (Table **9**, No. **19**) has also been prepared by reacting a suspension of (C₅H₅Fe(CO)₂)₂SnCl₂ in ether/benzene with C₂H₅MgBr for 2 h at reflux temperature. Recrystallization from methanol, 27.7% yield [3]. A 60% yield is indicated in [11], but without experimental detail.

The IR spectrum in CS₂ and CCl₄ is completely given from 828 to 3935 cm⁻¹ in [3]. v(CO) bands in tetrahydrofuran are at 1922, 1933, 1969, and 1978 cm⁻¹ [22].

References on p. 137

In [11] the compound is described as dark red. The compound is of low stability. Its odor resembles that of $(C_2H_5)_2SnCl_2$. It is soluble in petroleum ether, CH_2Cl_2, $CHCl_3$, CCl_4, and CS_2 [3].

$(C_5H_5Fe(CO)_2)_2Sn(C_4H_9\text{-n,i})_2$ (Table **9**, Nos. **20** and **21**) have probably been prepared by the Grignard method as described for No. 19. The increased dipole moment compared to No. 19 results probably from a decreased Fe–Sn–Fe angle due to the larger alkyl substituents [48, 53].

$(C_5H_5Fe(CO)_2)_2Sn(C_5H_5)_2$ (Table **9**, No. **23**). For a figure of the ^{1}H NMR spectrum in dimethoxyethane see [39]. In CS_2 solution at 30 °C, $\delta=6.12$ ppm has been measured for the $\sigma\text{-}C_5H_5$ groups, $J(C,H)=161\pm1$ Hz and $J(Sn,H)=13.6/14.2$ Hz [61]. The needle crystals belong to the orthorhombic system with $a=19.06\pm0.02$, $b=13.58\pm0.02$, and $c=8.70\pm0.01$ Å; space group $P2_12_12_1-D_2^4$. $Z=4$ gives $D_c=1.69$ g·cm^{-3}, $D_m=1.64$ g·cm^{-3} [13, 35]. The parameters b and c are exchanged in [13]. The molecular symmetry, which is show in **Fig. 38**, is close to C_2. The C_5H_5 rings joined to the Sn atom have the shape of an envelope with the flap folded at an angle of 27°. The C–C distances in these rings indicate a fairly complex delocalization of electrons [13, 25, 35], see also [37]. There are many shorter intramolecular nonbonding contacts than usual. The molecular packing in the crystal and intermolecular distances have been shown in a figure [35].

Fig. 38

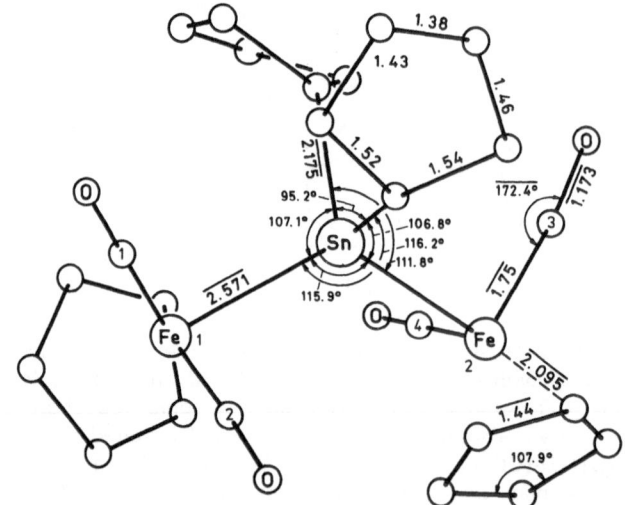

Molecular structure of $(C_5H_5Fe(CO)_2)_2Sn(C_5H_5)_2$ [13, 35].

Selected angles (°):

Sn–Fe(1)–C(1)	92.3	Sn–Fe(2)–C(3)	89.9
Sn–Fe(1)–C(2)	83.3	Sn–Fe(2)–C(4)	89.5
C(1)–Fe(1)–C(2)	94.9	C(3)–Fe(2)–C(4)	95.7

$(C_5H_5Fe(CO)_2)_2Sn(C_5H_4CH_3)_2$ (Table **9**, No. **24**). The preparation has not been mentioned. In the ^{1}H NMR spectrum protons of a Sn–CH grouping are not observed. A poorly resolved multiplet in the region of $\delta=5.5$ to 6.5 ppm is typical for an AA'BB' system. These results indicate a sandwich structure of this part of the molecule in solution [39].

$(C_5H_5Fe(CO)_2)_2Sn(C_6H_5)_2$ (Table 9, No. 25). The experimental dipole moment is compared with moments calculated for different configurations [48, 53]. More than four $v(CO)$ bands in the IR spectrum have been recorded in cyclohexane solution indicating the presence of rotational isomers: 1934, 1949, 1960, 1982, 1992, 1999, and 2010 cm^{-1} [43]. The compound crystallizes in the monoclinic system with a=34.18±0.08, b= 9.18±0.03, c=16.14±0.05 Å, and β=112.2±0.5°; space group C2/c−C_{2h}^6 or Cc−C_s^4. Z=8 gives $D_m \approx 1.8$ g·cm^{-3} [13].

Photolysis results in decarbonylation and formation of $C_5H_5Fe(CO)$-$(\mu$-CO$)(\mu$-Sn$(C_6H_5)_2)(CO)FeC_5H_5$ [43]. For the insertion of SO_2 see compound No. 11. Heating with the dichloro derivative No. 2 in benzene for 1 h gives small amounts of $(C_5H_5Fe(CO)_2)_2Sn(C_6H_5)Cl$ (compound No. 28) [58].

$(C_5H_5Fe(CO)_2)_2Sn(C_5H_5Mo(CO)_3)_2$ (Table 9, No. 26) has been prepared from $(C_5H_5Fe(CO)_2)_2SnCl_2$ and Na[$C_5H_5Mo(CO)_3$] (1:2 mole) in tetrahydrofuran at reflux for 24 h. After solvent removal in vacuum the residue is taken up in CH_2Cl_2/pentane under CO from which it crystallizes in the cold. Recrystallization from a minimum amount of CH_2Cl_2 in a CO atmosphere [5], see also [6].

Relatively large crystals of the compound appear deep red, while small crystals are red orange. The X-ray emission spectrum has bands corresponding to Sn at 12.41° ($K\beta_1$) and 14.0° to 14.12° (Kα, Kα$_1$), to Mo at 18.08° (Kβ, Kβ$_1$) and 20.29° to 20.41° (Kα, Kα$_1$, Kα$_2$), and to Fe at 51.73° (Kβ$_1$) and 57.47° (Kα$_1$) [5].

The solid is air-stable when dry, but in solution in the absence of CO it decomposes rapidly. It is insoluble in pentane, sparingly soluble in hexane and CCl$_4$, and soluble in CH_2Cl_2 and acetone [5]. Polarographic reduction occurs in $CH_3OCH_2CH_2OCH_3$ at $E_{1/2}$ −−1.8, −2.0, and −2.5 V (referred to 10^{-3} M Ag$^+$/Ag), but the last two steps are ill-defined [7].

$(C_5H_5Fe(CO)_2)_2Sn(Cl)C_6H_5$ (Table 9, No. 28) is prepared by heating equal amounts of $(C_5H_5Fe(CO)_2)_2SnCl_2$ and $(C_5H_5Fe(CO)_2)_2Sn(C_6H_5)_2$ at 155 °C for 1 h. The product is extracted with benzene and chromatographed on Al$_2$O$_3$ with benzene/petroleum ether. Recrystallization from toluene, 1.5% yield. The substance is soluble in benzene and well soluble in polar solvents like tetrahydrofuran and acetone [58].

$(C_5H_5Fe(CO)_2)_2Sn(Cl)C_5H_5Mo(CO)_3$ (Table 9, No. 29) is a secondary reaction product in the preparation of compound No. 26. It is recrystallized from CH_2Cl_2 under CO.

The substance crystallizes in the orthorhombic system with a=10.57±0.02, b= 12.14±0.02, and c=22.04±0.04 Å; space group P2$_1$2$_1$2$_1$−D$_2^4$, Z=4. The molecular structure is shown in **Fig. 39**, p. 136. The Sn-Mo bond length is considerably shorter than the expected distance based on the sum of the covalent radii (3.00 Å). The Sn-Fe bond lengths agree well with other distances between these metal atoms. The Sn-Cl bond is exceptionally long. The greatest amount of tetrahedral angle distortion at the Sn atom involves the less bulky Cl atom. The orientation of the C_5H_5 ligand on the Mo atom is distinctly different from that found in other C_5H_5Mo compounds. It is rotated about 36° from the orientation observed in $(C_5H_5Mo(CO)_3)_2$ [21].

Fig. 39

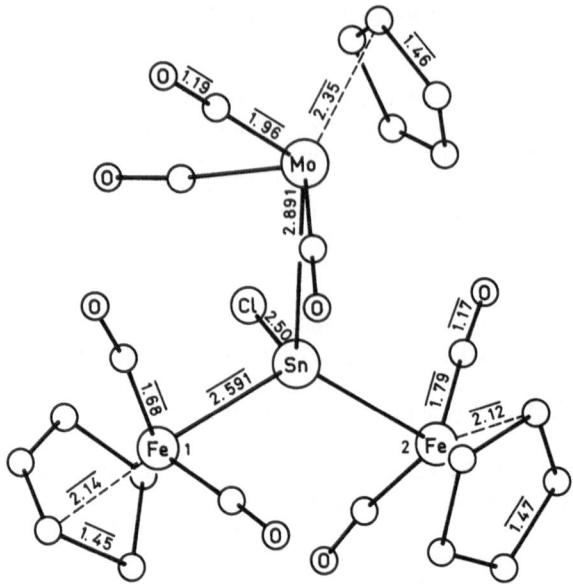

Molecular structure of $(C_5H_5Fe(CO)_2)_2Sn(Cl)C_5H_5Mo(CO)_3$ [21].

Bond angles (°):

Mo–Sn–Fe(1)	115.8(2)	Mo–C–O	173(3) mean
Mo–Sn–Fe(2)	117.9(2)		
Mo–Sn–Cl	98.4(3)	Sn–Fe(1)–CO	91(2) mean
Fe(1)–Sn–Fe(2)	115.6(2)	Fe(1)–C–O	176(4) mean
Fe(1)–Sn–Cl	107.4(3)		
Fe(2)–Sn–Cl	97.2(3)	Sn–Fe(2)–CO	87(2) mean
		Fe(2)–C–O	174(5) mean

$(C_5H_5Fe(CO)_2)_2Sn(Br)CH_3$ (Table **9**, No. **32**) appears to be formed together with $(CH_3)_3SnBr$ by a fast redistribution reaction of $C_5H_5Fe(CO)_2Sn(CH_3)_2Br$ [60].

$(C_5H_5Fe(CO)_2)_2Pb(CH_3)_2$ (Table **9**, No. **48**). The preparation by method III is carried out with $Na[C_5H_5Fe(CO)_2]$ and $(CH_3)_2PbCl_2$ (1.3:1 mole) in tetrahydrofuran at room temperature for 1 h. The substance can be recrystallized from n-hexane [1].

The black laminar crystals belong to the monoclinic system with a=9.76±0.02, b= 16.12±0.02, c=12.08±0.02 Å, and β=104.3°±1°; space group $P2_1/n-C_{2h}^5$. Z=4 gives D_c=2.15 and D_m=2.11 g·cm^{-3}. The molecular structure, see **Fig. 40**, corresponds largely to that of the Sn compound No. 18. The Pb–Fe bond lengths practically coincide with the sum of the tetrahedral radius of the Pb atom and the octahedral radius of the Fe atom. In the Sn compound the Sn–Fe bonds are somewhat shorter than the calculated covalent distances. This indicates that the metal–metal bond is weakened when Sn is replaced by Pb [24].

Also in chemical respects the Pb–Fe bond is not as strong as the Sn–Fe bond. The action of halides and HCl leads to a rupture of the Pb–Fe bond but not of the Pb–C bond [33]. Outstanding antiknock properties are claimed in [1] for the compound.

Fig. 40

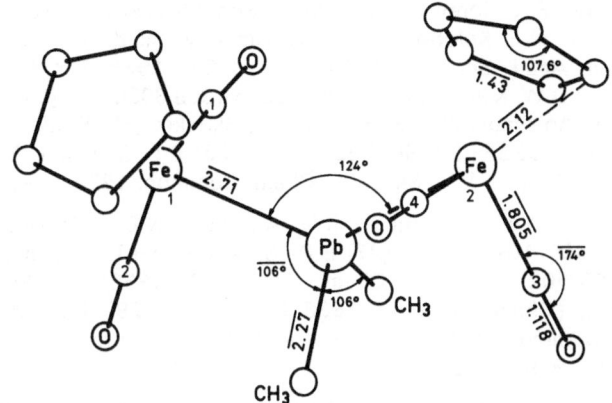

Molecular structure of $(C_5H_5Fe(CO)_2)_2Pb(CH_3)_2$ [24].

Angles (°) at the Fe atoms:

Pb–Fe(1)–C(1)	85		Pb–Fe(2)–C(3)	85
Pb–Fe(1)–C(2)	83		Pb–Fe(2)–C(4)	83
C(1)–Fe(1)–C(2)	93		C(3)–Fe(2)–C(4)	93

References:

[1] R.D. Gorsich, Ethyl Corp. (U.S. 3069449 [1962]; C.A. **58** [1963] 10241). – [2] T.P. Whaley, V. Norman, Ethyl Corp. (U.S. 3071493 [1963]; C.A. **58** [1953] 4235). – [3] F. Bonati, G. Wilkinson (J. Chem. Soc. **1964** 179/81). – [4] F. Bonati, S. Cenini, D. Morelli, R. Ugo (Inorg. Nucl. Chem. Letters **1** [1965] 107/8). – [5] S.V. Dighe, M. Orchin (J. Am. Chem. Soc. **87** [1965] 1146).

[6] S.V. Dighe (Diss. Univ. of Cincinnati 1965; Diss. Abstr. **26** [1965] 3627/8). – [7] R.E. Dessy, P.M. Weissman, R.L. Pohl (J. Am. Chem. Soc. **88** [1966] 5117/21). – [8] R.E. Edmondson, M.J. Newlands (Chem. Ind. [London] **1966** 1888/9). – [9] N. Flitcroft, D.A. Harbourne, J. Paul, P.M. Tucker, F.G.A. Stone (J. Chem. Soc. A **1966** 1130/3). – [10] J. Lewis, A.R. Manning, J.R. Miller, J.M. Wilson (J. Chem. Soc. A **1966** 1663/70).

[11] A.N. Nesmeyanov, K.N. Anisimov, N.E. Kolobova, V.V. Skripkin (Izv. Akad. Nauk SSSR Ser. Khim. **1966** 1292; Bull. Acad. Sci. USSR Div. Chem. Sci. **1966** 1248). – [12] D.J. Patmore, W.A.G. Graham (Inorg. Chem. **5** [1966] 1405/7). – [13] B.P. Bir'yukov, K.N. Anisimov, Yu.T. Struchkov, N.E. Kolobova, V.V. Skripkin (Zh. Strukt. Khim. **8** [1967] 556/7; J. Struct. Chem. [USSR] **8** [1967] 498/9). – [14] B.P. Bir'yukov, Yu.T. Struchkov, K.N. Anisimov, N.E. Kolobova, V.V. Skripkin (Chem. Commun. **1967** 750/1). – [15] F. Bonati, S. Cenini, R. Ugo (J. Chem. Soc. A **1967** 932/5).

[16] V.I. Gol'danskii, B.V. Borshagovskii, E.F. Makarov, R.A. Stukan, K.N. Anisimov, N.E. Kolobova, V.V. Skripkin (Teor. Eksperim. Khim. **3** [1967] 478/82; Theor. Exptl. Chem. [USSR] **3** [1967] 275/7). – [17] R.H. Herber (Symp. Faraday Soc. Nr. 1 [1967/68] 86/96). – [18] R.H. Herber, A. Hoffman (unpublished results according to [19]). – [19] R.H. Herber (Progr. Inorg. Chem. **8** [1967] 1/41, 34/5). – [20] J.E. O'Connor, E.R. Corey (Inorg. Chem. **6** [1967] 968/71).

[21] J.E. O'Connor, E.R. Corey (J. Am. Chem. Soc. **89** [1967] 3930/1). – [22] K.N. Anisimov, B.V. Lokshin, N.E. Kolobova, V.V. Skripkin (Izv. Akad. Nauk SSSR Ser. Khim. **1968** 1024/30; Bull. Acad. Sci. USSR Div. Chem. Sci. **1968** 978/81). – [23] D.M. Adams, J.N. Crosby, R.D.W. Kemmitt (J. Chem. Soc. A **1968** 3056). – [24] B.P. Bir'yukov, Yu.T. Struchkov, K.N. Anisimov, N.E. Kolobova, V.V. Skripkin (Zh. Strukt. Khim. **9** [1968] 922/5; J. Struct. Chem. [USSR] **9** [1968] 821/3). – [25] B.P. Bir'yukov, Yu.T. Struchkov, K.N. Anisimov, N.E. Kolobova, V.V. Skripkin (Chem. Commun. **1968** 1193/4).

[26] B.P. Bir'yukov, Yu.T. Struchkov, K.N. Anisimov, N.E. Kolobova, V.V. Skripkin (Chem. Commun. **1968** 159/60). – [27] B.P. Bir'yukov, Yu.T. Struchkov (Zh. Strukt. Khim. **9** [1968] 488/502; J. Struct. Chem. [USSR] **9** [1968] 412/25). – [28] R.F. Bryan, A.R. Manning (Chem. Commun. **1968** 1220/1). – [29] R.C. Edmondson, M.J. Newlands (Chem. Commun. **1968** 1219/20). – [30] D.E. Fenton, J.J. Zuckerman (J. Am. Chem. Soc. **90** [1968] 6226/8).

[31] R.J. Haines, A.L. du Preez (Chem. Commun. **1968** 1513/4). – [32] R.H. Herber, Y. Goscinny (Inorg. Chem. **7** [1968] 1293/8). – [33] A.N. Nesmeyanov, K.A. Anisimov, N.E. Kolobova, A.B. Antonova (Usp. Khim., in press according to [24]). – [34] P. Powell (Inorg. Chem. **7** [1968] 2458/9). – [35] B.P. Bir'yukov, Yu.T. Struchkov (Zh. Strukt. Khim. **10** [1969] 95/106; J. Struct. Chem. [USSR] **10** [1969] 86/96).

[36] S. Chandra, R.H. Herber (Mössbauer Eff. Methodol. Proc. Symp. **5** [1969/70] 45/64). – [37] Yu.T. Struchkov, V.G. Andrianov, B.P. Bir'yukov, A.I. Gusev, V.A. Semion (4th Intern. Conf. Organometal. Chem., Bristol 1969, A 15). – [38] P.F. Barrett, K.K.W. Sun (Can. J. Chem. **48** [1970] 3300/3). – [39] E.I. Fedin, L.A. Fedorov, R.B. Materikova (Zh. Strukt. Khim. **11** [1970] 174/92; J. Struct. Chem. [USSR] **11** [1970] 169/86). – [40] D.S. Field, M.J. Newlands (Abstr. Papers Chem. Inst. Can. Am. Chem. Soc. Joint Conf., Toronto 1970, INOR 30).

[41] S.R.A. Bird, J.D. Donaldson, A.F. Le C. Holding, B.J. Senior, M.J. Tricker (J. Chem. Soc. A **1971** 1616/21). – [42] H.C. Clark, B.K. Hunter (J. Organometal. Chem. **31** [1971] 227/32). – [43] A.J. Cleland, S.A. Fieldhouse, B.H. Freeland, R.J. O'Brien (J. Organometal. Chem. **32** [1971] C15/C18). – [44] R.C. Edmondson, D.S. Field, M.J. Newlands (Can. J. Chem. **49** [1971] 618/23). – [45] B.A. Goodman, R. Greatrex, N.N. Greenwood (J. Chem. Soc. A **1971** 1868/72).

[46] R.H. Herber (in: L. May, An Introduction to Mössbauer Spectroscopy, London 1971, p. 138/54, 150/2). – [47] B.V. Liengme (Inorg. Nucl. Chem. Letters **7** [1971] 1223/8). – [48] T.N. Mel'nikova, Yu.V. Kolodyazhnyi, O.A. Osipov, A.D. Garnovskii, N.E. Kolobova, K.N. Anisimov, B.V. Lokshin, V.V. Skripkin, W.N. Khandozhko (5th Intern. Conf. Organometal. Chem., Moscow 1971, Vol. 2, Abstr. No. 350). – [49] S. Cenini, B. Ratcliff, A. Fusi, A. Pasini (Gazz. Chim. Ital. **102** [1972] 141/63 in Engl.). – [50] T.C. Gibb, R. Greatrex, N.N. Greenwood (J. Chem. Soc. Dalton Trans. **1972** 238/40).

[51] P. Hackett, A.R. Manning (J. Chem. Soc. Dalton Trans. **1972** 1487/91). – [52] P. Hackett, A.R. Manning (J. Organometal. Chem. **34** [1972] C15/C17). – [53] Yu.V. Kolodyazhnyi, V.V. Skripkin, N.E. Kolobova, A.D. Garnovskii, B.V. Lokshin, O.A. Osipov, K.N. Anisimov, M.G. Gruntfest (Zh. Strukt. Khim. **13** [1972] 160/2; J. Struct. Chem. [USSR] **13** [1972] 148/50). – [54] R.M.G. Roberts (J. Organometal. Chem. **47** [1973] 359/66). – [55] C. Ungurenasu, G. Stiubianu, E. Streba (Syn. Reactiv. Inorg. Metal-org. Chem. **3** [1973] 211/4).

[56] B.J. Cole, J.D. Cotton, D. McWilliam (J. Organometal. Chem. **64** [1974] 223/7). – [57] G.M. Bancroft, K.D. Butler, T.K. Sham (J. Chem. Soc. Dalton Trans.

1975 1483/6). – [58] H.E. Sasse, M.L. Ziegler (Z. Naturforsch. **30b** [1975] 30/2). – [59] J.D. Cotton, A.M. Trewin (J. Organometal. Chem. **117** [1976] C7/C9). – [60] K. Triplet, M.D. Curtis (Inorg. Chem. **15** [1976] 431/3).

[61] Yu.K. Grishin, N.M. Sergeyev, Yu.A. Ustynyuk (J. Organometal. Chem. **34** [1972] 105/18).

2.5.2.2.3.3   Donor Substitution Products of $(C_5H_5Fe(CO)_2)_2Sn(X,R)_2$

Most of the compounds listed in Table 11 are monosubstituted derivatives of the type $C_5H_5Fe(CO)_2-Sn(X,R)_2-(CO)(^2D)FeC_5H_5$ with X=Cl or R=CH_3 (Nos. 1 to 11). Two disubstituted compounds of the type $C_5H_5Fe(^2D)(CO)-SnCl_2-(CO)(^2D)FeC_5H_5$ are described (Nos. 12 and 13).

The compounds are prepared by one or both of the following methods:

Method I: $SnCl_2 \cdot 2H_2O$ is added to $C_5H_5Fe(CO)(\mu-CO)_2(^2D)FeC_5H_5$, dissolved in $CH_3COOC_2H_5/CH_3OH$. After refluxing for 15 to 24 h the hot solution is filtered, evaporated to dryness, and the residue is recrystallized from an appropriate solvent [2].

Method II: $(C_5H_5Fe(CO)_2)_2SnCl_2$ and the donor molecule (1:1 mole) are refluxed in benzene or toluene for 15 to 24 h. Mixtures of benzene and dimethoxypropane are used for compounds No. 6, 10, and 13. Further work-up like method I [2].

Exact yields are not reported. When benzene is used as the solvent in method II, the reaction does not proceed to completion. In toluene the reactions are complete for smaller PR_3 and $P(OR)_3$ molecules (R=CH_3, C_2H_5) and give better yields for the larger ligands (R=C_3H_7 or C_4H_9).

The isomer shifts in the ^{57}Fe and ^{119}Sn Mössbauer spectra of the monosubstituted derivatives increase in the order $CO < P(OR)_3 < PR_3$. This correlates with a decreasing π-acceptor ability and increasing Lewis basicity of the ligands. The ^{119}Sn quadrupole splitting values increase with the bulkiness of the ligand and with the degree of CO substitution.

The monosubstituted complexes No. 5 to 8 produce IR spectra (in KBr) with three sharp CO bands, as predicted by C_1 symmetry. In the spectra of the other compounds, more than three $\nu(CO)$ bands appear or in KBr band broadening occurs. Either indicates rotational isomers.

The compounds are crystalline if not otherwise specified. They are air-stable as solids and in solutions of $CHCl_3$ and CH_2Cl_2. They are more soluble than the parent $(C_5H_5Fe(CO)_2)_2SnCl_2$, and the solubility increases with the length of the group R in PR_3 or $P(OR)_3$. Exceptions are compounds No. 10 and 13, whose solubility is substantially lower than that for the other compounds in this section [2].

Explanations to Table 11: The ^1H NMR spectra have been recorded in $CHCl_3$. The doublet at higher field belongs to the C_5H_5 ligand bonded to the same Fe atom as the donor ligand. Isomer shifts of the Mössbauer spectra, measured at 80 K, are referred to metallic Fe and $BaSnO_3$ at room temperature. All numerical values in Table 11 have only been reported in a dissertation [2]. Additional IR bands and Raman lines in the low-frequency region ($\nu(Sn-Cl)$ and $\nu(Sn-Fe)$ bands) are given in [2] but have not been included in the table.

Table 11
Donor substitution products of $(C_5H_5Fe(CO)_2)_2SnCl_2$ and $(C_5H_5Fe(CO)_2)_2Sn(CH_3)_2$.
Further information on compounds preceded by an asterisk is given at the end of the
table. For abbreviations and dimensions see p. 170.

No.	^2D ligand Method of preparation	Properties and further remarks Explanations on p. 139	Ref.
		Type $C_5H_5Fe(CO)_2-SnCl_2-(CO)(^2D)FeC_5H_5$:	
1	$P(C_2H_5)_3$ I, II	dark red (from pentane/ether) ^1H NMR: 4.69 (d, J(P,H) =1 to 2), 4.96 (s) $^{57}Fe-\gamma$: δ=0.183, Δ=1.695 $^{119}Sn-\gamma$: δ=2.10, Δ=2.60 IR (CHCl$_3$): 1922, 1951, 2001 (CS$_2$): 1914, 1925, 1947, 1996, 2007	[1, 2]
2	$P(C_3H_7-n)_3$ II	dark red (ill defined, from CH$_2$Cl$_2$/ligroin) ^1H NMR: 4.70 (d, J(P,H) =1 to 2), 4.99 (s) $^{57}Fe-\gamma$: δ=0.210, Δ=1.86 $^{119}Sn-\gamma$: δ=2.03, Δ=2.42 IR (CS$_2$): 1914, 1924, 1947, 1996, 2008 (KBr): 1916, 1935, 1981, 1989	[2]
3	$P(C_3H_7-i)_3$ I, II (?)	IR (CHCl$_3$): 1920, 1952, 2001 isolated as an oil	[2]
4	$P(C_4H_9-n)_3$ I, II	dark red (from hot CH$_3$OH or CH$_2$Cl$_2$/hexane) ^1H NMR: 4.66 (d, J(P,H) =1 to 2), 4.98 (s) $^{57}Fe-\gamma$: δ=0.171, Δ=1.71 $^{119}Sn-\gamma$: δ=2.09, Δ=2.37 IR (CHCl$_3$): 1919, 1950, 2002 (CS$_2$): 1913, 1928, 1949, 1959, 1997, 2009 (KBr): 1910, 1943, 1990	[2]
5	$P(OCH_3)_3$ I, II	orange; dec. >146° (from hot CH$_3$OH or CHCl$_3$/pentane) ^1H NMR: 4.8 (d, J(P,H) 1 to 2), 4.98 (s) $^{57}Fe-\gamma$: δ=0.130, Δ=1.52 $^{119}Sn-\gamma$: δ=1.84, Δ=2.28 IR (CHCl$_3$): 1940, 1962, 2010 (CS$_2$): 1935, 1960, 2004, 2011 (KBr): 1935, 1956, 2000	[2]
6	$P(OC_2H_5)_3$ I, II	red orange; dec. >125° (from hot CH$_3$OH or CHCl$_3$/pentane) ^1H NMR: 4.8 (d, J(P, H) =1 to 2), 4.98 (s) $^{57}Fe-\gamma$: δ=0.115, Δ=1.53 $^{119}Sn-\gamma$: δ=1.82, Δ=2.20 IR (CHCl$_3$): 1940, 1962, 2010 (CS$_2$): 1935, 1959, 2005, 2011 (KBr): 1924, 1943, 1999	[2]

References on p. 142

Table 11 [continued]

No.	^{2}D ligand Method of preparation	Properties and further remarks Explanations on p. 139	Ref.
7	P(OC$_3$H$_7$-i)$_3$ II	red orange (from hot CH$_3$OH or CH$_2$Cl$_2$/hexane) ^{1}H NMR: 4.8 (broad), 5.0 (s) ^{57}Fe-γ: $\delta=0.161$, $\Delta=1.70$ ^{119}Sn-γ: $\delta=2.19$, $\Delta=2.36$ IR (CHCl$_3$): 1957, 2010 (CS$_2$): 1940, 1953, 2000, 2010 (KBr): 1932, 1952, 2000	[1, 2]
8	P(OC$_4$H$_9$-n)$_3$ II	red brown or red orange, gummy solid (chromatography on SiO$_2$ with hexane/ether) ^{1}H NMR: 4.82 (s, broad), 5.0 (s, broad) ^{57}Fe-γ: $\delta=0.137$, $\Delta=1.48$ ^{119}Sn-γ: $\delta=1.77$, $\Delta=2.14$ IR (CHCl$_3$): 1940, 1963, 2011 (CS$_2$): 1940, 1956, 2003, 2010 (KBr): 1958, 1999.5, 2002	[2]
9	P(OC$_6$H$_5$)$_3$ I	red orange (from hot CH$_3$OH) ^{1}H NMR: 4.30, 4.97 (J(P,H) =1.3) ^{57}Fe-γ: $\delta=0.162$, $\Delta=1.72$ ^{119}Sn-γ: $\delta=1.98$, $\Delta=2.53$ IR (C$_6$H$_{12}$): 1944, 1952, 1969, 2007, 2017 (CHCl$_3$): 1950, 1970, 2012 (CS$_2$): 1945, 1966, 2007, 2014, 2024	[1, 2]
10	P(OCH$_2$)$_3$CCH$_3$ II	orange (from CHCl$_3$/pentane) ^{1}H NMR: 4.8 (s, broad), 4.96 (s, broad) ^{57}Fe-γ: $\delta=0.125$, $\Delta=1.50$ ^{119}Sn-γ: $\delta=1.78$, $\Delta=2.40$ IR (CHCl$_3$): 1960, 2010 (KBr): 1934, 1950, 1960, 2001	[2]

Type C$_5$H$_5$Fe(CO)$_2$-Sn(CH$_3$)$_2$-(CO)(^{2}D)FeC$_5$H$_5$:

*11	P(OCH$_3$)$_3$ see further information	yellow–orange needles IR (CS$_2$): 1915, 1970	[2]

Type C$_5$H$_5$Fe(^{2}D)(CO)-SnCl$_2$-(CO)(^{2}D)FeC$_5$H$_5$:

| *12 | P(OC$_2$H$_5$)$_3$ see further information | red orange ^{1}H NMR: 4.73 (d, J(P,H) =1 to 2) ^{57}Fe-γ: $\delta=0.117$, $\Delta=1.48$ ^{119}Sn-γ: $\delta=1.82$, $\Delta=2.29$ IR (CS$_2$): 1930 (very broad) (C$_6$H$_{12}$): 1937 (broad), (KBr): 1920, 1940 | [2] |

References on p. 142

Table 11 [continued]

No.	2D ligand Method of preparation	Properties and further remarks Explanations on p. 139	Ref.
13	$P(OCH_2)_3CCH_3$ II	orange yellow (ill defined, from $CHCl_3$/pentane) ^1H NMR: 4.79 (broad) IR ($CHCl_3$): 1945 (very broad) (KBr): 1943, 1953 preparation with excess $P(OCH_2)_3CCH_3$	[2]

* Further information:

$C_5H_5Fe(CO)_2-Sn(CH_3)_2-(CO)(P(OCH_3)_3)FeC_5H_5$ (Table **11**, No. **11**) has been prepared from $C_5H_5Fe(CO)_2-SnCl_2-(CO)(P(OCH_3)_3)FeC_5H_5$ and an excess of CH_3Li in tetrahydrofuran at -80 °C. The compound forms in a small yield [2].

$(C_5H_5Fe(CO)P(OC_2H_5)_3)_2SnCl_2$ (Table **11**, No. **12**) has been obtained from the monosubstituted complex No. 6 by further treatment with 2 moles of $P(OC_2H_5)_3$ in refluxing toluene for 32.5 h. Crystals are formed from ether/pentane at -78 °C [2].

References:

[1] R.J. Haines, A.L. du Preez (Chem. Commun. **1968** 1513). — [2] T.N. Pecoraro (Diss. Rutgers State Univ., New Brunswick, N.Y., 1975, 156 pp.; Diss. Abstr. Intern. B **36** [1975] 746).

2.5.2.2.4 Bridging Groups with Fe-B Bonds

The first example of a compound with two transition–metal atoms bonded to a pentaborane skeleton has recently been reported:

$C_5H_5Fe(CO)_2(\mu-B_5H_7)(CO)_2FeC_5H_5$ (Formula I) is formed in 11.0% yield by deprotonation of $C_5H_5Fe(CO)_2(2-B_5H_8)$ with KH in monoglyme at -40 °C/4 h and further reaction of the anion $[C_5H_5Fe(CO)_2(2-B_5H_7)]^-$ with $C_5H_5Fe(CO)_2I$ at -78 to -20 °C for 8 h with brief warming to room temperature. The compound is isolated by chromatography on SiO_2 with benzene and recrystallized from CH_2Cl_2.

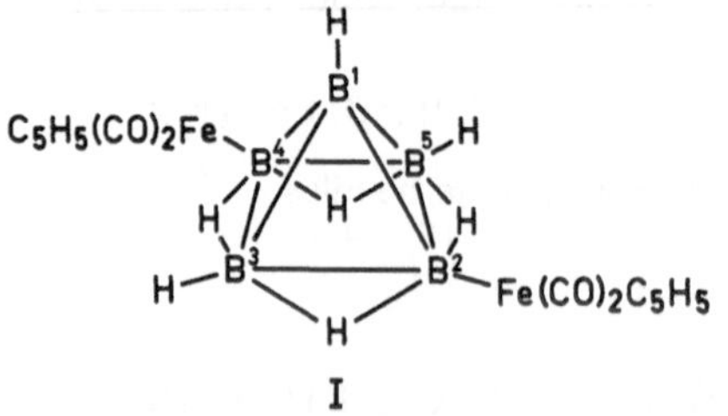

I

The yellow needles melt at 110 °C with decomposition. ^1H NMR spectrum (in CD_2Cl_2 at 28 °C): $\delta = -1.5$ (four bridging H), $+0.2$ (H-1), $+2.2$ (H-3,5), and $+4.9$ (C_5H_5) ppm. ^{11}B NMR spectrum (in CD_2Cl_2 at 28 °C): $\delta = -45.0$ (d, B-1, J(B,H) = 168), -8.6 (d, B-3,5, J(B,H) = 127), and $+1.4$ (s, B-2,4) ppm. The coupled and decoupled ^{11}B NMR spectra are shown as a figure. The resonance spectra are consistent with a basal–terminal substitution in the B_5H_7 skeleton with a trans (2,4) position of the Fe atoms as shown by Formula I. IR spectrum (KCl): ν(CO) bands at 1930 and 1988 cm^{-1}, ν(BH) bands at 2543 and 2580 cm^{-1}.

The high-resolution mass spectrum is given along with a fragmentation scheme. A unique feature is the favored formation of $[Fe_2(B_5H_5)(C_5H_5)_2]^+$, which constitutes the base peak and accounts for more than 15% of the total ion current. The solid can be handled in air without apparent change. It is very soluble in polar organic solvents.

Reference:

N.N. Greenwood, J.D. Kennedy, C.G. Savory, J. Staves, K.R. Trigwell (J. Chem. Soc. Dalton Trans. **1978** 237/44).

2.5.2.2.5   Compounds of the (^{5}LFe(CO)$_2$)$_2$M Type with M=Mg, Zn, Cd, and Hg

2.5.2.2.5.1   Compounds with Mg(^{2}D)$_n$ Central Groups

Structure proposals for the following compounds based on their spectra are shown by Formulas I and II. ^{2}D represents tetrahydrofuran (C$_4$H$_8$O) and pyridine (C$_5$H$_5$N). An X-ray study showed that an analogous Mo compound, $C_5H_5Mo(CO)_2$-CO-Mg(C$_5$H$_5$N)$_4$-OC-(CO)$_2$MoC$_5$H$_5$, has a structure like II [3].

I   II

$(C_5H_5Fe(CO)_2)_2Mg(OC_4H_8)_2$ and $(C_5H_5Fe(CO)_2)_2Mg(C_5H_5N)_2$ are prepared by reductive cleavage of $(C_5H_5Fe(CO)_2)_2$ with excess Mg/Hg (1% Mg). The former forms in tetrahydrofuran, while the latter forms in benzene in the presence of a slight excess of pyridine [1 to 5]. The reaction is run for 18 h at room temperature. The filtered solutions are concentrated under reduced pressure and the products precipitated by flooding with n-pentane. Typical yields are 90 to 95% [4, 5].

$(C_5H_5Fe(CO)_2)_2Mg(OC_4H_8)_2$ has alternatively been prepared by contacting C$_5$H$_5$Fe-(CO)$_2$I in tetrahydrofuran with Mg powder (200 mesh). After a short induction period, an exothermic reaction occurs for 15 to 20 min. Insoluble MgI$_2$·x C$_4$H$_8$O is removed by filtration and the complex precipitated by flooding with n-pentane, 96% yield [4]. The compounds can be recrystallized from benzene [4, 5].

The yellow diamagnetic compounds decompose without melting above 150 °C [1, 4]. Their ^{1}H NMR and IR spectra (δ in ppm, ν(CO) bands in cm^{-1}) are summarized below [4], see also [1].

C$_4$H$_8$O complex ^{1}H NMR (C$_6$H$_6$): 1.44, 3.91 (m's, C$_4$H$_8$O), 4.70 (s, C$_5$H$_5$)
IR (Nujol): 1772, 1854, 1925, 1962, 2018
(C$_4$H$_8$O): 1713, 1854, 1884, 1918, 1959, 2009;
(C$_6$H$_6$): 1854, 1921, 2015

C$_5$H$_5$N complex ^{1}H NMR (C$_4$H$_8$O): 4.25 (s, C$_5$H$_5$), 7.20, 7.54, 8.49 (m's, C$_5$H$_5$N)
IR (Nujol): 1834, 1905, 1921, 1966, 2024
(C$_6$H$_6$): 1847, 1917, 2018

The positions of the $v(CO)$ bands reveal that the polarity of the Mg–Fe bond is somewhat greater than that of the covalent Zn–Fe and Hg–Fe bonds but less than that of a purely ionic complex. The 1772 cm^{-1} band of solid $(C_5H_5Fe(CO)_2)_2Mg(C_4H_8O)_2$ most likely results from partial ionization in the solid state since this band is very close to the low–energy band of $[C_5H_5Fe(CO)_2]^-$. This band is absent in the solid state spectrum of the pyridine adduct, which does not appear to assume an ionic crystal lattice.

The complexity of the $v(CO)$ region in the tetrahydrofuran solution spectrum indicates the presence of several species. This complexity can be accounted for by an equilibrium with $(C_5H_5Fe(CO)_2)_2Mg(OC_4H_8)_4$. The strong bands at 1713 and 1884 cm^{-1} are assigned to this adduct. The band at 1713 cm^{-1} is likely associated with a Mg–O–C–Fe system. Thus the enhanced solvation of the Mg atom results in a changeover from a relatively covalent Mg–Fe interaction (e.g., in benzene) to a Mg–O–C–Fe interaction as shown in Formula II [4].

The two compounds remain unchanged after 24 h under a vaccum of 10^{-4} Torr at room temperature. They are soluble in benzene, tetrahydrofuran, and pyridine and remain monomeric in hydrocarbon and ether solvents. Conductivity measurements in tetrahydrofuran indicate that they are essentially nonionized in this solvent [1, 4, 5]. The compounds are extremely air–sensitive and are rapidly oxidized to MgO and $(C_5H_5Fe(CO)_2)_2$ [4]. They react cleanly with alkyl halides and phenol to produce alkyl and hydrido transition metal complexes [5]. The reactivity is dependent on the solvent. For example, the formation of $(C_5H_5Fe(CO)_2)_2Hg$ by the reaction with Hg(CN)$_2$ proceeds in benzene after 24 h to about 30% completion but is complete in several minutes upon addition of a small amount of tetrahydrofuran. This is probably related to the polar Mg–C–O–Fe interaction in tetrahydrofuran [4].

$(C_5H_5Fe(CO)_2)_2Mg(C_5H_5N)_4$ has been prepared like the previous compounds from $(C_5H_5Fe(CO)_2)_2$ and Mg/Hg in toluene containing a small excess of pyridine. A trace of MgCl$_2$ is necessary to promote the reaction. The yields are between 45 and 70% [3], see also [2]. The published articles do not make clear which conditions lead to the tetrakis pyridine compound and which conditions lead to the bis pyridine compound.

The IR spectrum in pyridine has $v(CO)$ bands at 1711, 1773 (weak), and 1857 cm^{-1}. The low–frequency band is assigned to the asymmetric $v(CO)$ of a doubly coordinated carbonyl group like that in structure II (D =pyridine).

The compound is sensitive to air and moisture [3]. The metal carbonyl unit retains substantial nucleophilic character. Compounds of this type are highly reactive, even in nonpolar solvents [2, 3], but particular applications of the title complex have not been reported.

References:

[1] G.B. McVicker, R.S. Matyas (J. Chem. Soc. Chem. Commun. 1972 972). –
[2] S.W. Ulmer (Diss. Cornell Univ. 1972; Diss. Abstr. Intern. B 33 [1973] 4718). –
[3] S.W. Ulmer, P.M. Skarstad, J.M. Burlitch, R.E. Hughes (J. Am. Chem. Soc. 95 [1973] 4469/71). – [4] G.B. McVicker (Inorg. Chem. 14 [1975] 2087/9). – [5] G.B. McVicker (Inorg. Syn. 16 [1976] 56/61).

2.5.2.2.5.2   Compounds with Zn, Cd, and Hg Central Atoms

The properties of four compounds of the $^5LFe(CO)_2-M-(CO)_2Fe^5L$ type are collected in Table 12. Their preparation is described under further information.

The solution IR spectra of the complexes with C_5H_5 ligands (Table 12, Nos. 1 to 3) are consistent with a structure of C_{2v} symmetry containing a linear Fe–M–Fe arrangement [10, 15]. The isomer shifts of the ^{57}Fe Mössbauer spectra show a significant and regular decrease in going from Hg to Zn, whereas the ν(CO) bands do not vary in a regular way. This indicates the metal-to-metal bonding is a combination of σ and π bonding [15].

The compounds are sufficiently volatile to give mass spectra which contain the molecular ions. They are readily soluble in tetrahydrofuran or acetone with an unexpected enhanced air-stability in acetone. In toluene the Cd derivative is less soluble than the Zn or Hg compound [15].

Table 12

Compounds of the $^5LFe(CO)_2-M-(CO)_2Fe^5L$ type with M = Zn, Cd, and Hg.
Further information on compounds preceded by an asterisk is given at the end of the table. For abbreviations and dimensions see p. 170.

No.	^5L ligand Metal M	Appearance Melting point	Spectra	Ref.
*1	C_5H_5 Zn	orange to red, prismatic crystals 158.5 to 160.0°	^{57}Fe-γ: $\delta=0.288$, $\Delta=1.733$ IR (KBr): 1860, 1880 (?), 1937, 1945, 1963 ($C_6H_5CH_3$): 1915, 1956, 1973 (HCON(CH$_3$)$_2$): 1869, 1926, 1937, 2016	[13, 15, 25]
*2	C_5H_5 Cd	red orange to dark red, octa-hedral crystals 177.2 to 179.8°	^{57}Fe-γ: $\delta=0.319$, $\Delta=1.700$ IR (KBr): 1858, 1891, 1901, 1940, 1953, 1963 ($C_6H_5CH_3$): 1914, 1952, 1971	[15, 25]
*3	C_5H_5 Hg	see further information decomposition at 142 to 150°	^{1}H NMR (CDCl$_3$): 4.77 (C_5H_5) ^{57}Fe-γ: $\delta=0.339$, $\Delta-1.633$ IR (Nujol): 1907, 1916, 1951, 1967, 1978 (C_6H_{12}): 1940, 1955, 1972, 2000 (CHCl$_3$): 1924, 1960, 1990	[3, 10, 11, 15, 19]
*4	$C_5(C_6H_5)_5$ Hg	orange prisms 262 to 264°	IR (KCl): 1931, 1970	[6]

* Further information:

$(C_5H_5Fe(CO)_2)_2Zn$ (Table 12, No. 1) has been prepared by stirring an excess of powdered Zn with $(C_5H_5Fe(CO)_2)_2$ in $CH_3OCH_2OCH_2OCH_3$ at reflux temperature for 67 h. Solvent removal by trap-to-trap vacuum distillation at room temperature gives the product in 93% yield [25]. The compound is also obtained by stirring $(C_5H_5Fe(CO)_2)_2$-Hg with Zn (1:10 mole) in tetrahydrofuran or toluene at room temperature for 1 h. After filtration the liquid is very slowly cooled to −80 °C, which gives the crystalline compound. It is recrystallized from toluene between 50 an −35 °C, 74 to 84% yield [15].

The solid is quite sensitive to air. Solutions decompose rapidly when exposed to traces of air [15]. In solutions of CH_3CN and $HCON(CH_3)_2$, the compound is completely associated. There is no evidence for the formation of $[C_5H_5Fe(CO)_2]^-$ when excess $[N(C_2H_5)_4]Br$ is added [12 to 14]. The reaction with $[C_5H_5Fe(CO)_2]^-$ gives $[(C_5H_5Fe-(CO)_2)_3Zn]^-$, which can be isolated as $[NR_4]^+$ or $[(C_6H_5)_3PNP(C_6H_5)_3]^+$ salts [27], but further details have not been reported. $C_5H_5Fe(CO)_2-Zn-Co(CO)_4$ is the major component in the equilibrium mixture obtained with $(Co(CO)_4)_2Zn$ in toluene at room temperature [15]. The reaction with ROH ($R=CH_3$ and C_2H_5) or with an equimolar quantity of either alcohol in toluene at room temperature leads to a product identified as $(C_5H_5Fe-(CO)_2ZnOR)_4$, and $C_5H_5Fe(CO)_2H$ has been detected simultaneously by the IR spectrum of the solution. Analogous treatment with H_2O or C_6H_5OH in a 1:1 mole ratio gives $(C_5H_5Fe(CO)_2ZnOR)_x$ ($R=H$ or C_6H_5) where x could not be determined. $C_5H_5Fe(CO)_2H$ and probably $Zn(OC_6H_5)_2$ are formed in the reaction with an excess of C_6H_5OH [20, 25].

$(C_5H_5Fe(CO)_2)_2Cd$ (Table 12, No. 2) has been obtained from $(C_5H_5Fe(CO)_2)_2$ and Cd (66% yield) and from $(C_5H_5Fe(CO)_2)_2Hg$ and Cd (72% yield) under conditions similar to those for the previous complex. When the second reaction is carried out in tetrahydrofuran, the crude product has a satisfactory melting point, 93% yield. The substance can be recrystallized from toluene [15].

Solutions of the compound are quite air–sensitive. Stirring with an excess of Hg in toluene leads quantitatively to $(C_5H_5Fe(CO)_2)_2Hg$ [15]. The compound can be refluxed in methanol for 2 d without evidence of the formation of either $(C_5H_5Fe(CO)_2CdOCH_3)_x$ or $C_5H_5Fe(CO)_2H$ [25]. It is mentioned without further details that $[(C_5H_5Fe(CO)_2)_3Cd]^-$ occurs in the reaction with $[C_5H_5Fe(CO)_2]^-$ [27]. The treatment with an equimolar amount of $(Co(CO)_4)_2Cd$ in toluene gives, after removal of the solvent, an orange–brown residue, whose mass spectrum shows the presence of $C_5H_5Fe(CO)_2-Cd-Co(CO)_4$. But the equilibrium mixture contains larger amounts of the symmetrical starting materials than in the case of the corresponding Zn and Hg derivatives [15].

$(C_5H_5Fe(CO)_2)_2Hg$ (Table 12, No. 3) has first been obtained by precipitation of $[C_5H_5Fe(CO)_2]^-$ in methanol with aqueous $Hg(CN)_2$ [1]. For a detailed description of this method in a modified form (85% crude yield) see [15]. Another method is the "symmetrization" of $C_5H_5Fe(CO)_2HgI$, which is added in acetone to aqueous alkaline Na_2SnO_2. After removal of Hg and extraction with benzene and chromatography on Al_2O_3 with petroleum ether/benzene the title compound is isolated in 73% yield [22]. The formation from $C_5H_5Fe(CO)_2HgCl$ and RLi ($R=n-C_4H_9$ and C_6F_5) proceeds via $C_5H_5Fe-(CO)_2HgR$, which disproportionates into the title compound and HgR_2. With $R=n-C_4H_9$ the reaction is carried out in ether/hexane with equimolar amounts of the starting materials. Filtration of LiCl followed by removal of the solvent and the volatile $Hg(C_4H_9-n)_2$ under vacuum gives an almost quantitative yield of $(C_5H_5Fe(CO)_2)_2Hg$ [17]. The complex is formed quantitatively from $(C_5H_5Fe(CO)_2)_2Cd$ and Hg in toluene at room temperature [15]. The direct insertion of Hg into $(C_5H_5Fe(CO)_2)_2$ takes place only in the presence of Hg_2I_2 when the suspension in benzene is stirred and irradiated for 16 h, 10% yield in addition to 5% $C_5H_5Fe(CO)_2I$ [19]. $(C_5H_5Fe(CO)_2)_2Hg$ has been identified among the products of several reactions with $Na[C_5H_5Fe(CO)_2]$ prepared from $(C_5H_5Fe(CO)_2)_2$ and Na/Hg, see [2, 3].

Due to the formation of the unstable $C_5H_5Fe(CO)_2HgCH_3$ and its decomposition, $(C_5H_5Fe(CO)_2)_2Hg$ occurs as one major product in reactions between $C_5H_5Fe-(CO)_2Sn(CH_3)_3$ or $(C_5H_5Fe(CO)_2)_2Sn(CH_3)_2$ and CH_3HgCl [26].

References on p. 148

$(C_5H_5Fe(CO)_2)_2Hg$ is best removed from other organoiron compounds by chromatography on Al_2O_3 with benzene [3]. It is polymorphic. From toluene at -35 °C a mixture of large dark red, octahedral crystals and much smaller yellow–orange, granular crystals is obtained. Recrystallization from methanol at -35 °C yields flacky, golden–yellow crystals [15]. It can also be recrystallized from hexane [9] and methylene chloride/hexane [3].

The $\nu(CO)$ region of the IR spectrum (in Nujol) is shown as figure in [10]. A trans configuration with C_{2h} symmetry has been suggested for the solid phase [10]. A strong band in the IR spectrum at 200 cm^{-1} is assigned to the asymmetric Fe–Hg–Fe vibration. A good Raman spectrum could not be obtained [8]. The compound has no electronic absorption above 200 nm [11].

The air–stable compound dissolves in the common organic solvents like benzene, ether, and acetone [1]. Air–saturated solutions in organic solvents deteriorate slowly when exposed to normal light [15]. Partial decomposition to $(C_5H_5Fe(CO)_2)_2$ and Hg occurs upon sublimation in a high vacuum at 80 to 90 °C [1]. The polarographic reduction in $CH_3OCH_2CH_2OCH_3$ at $E_{1/2} = -2.0$ V (referred to 10^{-3} M Ag$^+$/Ag) is a two–electron step giving $[C_5H_5Fe(CO)_2]^-$ and Hg [7, 16].

For exchange reactions with Zn and Cd metal see the two previous compounds. HCl in dioxane at 40 °C cleaves the compound completely in 30 min with the formation of $C_5H_5Fe(CO)_2Cl$ and $C_5H_5Fe(CO)_2HgCl$ [23]. Exchange reactions also take place with GeI_2, $SnCl_2$, $SnBr_2$, and In^I halides to give the compound types $(C_5H_5Fe(CO)_2)_2MX_2$ with M = Ge and Sn [4, 9, 21] and $(C_5H_5Fe(CO)_2)_2InX$ [18], see 2.5.2.2.3 and 2.5.2.2.6. Compounds of the type $C_5H_5Fe(CO)_2HgX$ are formed by exchange reactions with HgX_2 (X = Cl, Br, I, SCN, and CH_3COO) in acetone or methanol at room temperature [9, 11]. The equilibrium lies far towards these unsymmetrical products although $(C_5H_5Fe(CO)_2)_2Hg$ can be recovered by passing the equilibrium mixture in acetone through a Al_2O_3 column or by adding $P(C_6H_5)_3$ to precipitate $(P(C_6H_5)_3)_2HgX_2$ [11]. With HgR_2, no exchange to $C_5H_5Fe(CO)_2HgR$ occurs. Here the equilibrium is far towards the symmetrical starting materials [17]. The exchange with $(Co(CO)_4)_2Hg$ yields $C_5H_5Fe(CO)_2-Hg-Co(CO)_4$. With $(C_5H_5M(CO)_3)_2Hg$ (M = Mo or W), a roughly statistical distribution of complexes is found in solution [11]. An equimolecular mixture of $(C_5H_5Fe(CO)_2)_2Hg$ and $(C_5H_5Mo(CO)_3)_2Hg$ gave with $SnCl_2$ a poor yield of a compound which possibly was $C_5H_5Fe(CO)_2-SnCl_2-(CO)_3MoC_5H_5$ [9].

$Li[C_5H_5Fe(CO)_2]$ is formed from one mole of $(C_5H_5Fe(CO)_2)_2Hg$ and two moles of $n-C_4H_9Li$ in ether or tetrahydrofuran. However, a 1:1 mole ratio of the reactants gives a deep red solution probably containing $Li[(C_5H_5Fe(CO)_2)_2HgC_4H_9-n]$ as a major component [17]. The strong base $[C_5H_5Fe(CO)_2]^-$ similarly gives a complex anion, $[(C_5H_5Fe(CO)_2)_3Hg]^-$. Several other Lewis bases did not yield adducts [24, 27].

From a reaction with $CF_2=CF_2$ at 130 °C and 15 atm, small amounts of ferrocene and probably $C_5H_5Fe(CO)_2-CF_2=CF_2$ (4% yield) have been isolated [5].

$(C_5(C_6H_5)_5Fe(CO)_2)_2Hg$ (Table 12, No. 4) has been prepared by treating $C_5(C_6H_5)_5Fe(CO)_2Br$ with Na/Hg in tetrahydrofuran for 2 h. Filtration, solvent evaporation, and chromatography on Al_2O_3 with benzene gives a 30% yield of the compound which is crystallized from CH_2Cl_2. Its pyrolysis at 200 °C/1 Torr affords only 1,2,3,4,5–pentaphenylcyclopentadiene [6].

References on p. 148

References:

[1] E.O. Fischer, R. Böttcher (Z. Naturforsch. **10 b** [1955] 600/1). – [2] M.L.H. Green, P.L.I. Nagy (J. Chem. Soc. **1963** 189/97). – [3] R.B. King (J. Inorg. Nucl. Chem. **25** [1963] 1296/8). – [4] F. Bonati, S. Cenini, D. Morelli, R. Ugo (Inorg. Nucl. Chem. Letters **1** [1965] 107/8). – [5] M.L.H. Green, A.N. Stear (Z. Naturforsch. **20 b** [1965] 812).

[6] S. McVey, P.L. Pauson (J. Chem. Soc. **1965** 4312/8). – [7] R.E. Dessy, P.M. Weissman, R.L. Pohl (J. Am. Chem. Soc. **88** [1966] 5117/21). – [8] D.M. Adams, J.B. Cornell, J.L. Dawes, R.D.W. Kemmitt (Inorg. Nucl. Chem. Letters **3** [1967] 437/9). – [9] F. Bonati, S. Cenini, R. Ugo (J. Chem. Soc. A **1967** 932/5). – [10] R.D. Fischer, A. Vogler, K. Noack (J. Organometal. Chem. **7** [1967] 135/49).

[11] M.J. Mays, J.D. Robb (J. Chem. Soc. A **1968** 329/32). – [12] J.M. Burlitch (4th Intern. Conf. Organometal. Chem., Bristol 1969, A 9). – [13] J.M. Burlitch (J. Am. Chem. Soc. **91** [1969] 4562/3). – [14] J.M. Burlitch (J. Am. Chem. Soc. **91** [1969] 4563/4). – [15] J.M. Burlitch, A. Ferrari (Inorg. Chem. **9** [1970] 563/9).

[16] S.P. Gubin, L.I. Denisovich (Elektrokhim. Protseesy Uchastiem Org. Veshchestv **1970** 61/8 according to C.A. **74** [1971] No. 18800). – [17] S.C. Cohen, S.H. Sage, W.A. Baker, J.M. Burlitch, R.B. Petersen (J. Organometal. Chem. **27** [1971] C44/C46). – [18] A.T.T. Hsieh, M.J. Mays (Inorg. Nucl. Chem. Letters **7** [1971] 223/5). – [19] A.N. Nesmeyanov, L.G. Makarova, V.N. Vinogradova (Izv. Akad. Nauk SSSR Ser. Khim. **1971** 1984/7; Bull. Acad. Sci. USSR Div. Chem. Sci. **1971** 1869/72). – [20] J.M. Burlitch, S.E. Hayes (J. Organometal. Chem. **42** [1972] C13/C15).

[21] A.N. Nesmeyanov, L.G. Makarova, V.N. Vinogradova (Izv. Akad. Nauk SSSR Ser. Khim. **1972** 1449; Bull. Acad. Sci. USSR Div. Chem. Sci. **1972** 1406). – [22] A.N. Nesmeyanov, L.G. Makarova, V.N. Vinogradova (Izv. Akad. Nauk SSSR Ser. Khim. **1972** 2605; Bull. Acad. Sci. USSR Div. Chem. Sci. **1972** 2537/8). – [23] A.N. Nesmeyanov, L.G. Makarova, V.N. Vinogradova (Izv. Akad. Nauk SSSR Ser. Khim. **1972** 2798/9; Bull. Acad. Sci. USSR Div. Chem. Sci. **1972** 2725/7). – [24] H.L. Conder, W.R. Robinson (Inorg. Chem. **11** [1972] 1527/32). – [25] S.E. Hayes (Diss. Cornell Univ., Ithaca, N.Y., 1973; Diss. Abstr. Intern. B**34** [1973] 119).

[26] R.M.G. Roberts (J. Organometal. Chem. **47** [1973] 359/66). – [27] J.M. Burlitch, R.B. Petersen (Abstr. Papers Chem. Inst. Can. Am. Chem. Soc. Joint Conf., Toronto 1970, INOR 25).

2.5.2.2.6   Compounds of the $(C_5H_5Fe(CO)_2)_2MX$ Type with M =In and Tl

A review on "Organometallic Complexes Containing Bonds between Transition Metals and Group IIIB Metals" is given in [7].

$(C_5H_5Fe(CO)_2)_2InCl$ has been prepared by heating an equimolar mixture of $(C_5H_5Fe(CO)_2)_2$ and InCl in dioxane at reflux temperature for 6 h. The product obtained after solvent removal is dissolved in acetone and chromatographed on SiO_2 with tetrahydrofuran as eluent. Addition of petroleum ether and cooling to -30 °C gives the complex in 55% yield [6]. In the presence of an excess indium(I) halide, $C_5H_5Fe(CO)_2InX_2$ is obtained [4]. The compound can also be prepared from $Na[C_5H_5Fe(CO)_2]$ and anhydrous $InCl_3$ in tetrahydrofuran at room temperature or from $(C_5H_5Fe(CO)_2)_2Hg$ and InCl in tetrahydrofuran at room temperature. Yields for these methods have not been reported [6], see also [3, 4].

The compound forms yellow–brown crystals. The IR spectrum (in Nujol) has $v(CO)$ bands at 1912, 1924, 1963, and 1988 cm^{-1}. The available data do not allow a distinction between a planar three–coordinate In atom or a dimeric structure with four–coordinate In and two halogen bridges. In the mass spectrum a dimeric species has not been observed. The compound is essentially unaffected by exposure to air and moisture for short periods [6].

$(C_5H_5Fe(CO)_2)_2TlCl$ has only been obtained as an impure scarlet powder from the metathesis between $Na[C_5H_5Fe(CO)_2]$ and $TlCl_3$ in a 2:1 mole ratio. The reported analytical data approximate the above formulation [5, 7]. The compound has also been characterized by its IR spectrum [7], but the data have not been given.

$[(C_5H_5Fe(CO)_2)_2Tl]^+$. The existence of this cation is briefly mentioned in [1].

References:

[1] W.A.G. Graham (private communication, cited in [2]). – [2] J.M. Burlitch, R.B. Petersen, H.L. Condor, W.R. Robinson (J. Am. Chem. Soc. 92 [1970] 1783/4). – [3] A.T.T. Hsieh (Diss. Univ. of Cambridge 1971 according to [7]). – [4] A.T.T. Hsieh, M.J. Mays (Inorg. Nucl. Chem. Letters 7 [1971] 223/5. – [5] M.J. Mays, A.T.T. Hsieh (unpublished work cited in [7]).

[6] A.T.T. Hsieh, M.J. Mays (J. Organometal. Chem. 37 [1972] 9/14). – [7] A.T.T. Hsieh (Inorg. Chim. Acta 14 [1975] 87/104).

2.5.2.2.7　Other Compounds with Fe–Ni and Fe–Pt Bonds

Two compounds of the formula $(C_5H_5Fe(CO)_2)_2M(P(C_6H_5)_3)_2$ with M=Ni and Pt are reported to be formed from $(C_5H_5Fe(CO)_2)_2$ and the complex type $C_2H_4M(P(C_6H_5)_3)_2$ in benzene. The dark green solid (for M=Ni) and the dark orange product (for M=Pt) have not been further characterized by analyses or spectra [1].

$(C_5H_5Fe(CO)_2Pt(P(C_6H_5)_3)Cl)_2$ is the formulation of a pale brown crystalline material obtained by dropwise addition of $C_5H_5Fe(CO)_2Cl$ to $C_2H_4Pt(P(C_6H_5)_3)_2$ (1:1 mole) in refluxing benzene. The product deposits from the solution on standing at room temperature, 41% yield. The compound, insoluble in most organic solvents, has been characterized by elemental analysis and its IR spectrum, which shows frequencies associated with C_5H_5Fe, CO, Pt–P, and Pt–Cl [2]. However, the IR data in a table in [2] are those for the monomeric complex $C_5H_5Fe(CO)_2Pt(P(C_6H_5)_3)_2Cl$.

References:

[1] R.J. O'Brien, J.E. Guillet, the Governors of the University of Toronto (Fr. 1549337 [1966/68]; C.A. 71 [1969] No. 124677). – [2] M. Akhtar, H.C. Clark (J. Organometal. Chem. 22 [1970] 233/40).

2.5.2.2.8　Compounds of the $[C_5H_5Fe(CO)_2(\mu\text{-}H)(CO)_2FeC_5H_5]^+$　and $[C_5H_5Fe(CO)_2(\mu\text{-}H)(CO)(^2D)FeC_5H_5]^+$ Cations

The first title cation is formed on dissolving $(C_5H_5Fe(CO)_2)_2$ in 98% H_2SO_4 [4 to 6, 8], CF_3COOH [4, 5], other strong anhydrous acids like CH_3SO_3H, liquid HCl, HSO_3Cl, H_2PO_3F, $H_2PO_2F_2$ [5], and in BF_3/H_2O [4]. Solutions of $(C_5H_5Fe(CO)_2)_2$ in H_2SO_4 [1, 2] gave a van't Hoff i factor which appeared to indicate the presence of a $[(C_5H_5)_2Fe_2(CO)_4H_2]^{2+}$ cation [3].

The diamagnetic solutions of the cation in H_2SO_4, CF_3COOH, BF_3/H_2O, and $BF_3/H_2O/CF_3COOH$ are red brown [4]. Chemical shifts of the ^1H NMR spectrum (in H_2SO_4, HCl, CF_3COOH, and H_2SO_4/CH_3COOH) are $\delta = -26.3$ to -26.6 (singlet, Fe–H) and 5.23 to 5.34 (singlet, C_5H_5) ppm [4, 5, 8]. $\nu(CO)$ bands in the IR spectrum in various acid are at 2010 to 2012 and 2034 to 2040 cm^{-1} [5]. The band at 2068 cm^{-1} reported in [4] must be due to $[C_5H_5Fe(CO)_3]^+$ [5]. UV spectrum (in H_2SO_4): $\lambda_{max}(\varepsilon) = 287(3200)$, 340(2300), 426(1619?), and 565(206) [4]. For solutions in acetone/H_2SO_4 at 27 °C absorptions at $\lambda_{max}(\varepsilon) = 476(880)$ and 510(835) nm are given [8].

Conductometric titrations of the cation in liquid HCl with BCl_3 confirm the presence of a 1:1 electrolyte [5]. From changes in the ^1H NMR spectrum at different H_2SO_4 concentrations in CH_3COOH the equilibrium constant for $(C_5H_5Fe(CO)_2)_2 + H_2SO_4 \rightleftharpoons \{[(C_5H_5Fe(CO)_2)_2(\mu-H)]^+ HSO_4^-\}$ (ion pair) has been estimated to be $10^{-0.8}$ l·mol^{-1} at zero sulfuric acid concentration [6, 8].

Acidic solutions of the cation are stable at least for 10 min [4]. It is unstable in liquid HCl in the presence of light at room temperature and decomposes over a period of a few days, giving $(C_5H_5Fe(CO)_2)_2$ and tricarbonyl chlorides as the only identifiable products. The IR spectra of the acidic mulls soon show the $\nu(CO)$ bands of the $[C_5H_5Fe(CO)_3]^+$ ion, which increase in intensity with time. When the band at 2122 cm^{-1} was followed the reaction was found to be first-order in CH_3SO_3H, with a half-life of 15 min and $k = 4.6 \times 10^{-2}$ min^{-1} at about 40 °C [5]. The reaction is faster than with the cations $[C_5H_5Fe(CO)_2(\mu-X)(CO)_2FeC_5H_5]^+$ with X = Cl, Br, and I [7], see 2.5.2.2.9. Dilution with water and addition of KPF_6 affords $[C_5H_5Fe(CO)_3]PF_6$ [5]. Dilution of a concentrated H_2SO_4 solution gives $(C_5H_5Fe(CO)_2)_2$ [4]. H/D exchange does not occur on addition of DCl [5].

$[C_5H_5Fe(CO)_2(\mu-H)(CO)(^2D)FeC_5H_5]^+$ with $^2D = P(OCH_3)_3$ has been obtained as a green solution from $C_5H_5Fe(CO)(\mu-CO)_2(P(OCH_3)_3)FeC_5H_5$ in CH_3COOH and 98% H_2SO_4. ^1H NMR spectrum (in CH_3COOH/H_2SO_4): $\delta = -26.3$ (d, Fe–H, J(P,H) = 44 Hz), 3.82 (d, OCH_3, J(P,H) = 11.5 Hz), 4.95 (d, C_5H_5 on Fe(CO)P, J(P,H) = 1.5 Hz), and 5.21 (C_5H_5 on Fe(CO)$_2$) ppm. UV spectrum (HSO$_4^-$ salt in KBr at 80 K): $\lambda_{max} = 344$, 410, 429, $\approx$512, and 610 nm. The spectrum and the spectrum of the parent molecule are shown in a diagram [8].

$[C_5H_5Fe(CO)_2(\mu-H)(CO)_2FeC_5H_5]PF_6$ has been prepared by reaction of $(C_5H_5Fe(CO)_2)_2$ in anhydrous HF with PCl_5, removal of the excess of solvent in a stream of N_2 and extraction with hot CH_2Cl_2 to remove unreacted $(C_5H_5Fe(CO)_2)_2$. Washing with ether and drying in vacuum leaves the salt as a dark brown powder in 62% yield [4]. It has also been obtained as a red–brown solid by treating $(C_5H_5Fe(CO)_2)_2$ with PF_5 (1:3 mole) in anhydrous HCl at -84 °C overnight, removal of the volatiles in a vacuum and washing with CH_2Cl_2 [5].

^1H NMR spectrum (in SO_2): $\delta = -26.3$ (s, Fe–H) and 5.30 (C_5H_5) [4]. ^{57}Fe Mössbauer spectrum (4.2/300 K): $\delta = 0.55/0.47$ and $\Delta = 1.83/1.82$ mm·s^{-1} [5], see also [7]. IR spectrum (KCl?): $\nu(CO)$ bands at 2002 and 2032 cm^{-1} [5]. Other bands reported for this region in [4] are probably due to $[C_5H_5Fe(CO)_3]^+$ (at 2068 and 2138 cm^{-1}) and to $(C_5H_5Fe(CO)_2)_2$ (at 1767 cm^{-1}) [5]. Further assigned bands are: 555, 845 (PF$_6^-$), 1420 (C–C), and 3110 (C–H) cm^{-1} [5].

The salt is insoluble in ether, very sparingly soluble in CH_2Cl_2, and sparingly soluble in SO_2 [4]. In a KCl disk it decomposes almost completely within 18 h giving $(C_5H_5Fe(CO)_2)_2$ and $[C_5H_5Fe(CO)_3]^+$ [5]. Quantitative hydrolysis to $(C_5H_5Fe(CO)_2)_2$ has been observed in water/acetone [4].

$[C_5H_5Fe(CO)_2(\mu\text{-}D)(CO)_2FeC_5H_5]PF_6$ can be prepared like the previous salt with PF_5 in anhydrous DCl [5].

$[C_5H_5Fe(CO)_2(\mu\text{-}H)(CO)_2FeC_5H_5]BCl_4$ has been prepared like the PF_6^- salt from $(C_5H_5Fe(CO)_2)_2$ and BCl_3 (1:2 mole) in anhydrous HCl at $-84\,°C$. The red-brown solid gives a ^{57}Fe Mössbauer spectrum (4.2/300 K) with $\delta=0.60/0.49$ and $\Delta=1.90/1.87$ mm$\cdot$s^{-1}. The IR spectrum is the same as for the PF_6 salt except for the $\nu(BCl_4^-)$ at 659 and 691 cm^{-1}. The solid decomposes slowly [5].

$[C_5H_5Fe(CO)_2(\mu\text{-}H)(CO)(P(OCH_3)_3)FeC_5H_5]PF_6$. The preparation is not described in detail. The five $\nu(CO)$ bands in the IR spectrum (KBr) at 1969, 1991, 2023, 2050, and 2065 cm^{-1} suggest that at least two rotamers are present [7].

$[C_5H_5Fe(CO)_2(\mu\text{-}H)(CO)(P(OCH_3)_3)FeC_5H_5][B(C_6H_5)_4]$ has been prepared by adding the solution of the cation in CH_3COOH/H_2SO_4 to a stirred solution of $Na[B(C_6H_5)_4]$ in deoxygenated water. A green powder precipitates. It can be recrystallized from acetone/ether (4:3) by slow cooling to $-20\,°C$, 74% yield.

The salt melts at 154 to 159 °C with decomposition. The IR spectrum (in KBr) shows $\nu(CO)$ bands at 1955, 1984, and 2035 cm^{-1}. A sample stored at 4 °C for 5 months showed seven bands between 1730 und 2065 cm^{-1}, presumably due to isomerization to a mixture of rotamers as well as some decomposition to $C_5H_5Fe(CO)(\mu\text{-}CO)_2(P(OCH_3)_3)FeC_5H_5$. The UV spectrum (in KBr at 80 K) is shown as a diagram. The salt can be handled briefly in air and is stored under N_2 in the dark at $-20\,°C$ [8].

References:

[1] T.S. Piper, F.A. Cotton, G. Wilkinson (J. Inorg. Nucl. Chem. 1 [1955] 165/74). –
[2] B.F. Hallam, O.S. Mills, P.L. Pauson (J. Inorg. Nucl. Chem. 1 [1955] 313/6). –
[3] H.W. Sternberg, I. Wender (Intern. Conf. Coord. Chem., London 1959, Spec. Publ. No. 13, p. 35/55). – [4] A. Davison, W. McFarlane, L. Pratt, G. Wilkinson (J. Chem. Soc. 1962 3653/66). – [5] D.A. Symon, T.C. Waddington (J. Chem. Soc. A 1971 953/7).

[6] D.C. Harris (Diss. California Inst. of Technology 1973 according to Diss. Abstr. Intern. B 34 [1973/74] 1927). – [7] D.A. Symon, T.C. Waddington (J. Chem. Soc. Dalton Trans. 1974 78/81). – [8] D.C. Harris, H.B. Gray (Inorg. Chem. 14 [1975] 1215/7).

2.5.2.2.9　Compounds of the $[^5LFe(CO)_2(\mu\text{-}X)(CO)_2Fe^5L]^+$ Cations with X=Cl, Br, and I

The compounds collected in Table 13 contain C_5H_5 and $C_5H_4CH_3$ as ^5L ligands. A ^2D–^2D derivative of the iodine bridged cation (Formula I) has also been obtained

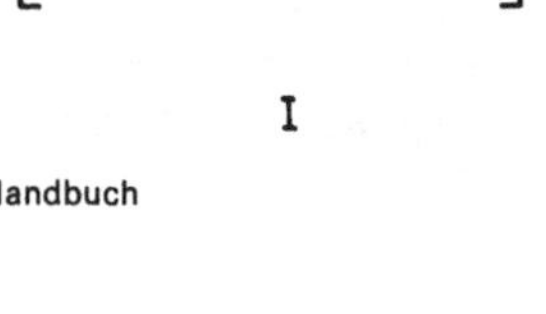

I

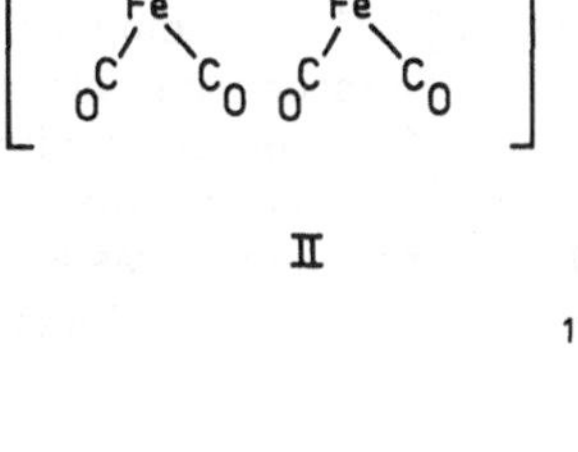

II

[4, 9], but it will be described among the compounds with twofold bridging groups in the next volume C 4. Another type with bridging cyclopentadienyl ligands (Formula II) contains formally a ^{10}L ligand and will also be treated later.

The compounds in Table 13 are prepared by the following methods:

Method I: Formation of the cations from $C_5H_5Fe(CO)_2X$ and concentrated H_2SO_4 at 0 °C, removal by pumping of the hydrogen halide evolved, dilution with H_2O at 0 °C and dropwise addition of HPF_6 (65% aqueous solution). Recrystallization from acetone/hexane [12].

Method II: Reaction of $C_5H_5Fe(CO)_2X$ with $AgPF_6$ in toluene for 30 min and filtration of the AgX. Solvent removal in a vacuum and crystallization of the residue from acetone/hexane [12].

Method III: Treatment of $C_5H_5Fe(CO)_2X$ with $BF_3 \cdot O(C_2H_5)_2$ at 40 to 70 °C for 1.5 to 2 h, addition of $NaBF_4$ at −196 °C, warming at room temperature with continuous shaking, and removal of the ether under vacuum. Recrystallization from acetone/ether [3, 14].

Method IV: Dropwise addition of Br_2 in CCl_4 or I_2 in benzene to $(C_5H_4RFe(CO)_2)_2$ (R = H and CH_3) in benzene (2:1 mole) and crystallization of the precipitated product from CH_2Cl_2/petroleum ether [5, 7].

Method V: Reaction of $(C_5H_4RFe(CO)_2)_2$ in C_6H_6/CH_3OH (12:1) in the presence of 2 moles $Na[B(C_6H_5)_4]$ with Cl_2 (1 mole) in CCl_4, Br_2 (2 moles) in CCl_4, or I_2 (2 moles) in C_6H_6. Recrystallization from CH_2Cl_2/petroleum ether [7].

Four compounds not listed in Table 13 have been mentioned in the literature without characterization.

$[(C_5H_5Fe(CO)_2)_2(\mu\text{-}I)]ClO_4$ and $[(C_5H_4CH_3Fe(CO)_2)_2(\mu\text{-}I)]ClO_4$ are prepared by method V in the presence of $LiClO_4$ and correspond in their properties to compounds No. 13 and 18 [7]. Spectroscopic evidence has also been obtained for the formation of $[(C_9H_7Fe(CO)_2)_2(\mu\text{-}I)]I_3$ (C_9H_7 = indenyl) from $(C_9H_7Fe(CO)_2)_2$ and I_2 in $CHCl_3$, but the salts are much less stable than those of the $[(C_5H_5Fe(CO)_2)_2(\mu\text{-}I)]^+$ cations [6].

The ^{57}Fe Mössbauer spectra show the equivalence of the two Fe atoms, i.e., symmetrical Fe–X–Fe bonding [8, 12, 13]. Almost constant isomer shifts are found for compounds No. 14, 16, and 18 [13]. For compounds No. 1, 7, and 11 a trend of decreasing isomer shift in the direction Cl > Br > I is observed, indicating that the s electron density at the Fe nuclei increases as the electronegativity of the halide decreases [12]. The cations formed by the reaction of $C_5H_5Fe(CO)_2X$ with concentrated H_2SO_4 show in the IR spectra the following ν(CO) bands: for X = Cl 2029, 2055, 2075 cm^{-1}, for X = Br 2023, 2059, 2071 cm^{-1}, and for X = I 2003, 2017, 2049, and 2063 cm^{-1}. Two main ν(CO) bands are expected, and each of these should be split by coupling with the other $C_5H_5Fe(CO)_2$ unit. But only the high–frequency band is split in all cases while the lower band is broad and unsymmetric [12].

The thermal stability increases in the series Cl, Br, and I, e.g., compounds No. 1, 7, and 11 [3, 12]. The decomposition in high vacuum gives $C_5H_5Fe(CO)_2X$ and $[C_5H_5Fe(CO)_3]^+$. Ferrocene has also been observed [1 to 3]. The decomposition on gentle heating was followed by monitoring the carbonyl region of the IR spectra. When X was chlorine or bromine, the only decomposition product identified was the $[C_5H_5Fe(CO)_3]^+$ ion; when X was iodine, $C_5H_5Fe(CO)_2I$ was also formed [12].

References on p. 156

The cations are readily susceptible to nucleophilic attack by halide ions with the formation of $^5LFe(CO)_2X$ (X=Cl, Br, I) in quantitative yield [5, 7]. In fact, the reaction of compound No. 13 with a tenfold excess of I^- in acetone has a half-life of about 30 s at 20 °C [7]. The reaction of $[(C_5H_5Fe(CO)_2)_2(\mu-I)]^+$ with Br^- as well as the reaction of $[(C_5H_5Fe(CO)_2)_2(\mu-Br)]^+$ with I^- in CH_2Cl_2 gives $C_5H_5Fe(CO)_2Br$ and $C_5H_5Fe(CO)_2I$ in equimolar amounts [5, 7]. The cations have also been observed to revert to $^5LFe(CO)_2X$ in CH_2Cl_2 solution in the presence of the corresponding halogen X_2, but the reaction rate is slow [5, 7]. Reactions with Lewis bases like $^2D=C_5H_5N$, $C_6H_5NH_2$, CH_3CN, and C_6H_5CN yield the $[C_5H_5Fe(CO)_2{}^2D]^+$ ions. For a given starting material like $[(C_5H_5Fe(CO)_2)_2(\mu-Br)]^+$, the reaction temperatures necessary increase in the ^2D series given above from room temperature to 50 °C. For a particular cation $[(C_5H_5Fe-(CO)_2)_2(\mu-X)]^+$ and a particular donor, e.g., CH_3CN, the reaction temperatures increase in the series X=Cl, Br, I from 25 to 70 °C [3], see also [1, 2]. Acrylonitrile also reacts as a base with compound No. 8 to give the $[C_5H_5Fe(CO)_2NCCH=CH_2]^+$ ion [3]. Formation of $C_5H_5Fe(CO)_2X$ compounds also occurs with the weaker bases acetone and tetrahydrofuran, but the corresponding $[C_5H_5Fe(CO)_2{}^2D]^+$ ions could not be isolated [2, 3].

Explanations to Table 13: All IR bands are $v(CO)$ vibrations. The electrical conductivity Λ, typical for 1:1 electrolytes, is given in $cm^2 \cdot \Omega^{-1} \cdot mol^{-1}$.

Table 13
Compounds of the $[^5LFe(CO)_2(\mu-X)(CO)_2Fe^5L]^+Y^-$ Type with X=Cl, Br, and I.
Further information on compounds preceded by an asterisk is given at the end of the table. For abbreviations and dimensions see p.170.

No.	Bridging atom X　Method of preparation	Anion Y^-	Properties and further remarks Explanations see above	Ref.
With $^5L=C_5H_5$:				
1	Cl I, II	PF_6	red crystals $^{57}Fe-\gamma$ (1.3 K): $\delta=0.67$, $\Delta=1.97$ (300 K): $\delta=0.58$, $\Delta=1.90$ IR (KBr): 2003, 2013, 2050, 2058 (CH_2Cl_2): 2025, 2064, 2071 UV (CH_2Cl_2): $\lambda_{max}(\varepsilon)=283$ (5750), 333 (4120), 390 (2630), 504 (475)	[12]
*2	Cl	$SbCl_6$	dark red crystals	[11]
*3	Cl III (35% yield)	BF_4	dark red crystals; dec. in air at 105° IR (Nujol): 2026, 2033, 2053, 2083 ($C_6H_5NO_2$): 2008, 2024, 2075	[3]
*4	Cl	BCl_4	red brown $^{57}Fe-\gamma$ (4.2 K): $\delta=0.60$, $\Delta=1.93$ (300 K): $\delta=0.50$, $\Delta=1.86$ IR (Nujol, 77 K): 2003, 2013, 2050, 2055	[8, 12]
5	Cl V (25 to 30% yield)	$B(C_6H_5)_4$	red ^1H NMR (CD_3COCD_3): 5.29 (C_5H_5) IR (CH_2Cl_2): 2022, 2061, 2071 $\Lambda=96$	[5, 7]

Table 13 [continued]

No.	Bridging atom X Method of preparation	Anion Y⁻	Properties and further remarks Explanations on p.153	Ref.
6	Br IV	Br₃	only mentioned, not characterized	[7]
*7	Br I, II	PF₆	dark red crystals; dec. at 80° in air, at 130° in vacuum ^{57}Fe-γ (1.3 K): δ=0.63, Δ=1.96 (300 K): δ=0.55, Δ=1.76 IR (Nujol): 2024, 2062, 2070 ($C_6H_5NO_2$): 2030, 2070, 2079 UV (CH_2Cl_2): $\lambda_{max}(\varepsilon)$=279 (5265), 342 (5000), 390 (3230), 505 (525)	[1, 3, 12]
*8	Br III (45% yield)	BF₄	dark red; dec. in air at 120° IR (Nujol): 2012, 2028, 2053, 2075 ($C_6H_5NO_2$): 2024, 2066, 2079	[3]
9	Br V	B(C_6H_5)₄	red ^{1}H NMR (CD_3COCD_3): 5.62 (C_5H_5) IR (CH_2Cl_2): 2022, 2056, 2069 Λ=108	[7]
*10	I IV	I₃	dark brown needles IR ($CHCl_3$): 2011, 2044, 2062	[5 to 7]
*11	I I, II	PF₆	dark red crystals; dec. at 111° in air, at 150° in vacuum ^{57}Fe-γ (1.3 K): δ=0.62, Δ=1.83 (300 K): δ=0.53, Δ=1.64 IR (Nujol): 2004, 2012, 2045 ($C_6H_5NO_2$): 2020, 2053, 2070 UV (CH_2Cl_2): 264 (18365), 347 (12800), 390 (10200), 490 (740)	[2, 3, 12]
*12	I III (65% yield)	BF₄	reddish-black crystals; dec. in air at 145° ^{1}H NMR (CD_3COCD_3): 5.56 (C_5H_5) IR (Nujol): 2008, 2066 ($C_6H_5NO_2$): 2020, 2053, 2070	[3, 10]
*13	I V (almost 100% yield)	B(C_6H_5)₄	brown ^{1}H NMR (CD_3COCD_3): 5.22 (C_5H_5) IR (CH_2Cl_2): 2014, 2049, 2062 Λ=107	[5, 7]

With $^5L=C_5H_4CH_3$:

No.	Bridging atom X Method of preparation	Anion Y⁻	Properties and further remarks	Ref.
14	Cl V (25 to 30% yield)	B(C_6H_5)₄	red ^{1}H NMR (CD_3COCD_3): 5.1 (very broad, C_5H_4) ^{57}Fe-γ (77 K): δ=0.24 (Fe), Δ=1.90 IR (CH_2Cl_2): 2016, 2054, 2063 Λ=91	[7, 13]

References on p. 156

Table 13 [continued]

No.	Bridging atom X Method of preparation	Anion Y$^-$	Properties and further remarks Explanations on p. 153	Ref.
15	Br III	Br$_3$	only mentioned, not characterized	[7]
16	Br V	B(C$_6$H$_5$)$_4$	red ^{1}H NMR (CD$_3$COCD$_3$): 5.0 (very broad, C$_5$H$_4$) ^{57}Fe-γ (77 K): δ=0.24 (Fe), Δ=1.86 IR (CH$_2$Cl$_2$): 2016, 2052, 2064 Λ=97	[7, 13]
17	I IV	I$_3$	only mentioned, no data given	[6, 7]
18	I V	B(C$_6$H$_5$)$_4$	brown ^{1}H NMR (CD$_3$COCD$_3$): 5.35 (C$_5$H$_4$) ^{57}Fe-γ (77 K): δ=0.23 (Fe), Δ=1.83 IR (CH$_2$Cl$_2$): 2009, 2043, 2057 Λ=100	[7, 13]

* Further information:

[(C$_5$H$_5$Fe(CO)$_2$)$_2$(μ-Cl)]SbCl$_6$ (Table **13**, No. **2**) has been obtained by the treatment of (C$_5$H$_5$Fe(CO)$_2$)$_2$ with SbCl$_5$ in CH$_2$Cl$_2$ at 0 °C. Recrystallization from CH$_2$Cl$_2$/ CHCl$_3$ [11].

[(C$_5$H$_5$Fe(CO)$_2$)$_2$(μ-X)]BF$_4$ (Table **13**, Nos. **3**, **8**, and **12**, X=Cl, Br, and I). The complete IR spectra (in Nujol) are given from 300 to 3125 cm^{-1}. Assignments were made for the vibrations of the C$_5$H$_5$ ligands. A broad band of compound No. **3** at 327 cm^{-1} may originate in a Fe–Cl vibration [3]. Bands of the BF$_4^-$ anion are between 1037 and 1075 cm^{-1} [3, 8]. The ν(CO) bands are also reported for solutions in acetone [3].

Chemical shifts of the ^{1}H NMR spectra (in acetone) are only given in Hz without indication of the frequency of measurement. The appearance of new signals over a period of 24 h shows that the cations decompose with decreasing rate in the series X=Cl, Br, and I [3].

[(C$_5$H$_5$Fe(CO)$_2$)$_2$(μ-I)]BF$_4$ has also been obtained from C$_5$H$_5$Fe(CO)$_2$I and AgBF$_4$ in benzene. It can be isolated from the red–brown precipitate by extraction with CH$_2$Cl$_2$ and is recrystallized from CH$_2$Cl$_2$/C$_6$H$_6$, 74% yield.

This compound crystallizes in the monoclinic system with a=15.602(2), b=9.607(2), c=12.373(2) Å, and β=104.86(1)°; space group P2/a–C$_{2h}^4$. Z=4 gives D$_c$=2.10 and D$_m$=2.08±0.02 g·cm^{-3}. The structure of the cation is shown in **Fig. 41**, p. 156. The transoid position of the C$_5$H$_5$ ligands and the large bond angle at the iodine atom suggest that the structure is essentially determined by intramolecular repulsion [10].

[(C$_5$H$_5$Fe(CO)$_2$)$_2$(μ-Cl)]BCl$_4$ (Table **13**, No. **4**) has been prepared from C$_5$H$_5$-Fe(CO)$_2$Cl and BCl$_3$ in liquid HCl at −84 °C. Bands of the anion in the IR spectrum are found at 661 and 691 cm^{-1} [8].

References on p. 156

Fig. 41

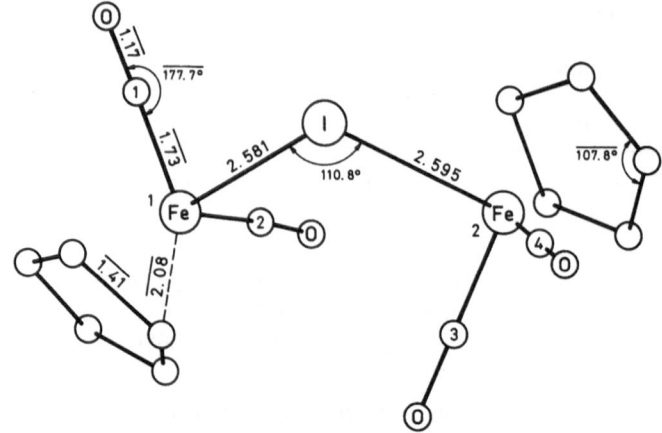

Molecular structure of the $[(C_5H_5Fe(CO)_2)_2(\mu\text{-}I)]^+$ cation [10].

Selected angles (°):

C(1)–Fe(1)–C(2)	94.9(8)	C(3)–Fe(2)–C(4)	95.8(7)
C(1)–Fe(1)–I	92.5(5)	C(3)–Fe(2)–I	94.5(6)
C(2)–Fe(1)–I	92.6(6)	C(4)–Fe(2)–I	90.0(6)

$[(C_5H_5Fe(CO)_2)_2(\mu\text{-}Br)]PF_6$ (Table **13**, No. **7**) has also been obtained in about 50% crude yield by the reaction of $C_5H_5Fe(CO)_2Br$ with $AlBr_3$ in liquid SO_2, addition of aqueous NH_4PF_6, and recrystallization from acetone/ether [1]. The complete IR spectrum is reported in [1]. See [3, 12] for the position of the $\nu(CO)$ bands in acetone and CH_2Cl_2. Changes of the ^1H NMR spectrum in acetone or nitrobenzene indicate slow decomposition of the salt. The dark red aqueous solution of the above preparation fades to orange, which may be due to a nucleophilic attack by water and possibly the formation of the $[C_5H_5Fe(CO)_2OH_2]^+$ ion [1].

$[(C_5H_5Fe(CO)_2)_2(\mu\text{-}I)]I_3$ (Table **13**, No. **10**) is also formed from $(C_5H_5Fe(CO)_2)_2$ and I_2 in $CHCl_3$ at 0 °C and crystallizes on cooling to −20 °C [6].

$[(C_5H_5Fe(CO)_2)_2(\mu\text{-}I)]PF_6$ (Table **13**, No. **11**) has been prepared in a melt of C_5H_5-$Fe(CO)_2I/AlCl_3$ at 60 °C (2 h) (1:3 mole). Careful hydrolysis with water under cooling and precipitation with NH_4PF_6 give the crude product in about 45% yield. Recrystallization from acetone/ether. The IR spectrum (in Nujol) is reported from 366 to 3125 cm^{-1} [2]. Bands of the anion are at 826 and 864 cm^{-1}. $\nu(CO)$ bands are also given for solutions in acetone [2, 3]. According to the ^1H NMR spectrum slow decomposition occurs in acetone solution [2].

$[(C_5H_5Fe(CO)_2)_2(\mu\text{-}I)][B(C_6H_5)_4]$ (Table **13**, No. **13**) is the only ionic product formed from $(C_5H_5Fe(CO)_2)_2$ and ICl or IBr in benzene in the presence of $Na[B(C_6H_5)_4]$. There was no evidence for the formation of the chloro- or bromo-bridged derivatives [5, 7]. The salt is recrystallized from CH_2Cl_2/petroleum ether [7].

References:

[1] E.O. Fischer, E. Moser (J. Organometal. Chem. **3** [1965] 16/24). – [2] E.O. Fischer, E. Moser (Z. Naturforsch. **20b** [1965] 184/5). – [3] E.O. Fischer, E. Moser (Z. Anorg. Allgem. Chem. **342** [1966] 156/64). – [4] R.J. Haines, A.L. du Preez, G.T.W.

Wittmann (Chem. Commun. **1968** 611/3). − [5] R.J. Haines, A.L. du Preez (J. Am. Chem. Soc. **91** [1969] 769/70).

[6] D.A. Brown, A.R. Manning, D.J. Thornhill (J. Chem. Soc. Chem. Commun. **1969** 338). − [7] R.J. Haines, A.L. du Preez (J. Chem. Soc. A **1970** 2341/6). − [8] D.A. Symon, T.C. Waddington (J. Chem. Soc. A **1971** 953/7). − [9] R.J. Haines, A.L. du Preez (Inorg. Chem. **11** [1972] 330/6). − [10] F.A. Cotton, B.A. Frenz, A.J. White (J. Organometal. Chem. **60** [1973] 147/52).

[11] W.R. Cullen, D.J. Patmore, J.R. Sams (Inorg. Chem. **12** [1973] 867/72). − [12] D.A. Symon, T.C. Waddington (J. Chem. Soc. Dalton Trans. **1974** 78/81). − [13] R. Greatrex, R.J. Haines (J. Organometal. Chem. **114** [1976] 199/211). − [14] E.O. Fischer, E. Moser (Inorg. Syn. **12** [1970] 35/43).

2.5.2.2.10   Compounds of the $[C_5H_5Fe(CO)_2(\mu\text{-}ER_n)(CO)_2FeC_5H_5]^+$ Cations with E = S, P, As, and Sb

The compounds listed in Table 14 are prepared by the following methods:

Method I: Reaction of $C_5H_5Fe(CO)_2Br$ with $(CH_3)_2P\text{-}P(CH_3)_2$ or $(CH_3)_2As\text{-}As(CH_3)_2$ (2:1 mole) in toluene at reflux temperature for 16 h. The precipitated solid is dissolved in water and treated with concentrated aqueous $NaClO_4$ or $Na[B(C_6H_5)_4]$. Recrystallization from aqueous acetone or acetone/hexane [1].

Method II: Treatment of $(C_5H_5Fe(CO)_2)_2$ with EX_3 (E = As, Sb, X = Cl, Br) in a two- to threefold excess in CH_2Cl_2 at room temperature for several hours (for SbX_3) or up to two weeks (for AsX_3). Recrystallization from CH_2Cl_2, $CH_2Cl_2/$ n-pentane, or acetone/ether [4].

Method III: Reaction between $C_5H_5Fe(CO)_2X$ (X = Cl, Br) and $C_5H_5Fe(CO)_2P(C_6H_5)_2$ in benzene [3], or $C_5H_5Fe(CO)_2Sb(CH_3)_2$ [7]. The former reaction is carried out in the presence of $Na[B(C_6H_5)_4]$ for 5 min to give the $[B(C_6H_5)_4]^-$ salt, which is recrystallized from acetone/methanol and acetone/petroleum ether [3].

Method IV: Formation of $[C_5H_5Fe(CO)_2OC(CH_3)_2]^+$ from $(C_5H_5Fe(CO)_2)_2$ and AgX (X = ClO_4, SbF_6, and BF_4) in a 1:2 mole ratio in acetone and further reaction of the cation with $C_5H_5Fe(CO)_2SR$ for 30 min [8].

The SR bridged cations (compounds No. 1 to 6) cannot be prepared from $C_5H_5Fe(CO)_2SR$ and $C_5H_5Fe(CO)_2X$ (X = halogen), the reaction analogous to method III [8]. C_5H_5Fe-$(CO)_2P(CH_3)_2$ may be an intermediate in the synthesis of compound No. 7 by method I [3].

The electrical conductivities of solutions of several salts have been measured [1 to 4, 6, 8]. The values indicate the presence of 1:1 electrolytes. The relatively small conductivities of compounds No. 16 and 19 could be due to association through halogen bridging [4].

The ^{57}Fe and ^{121}Sb Mössbauer spectra indicate that the fraction of π bonding in the Fe–Sb bond is greater than in the Fe–Sn bond of the isoelectronic $(C_5H_5Fe$-$(CO)_2)_2SnX_2$ compounds. The ^{121}Sb isomer shifts fall into the region between Sb^{III} and Sb^V [4, 6]. On the basis of the IR spectra, compounds No. 2 and 4 appear to occur as more than one conformer in solution [8].

The solids No. 7, 13, and 14 are stable in air [1], but No. 16 is sensitive to air and light [5].

Explanations to Table 14: μ represents the magnetic moment in B.M. measured by the Gouy method. Chemical shifts of the ^{121}Sb Mössbauer spectra are referred to $Ba^{121}SnO_3$ at 80 K. e^2qQ represents the quadrupole coupling constant in $mm \cdot s^{-1}$.

Table 14

Compounds of the $[C_5H_5Fe(CO)_2(\mu-ER_n)(CO)_2FeC_5H_5]Y$ Type, E=S, P, As, and Sb. Further information on compounds preceded by an asterisk is given at the end of the table. For abbreviations and dimensions see p. 170.

No.	Bridging group Method of preparation (yield in %)	Anion Y^-	Properties and further remarks Explanations see above	Ref.
With SR bridges:				
1	SC_2H_5 IV (−)	ClO_4	IR (CH_2Cl_2): 2008, 2051 conversion by $Na[B(C_6H_5)_4]$ in CH_3OH into No. 4	[8]
*2	SC_2H_5 IV (90)	SbF_6	red brown ^1H NMR (CD_3COCD_3): 1.50 (t, CH_3), 　2.50 (q, CH_2, J(H,H) =7.3), 5.55 (C_5H_5) IR (CH_2Cl_2): 2002, 2007, 2043, 2057	[8]
3	SC_2H_5 IV (−)	BF_4	IR (CH_2Cl_2): 2007, 2053	[8]
*4	SC_2H_5 IV (90)	$B(C_6H_5)_4$	purple red ^1H NMR (CD_3COCD_3): 1.20 (t, CH_3), 　2.40 (q, CH_2, J(H,H) =7.3), 　5.40 (C_5H_5), 7.0 (m, C_6H_5) IR (CH_2Cl_2): 2003, 2008, 2041, 2058	[8]
5	SC_4H_9-t IV (−)	SbF_6	IR (CH_2Cl_2): 2008, 2052	[8]
6	SC_4H_9-t IV (−)	BF_4	only mentioned, no data reported	[8]
With PR_2 bridges:				
*7	$P(CH_3)_2$ I (16)	$B(C_6H_5)_4$	yellow needles; m.p. 191 to 194° (dec.) ^1H NMR (CH_3SOCH_3): 1.89 (d, CH_3, 　J(P,H) =9.0), 5.46 (d, C_5H_5) IR (halocarbon): 1942, 1970, 1992, 2015, 　2026, 2038	[1]
*8	$P(C_6H_5)_2$ III (75)	$B(C_6H_5)_4$	yellow ^1H NMR (CD_2Cl_2): 4.55 (C_5H_5) IR (CH_2Cl_2): 1996, 2031, 2046	[3]
With AsR_2 bridges:				
*9	$AsCl_2$ II (−)	$AsCl_4$	−	[2]

References on p. 161

Table 14 [continued]

No.	Bridging group Method of preparation (yield in %)	Anion Y$^-$	Properties and further remarks Explanations on p. 158	Ref.
10	$AsCl_2$ II (−)	$FeCl_4$	orange–brown needles IR (CH_2Cl_2): 2030, 2057, 2071 $\mu = 5.9$	[4]
11	$AsBr_2$ II (−)	$AsBr_4$	−	[2]
12	$AsBr_2$ II (−)	$FeBr_4$	red crystals IR (CH_2Cl_2): 2028, 2054, 2069 $\mu = 6.2$	[4]

13	$As(CH_3)_2$
	I (31)
		ClO_4	orange crystals; dec. >150°
			^{1}H NMR (CH_3SOCH_3): 1.82 (CH_3),
			   5.48 (C_5H_5)
			IR (halocarbon): 1943, 1966, 1975, 2015,
			   2028; the complete spectrum is given
				[1]

14	$As(CH_3)_2$
	I (27)
		$B(C_6H_5)_4$	orange crystals; m.p. 211 to 213° (dec.)
			IR (halocarbon): 1943, 1969, 1973, 1975,
			   2018, 2028; the complete spectrum
			   is given
				[1]

With SbR_2 bridges:

No.	Bridging group Method of preparation (yield in %)	Anion Y$^-$	Properties and further remarks Explanations on p. 158	Ref.
*15	$SbCl_2$	PF_6	orange crystals ^{57}Fe-γ (80 K): $\delta = 0.40$, $\Delta = 1.86$ ^{121}Sb-γ (8.5 K): $\delta = -9.3$, $e^2qQ = +29.0 \pm 10$ IR (CH_2Cl_2): 2020, 2053, 2065	[4, 6]
*16	$SbCl_2$ II (−)	Sb_2Cl_7	orange cubes, red–brown plates ^{57}Fe-γ (80 K): $\delta = 0.40$, $\Delta = 1.81$ IR (CH_2Cl_2): 2020, 2054, 2065	[2, 4 to 6]
*17	$SbCl_2$	$Cr(SCN)_4$- $(NH_3)_2$	^{57}Fe-γ (80 K): $\delta = 0.40$, $\Delta = 1.83$ ^{121}Sb-γ (8.5 K): $\delta = -9.1$, $e^2qQ = +28.7 \pm 0.5$ IR (CH_3COCH_3): 2000, 2048, 2073	[4, 6]
18	$SbBr_2$ −	PF_6	^{57}Fe-γ (80 K): $\delta = 0.40$, $\Delta = 1.83$ ^{121}Sb-γ (8.5 K): $\delta = -9.6$, $e^2qQ = +26.6 \pm 0.4$	[6]
19	$SbBr_2$ II (−)	$SbBr_4$	orange needles	[2, 4]
20	$Sb(CH_3)_2$ III (−)	Br	m.p. 140 to 141° IR (CH_3COCH_3): 1981, 2018, 2035	[7]
21	$Sb(CF_3)_2$ −	PF_6	^{57}Fe-γ (80 K): $\delta = 0.42$, $\Delta = 1.80$	[6]

References on p. 161

Table 14 [continued]

No.	Bridging group Method of preparation (yield in %)	Anion Y⁻	Properties and further remarks Explanations on p. 158	Ref.
*22	$Sb(CF_3)_2$	$(CF_3)_2SbI_2$	yellow crystals IR (CH_2Cl_2): 2021, 2050, 2063	[2, 4]
23	$Sb(CF_3)_2$ –	$Cr(SCN)_4-$ $(NH_3)_2$	$^{121}Sb-\gamma$ (8.5 K): $\delta = -8.3$, $e^2qQ = +18.3 \pm 0.7$	[6]
*24	$Sb(C_6H_5)_2$	PF_6	orange crystals $^{57}Fe-\gamma$ (80 K): $\delta = 0.39$, $\Delta = 1.74$ $^{121}Sb-\gamma$ (8.5 K): $\delta = -7.0$, $e^2qQ = -7.0 \pm 0.4$ IR (CH_2Cl_2): 1985, 1994, 2027, 2045	[6]

* Further information:

$[(C_5H_5Fe(CO)_2)_2(\mu-SC_2H_5)]SbF_6$ (Table **14**, No. 2). The number of $\nu(CO)$ bands in the IR spectrum seems to indicate that more than one conformer is present in solution. UV irradiation leads to an increase of the intensity of the bands at 2002 an 2043 cm⁻¹ at the expense of the two other bands. UV irradiation in tetrahydrofuran for 2 h produces $[C_5H_5Fe(CO)(\mu-CO)(\mu-SC_2H_5)(CO)FeC_5H_5]SbF_6$ in 50% yield [8].

$[(C_5H_5Fe(CO)_2)_2(\mu-SC_2H_5)][B(C_6H_5)_4]$ (Table **14**, No. 4) has been obtained from compound No. 1 with $Na[B(C_6H_5)_4]$ in methanol. It decomposes in refluxing CH_2Cl_2 and gives after 1 h a 10% yield of $C_5H_5Fe(CO)_2C_6H_5$. UV irradiation in tetrahydrofuran for 2 h gives a 20% yield of $C_5H_5Fe(CO)_2C_6H_5$ and $[C_5H_5Fe(CO)(\mu-CO)(\mu-SC_2H_5)(CO)-FeC_5H_5][B(C_6H_5)_4]$ as a by-product [8].

$[(C_5H_5Fe(CO)_2)_2(\mu-P(CH_3)_2)][B(C_6H_5)_4]$ (Table **14**, No. 7). Attempts to prepare $C_5H_5Fe(CO)(\mu-P(CH_3)_2)(\mu-P(C_6H_5)_2)(CO)FeC_5H_5$ by reacting the complex with $(C_6H_5)_2P-P(C_6H_5)_2$ were unsuccessful [1]. The IR spectrum is completely given from 451 to 3124 cm⁻¹ [1].

$[(C_5H_5Fe(CO)_2)_2(\mu-P(C_6H_5)_2)][B(C_6H_5)_4]$ (Table **14**, No. 8). UV irradiation in tetrahydrofuran gives $(C_5H_5Fe(CO)_2)_2$ and not the CO bridged cation [3].

$[(C_5H_5Fe(CO)_2)_2(\mu-AsCl_2)]AsCl_4$ (Table **14**, No. 9). The preparation by method II first gives another crystalline product, whose analytical data is consistent with $[(C_5H_5Fe-(CO)_2)_2AsCl_2]_3[As_4Cl_{15}]$. On recrystallization from acetone/ether this is converted into compound No. 9 [2].

$[(C_5H_5Fe(CO)_2)_2(\mu-SbCl_2)]PF_6$ (Table **14**, No. 15) has been obtained from compound No. 16 and $NaPF_6$ in CH_2Cl_2 and is recrystallized from CH_2Cl_2 [4].

$[(C_5H_5Fe(CO)_2)_2(\mu-SbCl_2)][Sb_2Cl_7]$ (Table **14**, No. 16) crystallizes in the monoclinic system with a $=15.849(2)$, b$=11.968(2)$, c$=31.338(5)$ Å, and $\beta=92.33(1)°$; space group $P2_1/n-C_{2h}^5$. $Z=8$ gives $D_c=2.32$ and $D_m=2.2(1)$ g·cm⁻³. The asymmetric unit contains two cations and an anion grouping of formula $[Sb_4Cl_{14}]^{2-}$. These anions form ribbons parallel to the *ab* plane. The structure of the cation is shown in **Fig. 42** [5].

Fig. 42

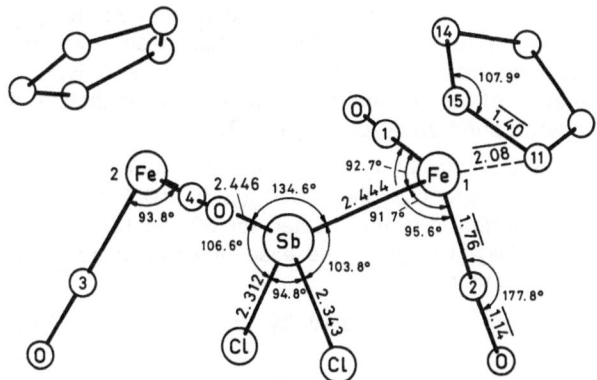

Structure of the cation in $[(C_5H_5Fe(CO)_2)_2(\mu-SbCl_2)][Sb_2Cl_7]$ [5].

Selected angles (°):

Sb–Fe(1)–C(1)	92.7(7)	Sb–Fe(2)–C(3)	93.0(7)
Sb–Fe(1)–C(2)	91.7(7)	Sb–Fe(2)–C(4)	91.1(7)
C(1)–Fe(1)–C(2)	95.6(9)	C(3)–Fe(2)–C(4)	93.8(9)

$[(C_5H_5Fe(CO)_2)_2(\mu-SbCl_2)][Cr(SCN)_4(NH_3)_2]$ (Table **14**, No. **17**) has been prepared from compound No. 16 and $NH_4[Cr(SCN)_4(NH_3)_2]\cdot H_2O$ in CH_2Cl_2 [4].

$[(C_5H_5Fe(CO)_2)_2(\mu-Sb(CF_3)_2)][(CF_3)_2SbI_2]$ (Table **14**, No. **22**) has been obtained from a reaction without solvent between $(C_5H_5Fe(CO)_2)_2$ and $(CF_3)_2SbI$ (1:1.9 mole) in an evacuated tube at 20 °C for 12 h. Recrystallization from CH_2Cl_2/n-pentane [4].

$[(C_5H_5Fe(CO)_2)_2(\mu-Sb(C_6H_5)_2)]PF_6$ (Table **14**, No. **24**) is formed from $Na[C_5H_5-Fe(CO)_2]$ and $(C_6H_5)_2SbCl_3$ in tetrahydrofuran at -80 °C in the presence of NH_4PF_6 and is recrystallized from CH_2Cl_2/n-pentane [6].

References:

[1] R.G. Hayter, L.F. Williams (Inorg. Chem. **3** [1964] 613/4). – [2] W.R. Cullen, D.J. Patmore, J.R. Sams, M.J. Newlands, L.K. Thompson (J. Chem. Soc. Chem. Commun. **1971** 952/3). – [3] R.J. Haines, A.L. du Preez, C.R. Nolte (J. Organometal. Chem. **55** [1973] 199/203). – [4] W.R. Cullen, D.J. Patmore, J.R. Sams (Inorg. Chem. **12** [1973] 867/72). – [5] F.W.B. Einstein, R.D.G. Stones (Inorg. Chem. **12** [1973] 1690/6).

[6] W.R. Cullen, D.J. Patmore, J.R. Sams, J.S. Scott (Inorg. Chem. **13** [1974] 649/55). – [7] W. Malisch, P. Panster (Angew. Chem. **86** [1974] 708/9; Angew. Chem. Intern. Ed. Engl. **13** [1974] 670). – [8] R.B. English, R.J. Haines, C.R. Nolte (J. Chem. Soc. Dalton Trans. **1975** 1030/3).

2.5.2.2.11   Compounds of the $C_5H_5Fe(CO)X(\mu-{}^2D-{}^2D)X(CO)FeC_5H_5$ Type

In the compounds described in this chapter and collected in Table 15 the bridging $^2D-^2D$ ligand is a diphosphine group. X represents a halogen atom except in No. 5 (X=$COCH_3$) and No. 10 (X=$SnCl_3$). The following methods have been used for preparation:

Method I: An excess of $C_5H_5Fe(CO)_2X$ (X=Cl, Br, and I) is allowed to react with the $^2D-^2D$ molecule in refluxing benzene for several hours [5, 7]. Compound

No. 2 has been obtained in a reaction at room temperature for 24 h [4]. Boiling methylene chloride (for 1 h) has been used for compound No. 11 [6]. Starting material for compound No. 5 is $C_5H_5Fe(CO)_2CH_3$ [1]. After evaporation of the solvent, the residue is either directly recrystallized or first chromatographed on Al_2O_3 [1, 4 to 8].

Method II: Equimolar amounts of I_2 in benzene are added dropwise to a solution of the complex type $C_5H_5Fe(\mu-CO)_2(\mu-{}^2D-{}^2D)FeC_5H_5$ in refluxing benzene, and the solution is further heated for 30 min [2, 7]. For the $^2D-^2D$ ligands $(C_6H_5)_2P(CH_2)_nP(C_6H_5)_2$ the ease of complex formation is reduced as n alters from 3 to 1 [2].

Preparative methods denoted in Table 15 by the term "special" are described under further information.

$C_5H_5Fe(CO)(Sn(CH_3)_3)(\mu-cis-(C_6H_5)_2PCH=CHP(C_6H_5)_2)(Sn(CH_3)_3)(CO)FeC_5H_5$ is possibly an intermediate in the photochemical reaction of $C_5H_5Fe(CO)_2Sn(CH_3)_3$ with $cis-(C_6H_5)_2PCH=CHP(C_6H_5)_2$. A single $\nu(CO)$ band at 1918 cm^{-1} appears in the IR spectrum of the solution. In the case of the irradiation of $C_5H_5Fe(CO)_2Si(CH_3)_3$ with $(C_6H_5)_2PCH_2P(C_6H_5)_2$, a dirty yellow, sparingly soluble solid was isolated. It showed a single $\nu(CO)$ band at 1900 cm^{-1} (in KBr). Analyses suggested it to be impure C_5H_5Fe-$(CO)(Si(CH_3)_3)(\mu-(C_6H_5)_2PCH_2P(C_6H_5)_2)(CO)FeC_5H_5$ [3].

Table 15
Compounds of the $C_5H_5Fe(CO)X(\mu-{}^2D-{}^2D)X(CO)FeC_5H_5$ Type.
Further information on compounds preceded by an asterisk is given at the end of the table. For abbreviations and dimensions see p. 170.

No.	$^2D-^2D$ ligand Method of preparation (yield in %)	X ligand	Properties and further remarks	Ref.
1	$(C_6H_5)_2PCH_2-$ $P(C_6H_5)_2$ II (−)	I	no properties reported	[2]
*2	$(CH_3)_2PCH_2CH_2-$ $P(CH_3)_2$ I (6.4)	I	dark green crystals; m.p. 205° ^{1}H NMR (CDCl$_3$): 4.66 (s, C$_5$H$_5$) IR (CH$_2$Cl$_2$): 1955	[4]
*3	$(C_6H_5)_2PCH_2CH_2-$ $P(C_6H_5)_2$ special	Br	green crystals ^{1}H NMR (CDCl$_3$): 2.42 (CH$_2$), 4.21 (C$_5$H$_5$), 7.20 (C$_6$H$_5$) IR: 1945 (KBr), 1950 (CH$_2$Cl$_2$), 1955 (C$_6$H$_6$)	[9]
*4	$(C_6H_5)_2PCH_2CH_2-$ $P(C_6H_5)_2$ I (≈65), II (55)	I	green crystals ^{1}H NMR (CDCl$_3$): 1.29, 1.47 (CH$_2$), 4.36 (C$_5$H$_5$), 7.45 (C$_6$H$_5$) IR: 1935 (KBr), 1942 (CH$_2$Cl$_2$), 1948 (C$_6$H$_6$)	[2, 7, 9]
5	$(C_6H_5)_2PCH_2CH_2-$ $P(C_6H_5)_2$ I (19.8)	COCH$_3$	orange crystals (from CHCl$_3$/octane) ^{1}H NMR (CDCl$_3$): 4.31 (d, C$_5$H$_5$, J(P,H) =1.5), 7.42 (m, C$_6$H$_5$) IR (CHCl$_3$): 1915; ν(C=O) at 1598 decomposes rapidly in CDCl$_3$	[1]

References on p. 164

Table 15 [continued]

No.	^{2}D–^{2}D ligand Method of preparation (yield in %)	X ligand	Properties and further remarks	Ref.
6	$(C_6H_5)_2PCH_2CH_2$- $CH_2P(C_6H_5)_2$ II (−)	I	no properties reported	[2]
*7	$(C_6H_5)_2PC\equiv C$- $P(C_6H_5)_2$ I (−)	Cl	green crystals; m.p. 155 to 156° ^{1}H NMR (CDCl$_3$): 4.53 (d, C$_5$H$_5$) ^{57}Fe-γ (77 K): δ=0.57, Δ=1.91 　(295 K): δ=0.51, Δ=1.87 IR: 1956, 1972 (Mull); 1973 (CHCl$_3$)	[5, 8]
*8	$(C_6H_5)_2PC\equiv C$- $P(C_6H_5)_2$ I (−)	Br	green plates; m.p. 173 to 174° ^{1}H NMR (CDCl$_3$): 4.52 (d, C$_5$H$_5$) ^{57}Fe-γ (77 K): δ=0.58, Δ=1.96 　(295 K): δ=0.50, Δ=1.91 IR: 1950, 1970 (Mull); 1970 (CHCl$_3$)	[5, 8]
*9	$(C_6H_5)_2PC\equiv C$- $P(C_6H_5)_2$ I (−)	I	green plates; m.p. 163 to 164° ^{1}H NMR (CDCl$_3$): 4.52 (d, C$_5$H$_5$) ^{57}Fe-γ (77 K): δ=0.56, Δ=1.88 　(295 K): δ=0.49, Δ=1.87 IR: 1945, 1972 (Mull); 1970 (CHCl$_3$)	[5, 8]
*10	$(C_6H_5)_2PC\equiv C$- $P(C_6H_5)_2$ special	SnCl$_3$	orange needles; m.p. 137 to 139° (dec.) ^{57}Fe-γ (77 K): δ=0.48, Δ=1.79 　(295 K): δ=0.40, Δ=1.77 ^{119}Sn-γ (77 K): δ=2.05 (BaSnO$_3$), Δ=1.99	[5, 8]
*11	$(C_6H_5)_2PCH_2C\equiv C$- $CH_2P(C_6H_5)_2$ I (1.6)	Br	green; m.p. 147 to 149° IR (KBr): 1940	[6]
12	$(C_2H_5O)_2PO$- $P(OC_2H_5)_2$ I (−)	I	no properties reported	[10]

* Further information:

$(C_5H_5Fe(CO)I)_2(CH_3)_2PCH_2CH_2P(CH_3)_2$ (Table 15, No. 2). The residue from the reaction by method I was chromatographed in CH_2Cl_2 on Al_2O_3. It was eluted and recrystallized with CH_2Cl_2/hexane. The complex was also obtained by UV irradiation of $C_5H_5Fe(CO)_2I$ and $(CH_3)_2PCH_2CH_2P(CH_3)_2$ in benzene at room temperature for 6 h. The complete IR spectrum from 659 to 3078 cm^{-1} is reported [4].

$(C_5H_5Fe(CO)Br)_2(C_6H_5)_2PCH_2CH_2P(C_6H_5)_2$ (Table 15, No. 3) has been prepared from $C_5H_5Fe(\mu$-CO)$_2(\mu$-$(C_6H_5)_2PCH_2CH_2P(C_6H_5)_2)FeC_5H_5$ and Br_2 (1:1 mole) in CH_2Cl_2 at −78 °C. Warming to room temperature, evaporation of the solvent, and chromatography on SiO_2 at 10 °C with tetrahydrofuran as eluent gives a 14% yield of the compound, which is recrystallized from CH_2Cl_2 by cooling to −78 °C.

UV irradiation in tetrahydrofuran at 10 °C for 3 h gives back the starting material in 10% yield. The reaction with N_2H_4 (0.6 mole) in tetrahydrofuran at 60 °C for 20 min

References on p. 164

leads to the cation $[C_5H_5Fe(CO)(C_6H_5)_2PCH_2CH_2P(C_6H_5)_2]^+$ which has been crystallized as PF_6^- salt. However, the same reaction in benzene gives the starting material and variable amounts of decomposition products. In the absence of N_2H_4, neither in tetrahydrofuran nor in benzene does a reaction occur [9].

$(C_5H_5Fe(CO)I)_2(C_6H_5)_2PCH_2CH_2P(C_6H_5)_2$ (Table 15, No. 4). Similar results as for compound No. 3 have been obtained on treatment with N_2H_4 in tetrahydrofuran or benzene. But the UV irradiation gives a higher yield (80%) of $C_5H_5Fe(\mu\text{-}CO)_2$-$(\mu\text{-}(C_6H_5)_2PCH_2CH_2P(C_6H_5)_2)FeC_5H_5$ [9].

$(C_5H_5Fe(CO)X)_2(C_6H_5)_2PC\equiv CP(C_6H_5)_2$ (Table 15, Nos. 7, 8, and 9, X = Cl, Br, and I) are recrystallized from $CHCl_3/CH_3OH$. The coupling constants for the C_5H_5 doublets in the ^1H NMR spectrum are J(P,H) =1.9 Hz for all three complexes. The ^{57}Fe Mössbauer parameters do not vary significantly in this series.

Compound No. 7 decomposes even in thoroughly degassed chloroform. Compound No. 8 reacts with $P(OCH_3)_3$ in refluxing benzene to give $C_5H_5Fe(P(OCH_3)_3)_2Br$ and $(C_6H_5)_2PC\equiv CP(C_6H_5)_2$. Its reaction with $SnCl_2$ has been used for the preparation of compound No. 10 [5, 8].

$(C_5H_5Fe(CO)SnCl_3)_2(C_6H_5)_2PC\equiv CP(C_6H_5)_2$ (Table 15, No. 10) has been prepared from compound No. 8 in benzene and an excess of $SnCl_2$ in methanol. After 6 h at reflux temperature and filtration, the product crystallizes on slow removal of solvent. The compound is not sufficiently soluble in benzene, acetone, or chloroform for a ^1H NMR spectrum to be recorded. The ^{57}Fe Mössbauer parameters are significantly smaller than for compounds No. 7 to 9. An interaction of the $SnCl_3$ and $P(C_6H_5)_2$ groups may decrease the Cl–Sn–Cl and C–P–C angles, thus increasing the s contribution to the Sn–Fe and P–Fe bonds [5, 8].

$(C_5H_5Fe(CO)Br)_2(C_6H_5)_2PCH_2C\equiv CCH_2P(C_6H_5)_2$ (Table 15, No. 11) has been prepared by method I in boiling CH_2Cl_2 (1 h). The second product is the $[((C_5H_5Fe(CO)_2)_2$-$(C_6H_5)_2PCH_2C\equiv CCH_2P(C_6H_5)_2]^{2+}$ cation (2.5% yield). The complex is eluted from Al_2O_3 with acetone. The solubilities in $CDCl_3$ and CD_3COCD_3 are too low for a ^1H NMR spectrum to be recorded. The IR spectrum (in KBr) is completely given from 698 to 3068 cm^{-1} [6].

References:

[1] K.W. Barnett (Diss. Univ. of Wisconsin, Madison, 1967; Diss. Abstr. B 28 [1967/68] 3203). – [2] R.J. Haines, A.L. du Preez, G.T.W. Wittmann (Chem. Commun. 1968 611/3). – [3] R.B. King, K.H. Pannell (Inorg. Chem. 7 [1968] 1510/3). – [4] R.B. King, L.W. Houk, K.H. Pannell (Inorg. Chem. 8 [1969] 1042/8). – [5] A.J. Carty, A. Efrati, T.W. Ng, T. Birchall (Inorg. Chem. 9 [1970] 1263/8).

[6] R.B. King, A. Efrati (Inorg. Chim. Acta 4 [1970] 123/8). – [7] R.J. Haines, A.L. du Preez (Inorg. Chem. 11 [1972] 330/6). – [8] T.W. Ng (Diss. Univ. of Waterloo, Canada, 1973; Diss. Abstr. Intern. B 34 [1973/74] 4865/6). – [9] D. Sellmann, E. Kleinschmidt (Z. Naturforsch. 32b [1977] 1010/4). – [10] A.L. du Preez, I.L. Marais, R.J. Haines, A. Pidcock, M. Safari (J. Organometal. Chem. 141 [1977] C10/C12).

2.5.2.2.12 Compounds of the $[C_5H_5Fe(CO)_2(\mu\text{-}^2D\text{-}^2D)(CO)_2FeC_5H_5]^{2+}$ and $[C_5H_5Fe(CO)(^2D')(\mu\text{-}^2D\text{-}^2D)(^2D')(CO)FeC_5H_5]^{2+}$ Cations

The compounds in Table 16 except for Nos. 1 and 2 contain ditertiary phosphines as bridging ligands. $^2D'$ substituted derivatives with $^2D'$=CH_3CN (Nos. 17 to 19) have been observed in the electrochemical oxidation of $C_5H_5Fe(\mu-CO)_2(\mu-(C_6H_5)_2P(CH_2)_n-P(C_6H_5)_2)FeC_5H_5$ complexes.

The compounds listed in Table 16 are prepared by the following methods. Warning: the solutions and salts containing the ClO_4^- anion may be explosive.

Method I: The acetone intermediate $[C_5H_5Fe(CO)_2OC(CH_3)_2]ClO_4$ is prepared from $(C_5H_5Fe(CO)_2)_2$ and $Fe(ClO_4)_3$ (1:3.3 mole) in deaerated acetone. To a solution of this intermediate in about eightfold excess is added the $^2D-^2D$ molecule in benzene. After 2 h the solution is added dropwise to ether giving the perchlorate salts as yellow powders. It is reprecipitated from a solution in CH_3CN with ether [6].

Method II: Solutions of the salts No. 3, 5, 9, 11, and 12 in acetonitrile or acetone are treated with large excess of aqueous NH_4PF_6, KPF_6, or $Na[B(C_6H_5)_4]$ [1, 3, 6].

Method III: $C_5H_5Fe(CO)_2Cl$ and $(C_6H_5)_2PCH_2CH_2P(C_6H_5)_2$ (2:1 mole) are reacted in benzene at room temperatures for 15 h. The product separates from the solution and is washed with benzene [8]. The preparation of No. 7 was carried out with equimolar amounts of the starting materials in the presence of $AlCl_3$ in refluxing benzene for 3 h [1].

Method IV: The compound $(C_5H_5Fe(CO)(\mu-CO)_2FeC_5H_5)_2(C_6H_5)_2PC\equiv CP(C_6H_5)_2$ is cleaved with Br_2 or I_2 by heating in CH_2Cl_2 at reflux temperature for 6 h. The crude product precipitates and is recrystallized from boiling acetone/ethanol. Compound No. 11 has been prepared by bubbling Cl_2 through the solution at room temperature. Rapid removal of the solvent in a vacuum and addition of n-heptane precipitates the $[FeCl_4]^-$ salt [3].

Method V: Anhydrous $CoCl_2$ in ethanol is treated with 1.4 moles of $[C_5H_5Fe(CO)_2(C_6H_5)_2PCH_2CH_2P(C_6H_5)_2]PF_6$ in acetone at room temperature for 2 h. The products precipitate on addition of ether. They are reprecipitated from a solution in CH_2Cl_2 with ether [6].

Table 16

Compounds of the $[C_5H_5Fe(CO)_2(\mu-^2D-^2D)(CO)_2FeC_5H_5]^{2+}$ and $[C_5H_5Fe(CO)(NCCH_3)(\mu-^2D-^2D)(CH_3CN)(CO)FeC_5H_5]^{2+}$ cations.

Further information on compounds preceded by an asterisk is given at the end of the table. For abbreviations and dimensions see p. 170.

No.	$^2D-^2D$ ligand Method of preparation (yield in %)	Anion	Properties and further remarks	Ref.

Type $[C_5H_5Fe(CO)_2(\mu-^2D-^2D)(CO)_2FeC_5H_5]^{2+}$:

No.	$^2D-^2D$ ligand	Anion	Properties and further remarks	Ref.
1	$CH_3SCH_2CH_2SCH_3$ I (81)	ClO_4	golden precipitate ^1H NMR (CD_3CN): 5.48 (C_5H_5) IR (Nujol): 2033, 2071	[4, 6]
2	(N—N ring structure) I (38)	ClO_4	golden precipitate ^1H NMR (CD_3CN): 5.89 (C_5H_5) IR (Nujol): 2031, 2067	[4, 6]

References on p. 168

Table 16 [continued]

No.	$^2D-^2D$ ligand Method of preparation (yield in %)	Anion	Properties and further remarks	Ref.

3	$(C_6H_5)_2PCH_2-$
	$P(C_6H_5)_2$
	I (−)
		ClO_4	yellow solid
			the salt is explosive when dry
			   and should not be isolated.
			   It is converted into No. 4
				[6]

4	$(C_6H_5)_2PCH_2-$ $P(C_6H_5)_2$ II (39)	$B(C_6H_5)_4$	orange solid (from acetone/ether) ^1H NMR (CD_3CN): 5.01 (C_5H_5), 6.93, 7.07, 7.55 (m) (C_6H_5) IR (CH_2Cl_2): 2015, 2055	[6]
5	$(C_6H_5)_2PCH_2CH_2-$ $P(C_6H_5)_2$ III (≈75)	Cl	yellow crystals IR (CH_2Cl_2): 2008, 2053	[8]
6	$(C_6H_5)_2PCH_2CH_2-$ $P(C_6H_5)_2$ I (83)	ClO_4	^1H NMR (CD_3CN): 5.12 (C_5H_5), 7.44 (C_6H_5) IR (Nujol): 1998, 2050	[4, 6]
7	$(C_6H_5)_2PCH_2CH_2-$ $P(C_6H_5)_2$ III and II (35)	PF_6	yellow crystals (from acetone/ether) IR (Nujol): 1995, 2040; bands between 557 and 1440 are also given	[1]
8	$(C_6H_5)_2PCH_2CH_2-$ $CH_2P(C_6H_5)_2$ I (87)	ClO_4	yellow solid (from acetonitrile/ether) ^1H NMR (CD_3CN): 5.13, 5.15 (C_5H_5), 7.57 (C_6H_5) IR (CH_2Cl_2): 2009, 2050	[6]

*9	$(C_6H_5)_2PC≡C-$
	$P(C_6H_5)_2$
	IV (−)
		I_3	red−brown needles; m.p. 167 to 169°
			^1H NMR (CD_3COCD_3): 5.79 (d, C_5H_5,
			   J(P,H) =1.9)
			$^{57}Fe-\gamma$ (77 K): δ=0.36, Δ=1.80
			   (295 K): Δ=0.31, Δ=1.79
			IR (Mull): 1985, 1990, 2021, 2059
				[3, 9]

*10	$(C_6H_5)_2PC≡C-$
	$P(C_6H_5)_2$
	II (−)
		PF_6	yellow plates; m.p. 245 to 247°
			^1H NMR (CD_3COCD_3): 5.78 (d, C_5H_5,
			   J(P,H) =1.9)
			$^{57}Fe-\gamma$ (77 K): δ=0.35, Δ=1.82
			   (295 K): δ=0.28, Δ=1.80
			IR (Mull): 1984, 2018, 2058
				[3, 9]

*11	$(C_6H_5)_2PC≡C-$ $P(C_6H_5)_2$ IV (−)	$FeCl_4$	yellow blocks; m.p. 189 to 192° $^{57}Fe-\gamma$ (77 K): δ=0.50, 0.35, Δ=1.80 (295 K): δ=0.28, Δ=1.74 IR (Mull): 1989, 2018, 2058	[3, 9]
*12	$(C_6H_5)_2PC≡C-$ $P(C_6H_5)_2$ IV (−)	$FeBr_4$	red crystals; m.p. 171 to 173° $^{57}Fe-\gamma$ (77 K): δ=0.62, 0.36, Δ=1.79 IR (Mull): 1980, 2018, 2053	[3, 9]

References on p. 168

Table 16 [continued]

No.	^{2}D–^{2}D ligand Method of preparation (yield in %)	Anion	Properties and further remarks	Ref.
*13	$(C_6H_5)_2PCH_2C\equiv C$- $CH_2P(C_6H_5)_2$ special	PF_6	tan crystals; m.p. 244 to 246° ^{1}H NMR (CDCl$_3$): 5.03 (d, CH$_2$), 5.62 (C$_5$H$_5$), 7.58, 7.63 (C$_6$H$_5$) IR (KBr): 2008, 2050	[2]
14	$(C_2H_5O)_2PO$- $P(OC_2H_5)_2$ III (−)	$B(C_6H_5)_4$	no data reported	[10]
*15	$[(C_6H_5)_2PCH_2CH_2$- $P(C_6H_5)_2]_2CoCl_2$ V (53)	PF_6	turquoise solid IR (CH$_2$Cl$_2$): 2011, 2061 UV (CH$_2$Cl$_2$): λ_{max}=340, 595, 640, 650, 715	[6]
*16	$[(C_6H_5)_2PCH_2CH_2$- $CH_2P(C_6H_5)_2]_2$- $CoCl_2$ V (48)	PF_6	IR (CH$_2$Cl$_2$): 2011, 2056 UV (CH$_2$Cl$_2$): λ_{max}=340, 597, 640, 660, 690	[6]

Type $[C_5H_5Fe(CO)(NCCH_3)(\mu\text{-}^2D\text{-}^2D)(CH_3CN)(CO)FeC_5H_5]^{2+}$:

No.	^{2}D–^{2}D ligand	Anion	Properties and further remarks	Ref.
17	$(C_6H_5)_2PCH_2$- $P(C_6H_5)_2$	?	obtained like No. 18, no data given	[5]

*18	$(C_6H_5)_2PCH_2CH_2$-
	$P(C_6H_5)_2$
	special
		PF_6	orange solid
			^{1}H NMR (CD$_3$COCD$_3$): 2.49 (CH$_2$), 2.86
			   (CH$_3$CN), 4.83 (C$_5$H$_5$), 7.48 (C$_6$H$_5$)
			IR (CH$_3$CN): 1982, ν(CN) at 2189 in KBr
			UV (CH$_3$CN): $\lambda_{max}(\varepsilon)$ =328 ($\approx$1672),
			   398 ($\approx$1700)
				[7]

19	$(C_6H_5)_2PCH_2CH_2$- $CH_2P(C_6H_5)_2$	?	obtained like No. 18, no data given	[5]

* Further information:

$[(C_5H_5Fe(CO)_2)_2(C_6H_5)_2PC\equiv CP(C_6H_5)_2]X_2$ (Table 16, Nos. 9 to 12, X=I$_3^-$, PF$_6^-$, FeCl$_4^-$, and FeBr$_4^-$). The C$_5$H$_5$ chemical shifts in the ^{1}H NMR spectra of Nos. 9 and 10 are among the lowest recorded for π–cyclopentadienyliron compounds, owing to the presence of a dipositive charge on the cation. Compounds No. 11 and 12 are paramagnetic due to the anions. Their ^{57}Fe Mössbauer spectra show two kinds of Fe atoms, and the one with the higher isomer shift is attributed to the FeX$_4^-$ anions. Compounds No. 9, 11, and 12 react with KPF$_6$ as described in method II to give the PF$_6^-$ salt [3, 9].

$[(C_5H_5Fe(CO)_2)_2(C_6H_5)_2PCH_2C\equiv CCH_2P(C_6H_5)_2][PF_6]_2$ (Table 16, No. 13) has been prepared from C$_5$H$_5$Fe(CO)$_2$Br and (C$_6$H$_5$)$_2$PCH$_2$C≡CCH$_2$P(C$_6$H$_5$)$_2$ (1:1.2 mole) which are refluxed first in CH$_2$Cl$_2$ (60 min) and then in toluene (45 min). The brown precipitate was dissolved in acetone/ethanol and treated with an excess of NH$_4$PF$_6$, 2.5% yield [2].

$[(C_5H_5Fe(CO)_2(C_6H_5)_2P(CH_2)_nP(C_6H_5)_2)_2CoCl_2][PF_6]_2$ (Table **16**, Nos. **15** and **16**, $n=2$ and 3). The ^{1}H NMR spectra show extensive paramagnetic shifts. The pattern of the UV spectrum indicates a tetrahedral geometry at the Co atom. There are no bands which could be attributed to an interaction between the metal centers. The salts are readily hydrolyzed in acetone by a few drops of water, giving Co^{2+} and $[C_5H_5Fe(CO)_2(C_6H_5)_2P(CH_2)_nP(C_6H_5)_2]^+$ [6].

$[(C_5H_5Fe(CO)NCCH_3)_2(C_6H_5)_2CH_2CH_2P(C_6H_5)_2][PF_6]_2$ (Table **16**, No. **18**) has been prepared by the electrochemical two-electron oxidation of $C_5H_5Fe(\mu-CO)_2$- $(\mu-(C_6H_5)_2PCH_2CH_2P(C_6H_5)_2)FeC_5H_5$ suspended in CH_3CN which contains NH_4PF_6 (0.1 M). The electrolysis on a Pt electrode at $+1.20$ V (referred to the SCE) at 75 °C for 80 min gives quantitative conversion. Evaporation of the solvent and washing with water and benzene gives the crude product, which is recrystallized from dichloromethane/ pentane, 85% yield. On the basis of IR and UV spectra, the dication also forms along with $C_5H_5Fe(\mu-CO)_2(\mu-(C_6H_5)_2PCH_2CH_2P(C_6H_5)_2)FeC_5H_5$ by disproportionation of the monocation of the latter in CH_3CN. However, the two disproportionation products cannot be separated by chromatography on SiO_2.

The IR and UV spectra have been recorded in CH_3CN in the presence of $[N(C_4H_9-n)_4]PF_6$. A further UV absorption at $\lambda_{max}=510$ nm ($\varepsilon\approx460$) is observed in a solution of CH_2Cl_2 [7], see also [5].

References:

[1] P.M. Treichel, R.L. Shubkin, K.W. Barnett, D. Reichard (Inorg. Chem. **5** [1966] 1177/81). – [2] R.B. King, A. Efrati (Inorg. Chim. Acta **4** [1970] 123/8). – [3] A.J. Carty, A. Efrati, T.W. Ng, T. Birchall (Inorg. Chem. **9** [1970] 1263/8). – [4] M.L. Brown, T.J. Meyer, N. Winterton (J. Chem. Soc. Chem. Commun. **1971** 309). – [5] J.A. Ferguson (J. Chem. Soc. Chem. Commun. **1971** 1544/5).

[6] M.L. Brown, J.L. Cramer, J.A. Ferguson, T.J. Meyer, N. Winterton (J. Am. Chem. Soc. **94** [1972] 8707/10). – [7] J.A. Ferguson, T.J. Meyer (Inorg. Chem. **11** [1972] 631/6). – [8] R.J. Haines, A.L. du Preez (Inorg. Chem. **11** [1972] 330/6). – [9] T.W. Ng (Diss. Univ. of Waterloo, Canada, 1973; Diss. Abstr. Intern. B **34** [1974] 4865/6).

2.5.2.2.13 Compounds of the $C_5H_5Fe(CO)(H)(\mu-PR_2)(CO)FeC_5H_5$ Type

Among the carbonyl compounds with single bridging groups the title complexes are unique in that they must have a Fe–Fe bond in addition to the bridging PR_2 group. The presence of a Fe–H bond is clearly indicated by the ^{1}H NMR spectra, but the location of the H atom is uncertain. It may undergo rapid exchange between equivalent positions at the Fe atoms (Formula I), or it may be a part of a three-center bond, possibly also involving the electrons of the Fe–Fe bond (Formula II).

The diamagnetic complexes are prepared from $Na[C_5H_5Fe(CO)_2]$ and R_2PCl ($R=CH_3$ and C_6H_5) in equimolar amounts by reacting the components in tetrahydrofuran for 1 to 2 h, removing the solvent, and heating in toluene at reflux temperature for 16 h.

The crude products also contain the cis and trans isomers of the $C_5H_5Fe(CO)(\mu-PR_2)_2$-$(CO)FeC_5H_5$ compounds.

$C_5H_5Fe(CO)(H)(\mu-P(CH_3)_2)(CO)FeC_5H_5$ has been isolated by chromatography on Al_2O_3 with hexane/benzene (30%) and sublimation at 90 °C/0.2 Torr, 13% yield. The dark brown crystals melt at 137 to 140 °C with decomposition. ^1H NMR spectrum (CS_2): $\delta = 1.80$ and 1.99 (d's, CH_3, $J(P,H) \approx 0.5$ Hz), and 4.37 (d, C_5H_5, $J(P,H) = 1.1$ Hz) ppm; in benzene solution the Fe–H resonance occurs as doublet at $\delta = -18.73$ ppm with $J(P,H) = 43.5$ Hz. The nonequivalence of the CH_3 groups has been discussed. IR spectrum (Nujol or halocarbons): $\nu(CO)$ bands at 1850, 1860, 1898, and 1915 cm^{-1}. Other bands from 498 to 2980 cm^{-1} are given.

$C_5H_5Fe(CO)(H)(\mu-P(C_6H_5)_2)(CO)FeC_5H_5$ has been separated from other products by chromatography on Al_2O_3 with hexane/benzene (30%) and crystallization from hexane with cooling, 5% yield. The dark brown crystals melt at 153 to 157 °C. ^1H NMR spectrum (CS_2): $\delta = 4.20$ (d, C_5H_5, $J(P,H) = 1.3$ Hz), 7.17 and 7.5 (m's, C_6H_5). In benzene Fe–H gives a doublet at $\delta = -18.67$ ppm with $J(P,H) = 41.0$ Hz. IR spectrum (Nujol and halocarbon): $\nu(CO)$ bands at 1863, 1872, 1903, 1911, 1922, and 1932 cm^{-1}. The two bands on the high frequency side may be associated with the Fe–H stretching vibration. Other IR bands are given from 482 to 3060 cm^{-1}.

Reference:

R.G. Hayter (J. Am. Chem. Soc. 85 [1963] 3120/4).

Remarks on abbreviations and dimensions

Many compounds in this volume are presented in tables in which numerous abbreviations are used and the dimensions are omitted for the sake of conciseness. This necessitates the following clarification.

Temperatures are given in °C, otherwise K stands for Kelvin. Abbreviations used with temperatures are m.p. for melting point and dec.p. for decomposition point.

NMR represents nuclear magnetic resonance. Chemical shifts are given as δ values in ppm; reference substances and signs for δ are shown in the scheme below:

		Increasing field	
		$\delta = 0$ for	
^1H	+	$Si(CH_3)_4$	−
^{11}B	+	$BF_3 \cdot O(C_2H_5)_2$	−
^{13}C	+	$Si(CH_3)_4$	−
^{19}F	−	$CFCl_3$	+
^{31}P	−	H_3PO_4	+

coupling constants J in Hz given as J(A,B) or as J(1,3) referring to labelled structural formulas

Multiplicities of the signals are abbreviated as s, d, t, q (singlet to quartet), quint, sext, sept (quintet to septet), and m (multiplet); terms like dd (double doublet) and t's (triplets) are also used. Assignments referring to labelled structural formulas are given in the form C-4, H-3,5.

^{57}Fe Mössbauer spectra are represented by $^{57}Fe-\gamma$: both the isomer shift δ (relative to sodium nitroprusside at room temperature) and the quadrupole splitting Δ are given in $mm \cdot s^{-1}$; the experimental error has generally been omitted. Other reference substances for δ are indicated after the numerical value, e.g., $\delta = 0.23$ (Fe).

Optical spectra are labelled as IR (infrared), R (Raman), and UV (electronic spectrum including the visible region). IR bands and Raman lines are given in cm^{-1}; the assigned bands are usually labelled with the symbols v for stretching vibration and δ for deformation vibration whereas unlabelled bands belong to CO stretching vibrations; p or np indicates polarized and depolarized Raman lines, respectively. The UV absorption maxima (λ_{max}) are given in nm followed by the extinction coefficient ε ($l \cdot cm^{-1} \cdot mol^{-1}$) or lg ε in parentheses, respectively; sh means shoulder.

Solvents or the physical state of the sample and the temperature (in °C or K) are given in parentheses immediately after the spectral symbol, e.g., R (solid), ^{13}C NMR (C_6D_6, 50 °C). Common solvents are given by their formula (C_6H_{12}=cyclohexane) except THF which represents tetrahydrofuran.

Figures give only selected parameters. Barred bond lengths (in Å) or angles are mean values for parameters of the same type.

Empirical Formula Index

In the following index the compounds are listed in the order of increasing carbon content. Empirical formulas of ionic compounds are given in brackets; ions as well as components of solvates and adducts are separated by a period.

Page references are printed in ordinary types, table numbers in bold-face, and compound numbers of the tables in italics.

Ligand Formula Index

Ligands containing carbon atoms (except CO, CN, CNO, and CNS) are used in the following Ligand Formula Index to locate a compound in the order of increasing carbon content of the respective ligand. The number of identical ligands in a compound and the nature of bonding are not taken into consideration, so that several compounds may be listed at one position, if need be. On the other hand, compounds having two or more different types of carbon–containing ligands occur at two or more positions, respectively. The following examples illustrate the arrangement.

$(CO)_4Fe(\mu-CH_2=CHCH=CH_2)Fe(CO)_4$:

| C_4H_6 | – | – | CO | – |

$C_5H_5Fe(CO)_2(\mu-Sn(CH_3)_2)(CO)(P(OCH_3)_3)FeC_5H_5$:

C_2H_6Sn	$C_3H_9O_3P$	C_5H_5	CO	–
$C_3H_9O_3P$	C_2H_6Sn	C_5H_5	CO	–
C_5H_5	C_2H_6Sn	$C_3H_9O_3P$	CO	–

$[C_5H_5Fe(CO)_2(\mu-H)(CO)(P(OCH_3)_3)FeC_5H_5][B(C_6H_5)_4]$:

| $C_3H_9O_3P$ | C_5H_5 | – | CO | H |
| C_5H_5 | $C_3H_9O_3P$ | – | CO | H |

In view of the very large number of carbonyl compounds, CO is not included in the first column; it is given in column 4 along with other non–organic ligands, such as H, NO, $SnCl_2$ etc., in columns 3 and 5.

Page references are printed in ordinary types, table numbers in bold–face, and compound numbers of the tables in italics.

CH_3	C_5H_5	–	–	NO	98
CH_3BrSn	C_5H_5	–	CO	–	127, **9**, *32*
CH_3ClSi	C_5H_5	–	CO	–	115, **8**, *5*
CH_3F_2NP	$CH_3F_4NP_2$	C_5H_5	–	PF_2	103/4
CH_3F_2NP	C_5H_5	$CH_3F_4NP_2$	–	PF_2	103/4
$CH_3F_4NP_2$	CH_3F_2NP	C_5H_5	–	PF_2	103/4
$CH_3F_4NP_2$	C_5H_5	CH_3F_2NP	–	PF_2	103/4
CH_3S	C_5H_5	–	–	S_2	99
CH_3S	C_5H_5	–	CO	–	82
CH_3S	C_7H_8	–	CO	–	78
CH_3Sb	C_5H_5	–	CO	–	110, **7**, *5*
CH_4Si	C_5H_5	–	CO	–	115, **8**, *4*
C_2F_6P	C_2F_6PO	C_5H_5	CO	–	108/9
C_2F_6P	C_5H_5	–	CO	–	84, **5**, *2*
C_2F_6P	C_5H_5	–	CO	NO	90
C_2F_6PO	C_5H_5	C_2F_6P	CO	–	108/9
C_2F_6Sb	C_5H_5	–	CO	–	159, **14**, *21;* 160, **14**, *22, 23*
$C_2H_2O_4Sn$	C_5H_5	–	CO	–	125, **9**, *17*
$C_2H_2S_2$	C_5H_5	–	CO	–	107

C_2H_3N	C_2H_5S	C_5H_5	–	–	102
C_2H_3N	C_5H_5	C_2H_5S	–	–	102
C_2H_3N	C_5H_5	$C_{25}H_{22}P_2$	CO	–	167, **16**, *17*
C_2H_3N	C_5H_5	$C_{26}H_{24}P_2$	CO	–	167, **16**, *18*
C_2H_3N	C_5H_5	$C_{27}H_{26}P_2$	CO	–	167, **16**, *19*
C_2H_3N	$C_{25}H_{22}P_2$	C_5H_5	CO	–	167, **16**, *17*
C_2H_3N	$C_{26}H_{24}P_2$	C_5H_5	CO	–	167, **16**, *18*
C_2H_3N	$C_{27}H_{26}P_2$	C_5H_5	CO	–	167, **16**, *19*
C_2H_3O	C_5H_5	$C_{26}H_{24}P_2$	CO	–	162, **15**, *5*
C_2H_3O	$C_{26}H_{24}P_2$	C_5H_5	CO	–	162, **15**, *5*
$C_2H_4S_2Sn$	C_5H_5	–	CO	–	124, **9**, *9*
C_2H_5S	C_2H_3N	C_5H_5	–	–	102
C_2H_5S	C_5H_5	–	–	CN	102
C_2H_5S	C_5H_5	–	–	S_2	99/100; 101
C_2H_5S	C_5H_5	–	–	SCN	102
C_2H_5S	C_5H_5	–	CO	–	158, **14**, *1/4*
C_2H_5S	C_5H_5	C_2H_3N	–	–	102
C_2H_6As	C_5H_5	–	CO	–	84, **5**, *4*; 159, **14**, *13, 14*
C_2H_6Ge	C_5H_5	–	CO	–	116, **8**, *19*
$C_2H_6O_2Ge$	C_5H_5	–	CO	–	116, **8**, *15*
C_2H_6P	C_5H_5	–	CO	–	83, **5**, *1*; 88, **6**, *1*; 158, **14**, *7*
C_2H_6P	C_5H_5	–	CO	–	169
C_2H_6Pb	C_5H_5	–	CO	–	128, **9**, *48*
C_2H_6Sb	C_5H_5	–	CO	–	159, **14**, *20*
C_2H_6Sn	$C_3H_9O_3P$	C_5H_5	CO	–	141, **11**, *11*
C_2H_6Sn	C_5H_5	–	CO	–	125, **9**, *18*
C_2H_6Sn	C_5H_5	$C_3H_5O_3P$	CO	–	141, **11**, *11*
C_3H_5ClSn	C_5H_5	–	CO	–	126, **9**, *27*
$C_3H_6O_4S_2$	C_5H_5	–	CO	–	107
$C_3H_9O_3P$	C_2H_6Sn	C_5H_5	CO	–	141, **11**, *11*
$C_3H_9O_3P$	C_5H_5	–	CO	H	151
$C_3H_9O_3P$	C_5H_5	–	CO	$SnCl_2$	140, **11**, *5*
$C_3H_9O_3P$	C_5H_5	C_2H_6Sn	CO	–	141, **11**, *11*
$C_3H_9O_3P$	C_5H_5	$C_{12}H_{10}P$	CO	–	84, **5**, *7*; 85, **5**, *12*; 88, **6**, *5*
$C_3H_9O_3P$	$C_{12}H_{10}P$	C_5H_5	CO	–	84, **5**, *7*; 85, **5**, *12*; 88, **6**, *5*
C_4H_4					
$H_2C=C=C=CH_2$	–	–	CO	–	13, **2**, *1*
$H_2C=C=C=CH_2$	$C_{12}H_{27}P$	–	CO	–	22, **2a**, *1*; 23
$H_2C=C=C=CH_2$	$C_{18}H_{15}O_3P$	–	CO	–	22, **2a**, *3*
$H_2C=C=C=CH_2$	$C_{18}H_{15}P$	–	CO	–	22, **2a**, *2*
$-CH=CH-CH=CH-$	–	–	CO	–	28, **3**, *1*
$-CH=CH-CH=CH-$	$C_{18}H_{15}P$	–	CO	–	60
Cyclobutadiene	–	–	CO	–	79

$C_4H_4N_2$	C_5H_5	—	CO	—	165, **16**, *2*
C_4H_4O	—	—	CO	—	39, **3**, *76*
$C_4H_4O_2$	—	—	CO	—	33, **3**, *33*
C_4H_6	—	—	CO	—	1, **1**, *1*
$C_4H_6GeO_4$	C_5H_5	—	CO	—	116, **8**, *16*
$C_4H_6O_4Sn$	C_5H_5	—	CO	—	125, **9**, *13*
C_4H_9S	C_5H_5	—	CO	—	87;
					158, **14**, *5, 6*
$C_4H_{10}Ge$	C_5H_5	—	CO	—	117, **8**, *20*
$C_4H_{10}GeS_2$	C_5H_5	—	CO	—	116, **8**, *18*
$C_4H_{10}SSn$	C_5H_5	—	CO	—	124, **9**, *7*
$C_4H_{10}S_2$	C_5H_5	—	CO	—	165, **16**, *1*
$C_4H_{10}Sn$	C_5H_5	—	CO	—	125, **9**, *19*
$C_4H_{12}Ge_2$	C_5H_5	—	CO	—	117, **8**, *25*
$C_4H_{12}Ge_2O$	C_5H_5	—	CO	—	117, **8**, *27*
$C_4H_{12}Si_2$	C_5H_5	—	CO	—	115, **8**, *7*
$C_4N_2S_2$	C_5H_5	—	CO	—	107
C_5BrCrO_5Sb	C_5H_5	—	CO	—	110, **7**, *7*
C_5ClMnO_5Sn	C_5H_5	—	CO	—	126, **9**, *30*
C_5ClO_5ReSn	C_5H_5	—	CO	—	126, **9**, *31*
C_5CrFO_5Sb	C_5H_5	—	CO	—	110, **7**, *6*
C_5F_6	—	—	CO	—	3, **1**, *11*
C_5H_5	—	—	—	NO	97/8
C_5H_5	—	—	CO	$AsBr_2$	159, **14**, *11, 12*
C_5H_5	—	—	CO	$AsCl_2$	158, **14**, *9;*
					159, **14**, *10*
C_5H_5	—	—	CO	B_5H_7	142/3
C_5H_5	—	—	CO	Br	154, **13**, *6/9*
C_5H_5	—	—	CO	Cd	145, **12**, *2*
C_5H_5	—	—	CO	Cl	153, **13**, *1/5*
C_5H_5	—	—	CO	D	151
C_5H_5	—	—	CO	$GeBr_2$	116, **8**, *13*
C_5H_5	—	—	CO	$GeCl_2$	116, **8**, *12*
C_5H_5	—	—	CO	GeF_2	115, **8**, *11*
C_5H_5	—	—	CO	GeH_2	115, **8**, *10*
C_5H_5	—	—	CO	GeI_2	116, **8**, *14*
C_5H_5	—	—	CO	$Ge(NCS)_2$	116, **8**, *17*
C_5H_5	—	—	CO	H	149/51
C_5H_5	—	—	CO	Hg	145, **12**, *3*
C_5H_5	—	—	CO	InCl	148/9
C_5H_5	—	—	CO	I	152;
					154, **13**, *10/13*
C_5H_5	—	—	CO	SO_2	105/7
C_5H_5	—	—	CO	S_3	105
C_5H_5	—	—	CO	SbBr	109, **7**, *4*
C_5H_5	—	—	CO	$SbBr_2$	159, **14**, *18, 19*
C_5H_5	—	—	CO	SbCl	109
C_5H_5	—	—	CO	$SbCl_2$	159, **14**, *15/17*
C_5H_5	—	—	CO	SbI	109
C_5H_5	—	—	CO	$SiCl_2$	115, **8**, *3*

C_5H_5					
C_5H_5	–	–	CO	SiHCl	115, **8**, *2*
C_5H_5	–	–	CO	SiH_2	114, **8**, *1*
C_5H_5	–	–	CO	$SnBr_2$	124, **9**, *3*
C_5H_5	–	–	CO	$SnCl_2$	123, **9**, *2*
C_5H_5	–	–	CO	SnF_2	123, **9**, *1*
C_5H_5	–	–	CO	SnI_2	124, **9**, *4*
C_5H_5	–	–	CO	$Sn(NCS)_2$	125, **9**, *16*
C_5H_5	–	–	CO	$Sn(NO_2)_2$	125, **9**, *14*
C_5H_5	–	–	CO	$Sn(NO_3)_2$	125, **9**, *15*
C_5H_5	–	–	CO	$Sn(OH)_2$	125, **9**, *12*
C_5H_5	–	–	CO	SnS	124, **9**, *5*
C_5H_5	–	–	CO	SnS_4	124, **9**, *6*
C_5H_5	–	–	CO	Tl	149
C_5H_5	–	–	CO	TlCl	149
C_5H_5	–	–	CO	Zn	145, **12**, *1*
C_5H_5	CH_3	–	–	NO	98
C_5H_5	CH_3BrSn	–	CO	–	127, **9**, *32*
C_5H_5	CH_3ClSi	–	CO	–	115, **8**, *5*
C_5H_5	CH_3F_2NP	$CH_3F_4NP_2$	–	PF_2	103/4
C_5H_5	$CH_3F_4NP_2$	CH_3F_2NP	–	PF_2	103/4
C_5H_5	CH_3S	–	CO	–	82
C_5H_5	CH_3S	–	–	S_2	99
C_5H_5	CH_3Sb	–	CO	–	110, **7**, *5*
C_5H_5	CH_4Si	–	CO	–	115, **8**, *4*
C_5H_5	C_2F_6P	–	CO	–	84, **5**, *2*
C_5H_5	C_2F_6P	–	CO	NO	90
C_5H_5	C_2F_6P	C_2F_6PO	CO	–	108/9
C_5H_5	C_2F_6PO	C_2F_6P	CO	–	108/9
C_5H_5	C_2F_6Sb	–	CO	–	159, **14**, *21*; 160, **14**, *22, 23*
C_5H_5	$C_2H_2O_4Sn$	–	CO	–	125, **9**, *17*
C_5H_5	$C_2H_2S_2$	–	CO	–	107
C_5H_5	C_2H_3N	C_2H_5S	–	–	102
C_5H_5	C_2H_3N	$C_{25}H_{22}P_2$	CO	–	167, **16**, *17*
C_5H_5	C_2H_3N	$C_{26}H_{24}P_2$	CO	–	167, **16**, *18*
C_5H_5	C_2H_3N	$C_{27}H_{26}P_2$	CO	–	167, **16**, *19*
C_5H_5	C_2H_3O	$C_{26}H_{24}P_2$	CO	–	162, **15**, *5*
C_5H_5	$C_2H_4S_2Sn$	–	CO	–	124, **9**, *9*
C_5H_5	C_2H_5S	–	–	CN	102
C_5H_5	C_2H_5S	–	–	SCN	102
C_5H_5	C_2H_5S	–	–	S_2	99/101
C_5H_5	C_2H_5S	–	CO	–	158, **14**, *1/4*
C_5H_5	C_2H_5S	C_2H_3N	–	–	102
C_5H_5	C_2H_6As	–	CO	–	84, **5**, *4*; 159, **14**, *13, 14*
C_5H_5	C_2H_6Ge	–	CO	–	116, **8**, *19*
C_5H_5	$C_2H_6GeO_2$	–	CO	–	116, **8**, *15*
C_5H_5	C_2H_6P	–	CO	–	83, **5**, *1*; 88, **6**, *1*; 158, **14**, *7*
C_5H_5	C_2H_6P	–	CO	H	169

C_5H_5	C_2H_6Pb	–	CO	–	128, **9**, *48*
C_5H_5	C_2H_6Sb	–	CO	–	159, **14**, *20*
C_5H_5	C_2H_6Sn	–	CO	–	125, **9**, *18*
C_5H_5	C_2H_6Sn	$C_3H_9O_3P$	CO	–	141, **11**, *11*
C_5H_5	C_3H_5ClSn	–	CO	–	126, **9**, *27*
C_5H_5	$C_3H_6O_4S_2$	–	CO	–	107
C_5H_5	$C_3H_9O_3P$	–	CO	H	151
C_5H_5	$C_3H_9O_3P$	–	CO	$SnCl_2$	140, **11**, *5*
C_5H_5	$C_3H_9O_3P$	C_2H_6Sn	CO	–	141, **11**, *11*
C_5H_5	$C_3H_9O_3P$	$C_{12}H_{10}P$	CO	–	84, **5**, *7;* 85, **5**, *12;* 88, **6**, *5*
C_5H_5	$C_4H_6O_4Ge$	–	CO	–	116, **8**, *16*
C_5H_5	$C_4H_6O_4Sn$	–	CO	–	125, **9**, *13*
C_5H_5	C_4H_9S	–	CO	–	87; 158, **14**, *5, 6*
C_5H_5	$C_4H_{10}Ge$	–	CO	–	117, **8**, *20*
C_5H_5	$C_4H_{10}GeS_2$	–	CO	–	116, **8**, *18*
C_5H_5	$C_4H_{10}S_2Sn$	–	CO	–	124, **9**, *7*
C_5H_5	$C_4H_{10}S_2$	–	CO	–	165, **16**, *1*
C_5H_5	$C_4H_{10}Sn$	–	CO	–	125, **9**, *19*
C_5H_5	$C_4H_{12}Ge_2$	–	CO	–	117, **8**, *25*
C_5H_5	$C_4H_{12}Ge_2O$	–	CO	–	117, **8**, *27*
C_5H_5	$C_4H_{12}Si_2$	–	CO	–	115, **8**, *7*
C_5H_5	$C_4N_2S_2$	–	CO	–	107
C_5H_5	C_5BrCrO_5Sb	–	CO	–	110, **7**, *7*
C_5H_5	C_5ClMnO_5Sn	–	CO	–	126, **9**, *30*
C_5H_5	C_5ClO_5ReSn	–	CO	–	126, **9**, *31*
C_5H_5	C_5CrFO_5Sb	–	CO	–	110, **7**, *6*
C_5H_5	$C_5H_9O_3P$	–	CO	$SnCl_2$	141, **10**, *10;* 142, **10**, *13*
C_5H_5	C_6H_5ClSn	–	CO	–	126, **9**, *28*
C_5H_5	C_6H_5S	–	CO	–	82
C_5H_5	C_6H_7	–	CO	$SnBr_2$	127, **9**, *42*
C_5H_5	C_6H_7	–	CO	$SnCl_2$	127, **9**, *41*
C_5H_5	C_6H_7	–	–	SnI_2	127, **9**, *43*
C_5H_5	C_6H_7	$C_{15}H_{27}ClCo\text{-}O_3PSn$	CO	–	127, **9**, *44*
C_5H_5	$C_6H_{10}Ge$	–	CO	–	117, **8**, *21*
C_5H_5	$C_6H_{10}Sn$	–	CO	–	126, **9**, *22*
C_5H_5	$C_6H_{15}O_3P$	–	CO	$SnCl_2$	140, **11**, *6;* 141, **11**, *12*
C_5H_5	$C_6H_{15}O_3P$	$C_{12}H_{10}P$	CO	–	88, **6**, *6*
C_5H_5	$C_6H_{15}P$	–	CO	$SnCl_2$	140, **11**, *1*
C_5H_5	$C_6H_{15}P$	$C_{12}H_{10}P$	CO	–	84, **5**, *6;* 85, **5**, *11;* 88, **6**, *3*
C_5H_5	$C_6H_{16}P_2$	–	CO	I	162, **15**, *2*
C_5H_5	$C_6H_{16}P_2$	–	–	N_2	103
C_5H_5	$C_6H_{16}Si_2$	–	CO	–	115, **8**, *9*
C_5H_5	$C_6H_{18}Si_3$	–	CO	–	115, **8**, *8*

C_5H_5	C_7H_5Br-MnO_2Sb	–	CO	–	110, **7**, *8*
C_5H_5	$C_7H_6S_2Sn$	–	CO	–	124, **9**, *10*
C_5H_5	C_7H_7S	–	–	S_2	99
C_5H_5	C_8H_5ClMo-O_3Sn	–	CO	–	126, **9**, *29*
C_5H_5	$C_8H_{12}As_2CrO_4$	–	CO	–	109, **7**, *1*
C_5H_5	$C_8H_{12}As_2$-MoO_4	–	CO	–	109, **7**, *2*
C_5H_5	$C_8H_{12}As_2$-O_4W	–	CO	–	109, **7**, *3*
C_5H_5	$C_8H_{16}MgO_2$	–	CO	–	143/4
C_5H_5	$C_8H_{18}Ge$	–	CO	–	117, **8**, *22*
C_5H_5	$C_8H_{18}Sn$	–	CO	–	126, **9**, *20, 21*
C_5H_5	$C_8H_{20}Ge_2$	–	CO	–	117, **8**, *26*
C_5H_5	$C_8H_{20}O_5P_2$	–	CO	–	167, **16**, *14*
C_5H_5	$C_8H_{20}O_5P_2$	–	CO	I	163, **15**, *12*
C_5H_5	$C_8Mn_2O_8S_2$	–	CO	–	107/8
C_5H_5	C_9H_7	–	CO	$SnBr_2$	128, **9**, *46*
C_5H_5	C_9H_7	–	CO	$SnCl_2$	128, **9**, *45*
C_5H_5	C_9H_7	–	CO	SnI_2	128, **9**, *47*
C_5H_5	$C_9H_{21}O_3P$	–	CO	$SnCl_2$	141, **11**, *7*
C_5H_5	$C_9H_{21}O_3P$	$C_{12}H_{10}P$	CO	–	84, **5**, *8;* 85, **5**, *13;* 88, **6**, *7*
C_5H_5	$C_9H_{21}P$	–	CO	$SnCl_2$	140, **11**, *2, 3*
C_5H_5	$C_{10}H_{10}Ge$	–	CO	–	117, **8**, *23*
C_5H_5	$C_{10}H_{10}MgN_2$	–	CO	–	143/4
C_5H_5	$C_{10}H_{10}Sn$	–	CO	–	126, **9**, *23*
C_5H_5	$C_{12}H_{10}As$	–	CO	–	84, **5**, *5*
C_5H_5	$C_{12}H_{10}Ge$	–	CO	–	117, **8**, *24*
C_5H_5	$C_{12}H_{10}O_4S_2Sn$	–	CO	–	124, **9**, *11*
C_5H_5	$C_{12}H_{10}P$	–	CO	–	84, **5**, *3;* 88, **6**, *2;* 158, **14**, *8*
C_5H_5	$C_{12}H_{10}P$	–	CO	H	169
C_5H_5	$C_{12}H_{10}P$	$C_3H_9O_3P$	CO	–	84, **5**, *7;* 85, **5**, *12;* 88, **6**, *5*
C_5H_5	$C_{12}H_{10}P$	$C_6H_{15}O_3P$	CO	–	88, **6**, *6*
C_5H_5	$C_{12}H_{10}P$	$C_6H_{15}P$	CO	–	84, **5**, *6;* 85, **5**, *11;* 88, **6**, *3*
C_5H_5	$C_{12}H_{10}P$	$C_9H_{21}O_3P$	CO	–	85, **5**, *13;* 88, **6**, *7*
C_5H_5	$C_{12}H_{10}P$	$C_9H_{21}P$	CO	–	84, **5**, *8*
C_5H_5	$C_{12}H_{10}P$	$C_{18}H_{15}O_3P$	CO	–	84, **5**, *9;* 85, **5**, *14;* 88, **6**, *8*
C_5H_5	$C_{12}H_{10}P$	$C_{18}H_{15}P$	CO	–	88, **6**, *4*
C_5H_5	$C_{12}H_{10}P$	$C_{25}H_{22}P_2$	CO	–	85, **5**, *15;* 88, **6**, *9*

C_5H_5	$C_{12}H_{10}P$	$C_{26}H_{24}P_2$	CO	–	84, **5**, _10_
C_5H_5	$C_{12}H_{10}S_2Sn$	–	CO	–	124, **9**, _8_
C_5H_5	$C_{12}H_{10}Sb$	–	CO	–	160, **14**, _24_
C_5H_5	$C_{12}H_{10}Si$	–	CO	–	115, **8**, _6_
C_5H_5	$C_{12}H_{10}Sn$	–	CO	–	126, **9**, _25_
C_5H_5	$C_{12}H_{11}P$	$C_{24}H_{20}P_2$	–	–	103
C_5H_5	$C_{12}H_{14}Sn$	–	CO	–	126, **9**, _24_
C_5H_5	$C_{12}H_{27}O_3P$	–	CO	$SnCl_2$	141, **11**, _8_
C_5H_5	$C_{12}H_{27}P$	–	CO	$SnCl_2$	140, **11**, _4_
C_5H_5	$C_{15}H_{27}ClCo\text{-}O_3PSn$	C_6H_7	CO	–	127, **9**, _44_
C_5H_5	$C_{16}H_{10}Mo_2\text{-}O_6Sn$	–	CO	–	126, **9**, _26_
C_5H_5	$C_{18}H_{15}ClPPt$	–	CO	–	149
C_5H_5	$C_{18}H_{15}O_3P$	–	CO	$SnCl_2$	141, **11**, _9_
C_5H_5	$C_{18}H_{15}O_3P$	$C_{12}H_{10}P$	CO	–	84, **5**, _9;_ 85, **5**, _14;_ 88, **6**, _8_
C_5H_5	$C_{18}H_{15}P$	$C_{12}H_{10}P$	CO	–	88, **6**, _4_
C_5H_5	$C_{20}H_{20}MgN_4$	–	CO	–	144
C_5H_5	$C_{24}H_{20}P_2$	$C_{12}H_{11}P$	–	–	103
C_5H_5	$C_{25}H_{22}P_2$	–	CO	–	166, **16**, _3, 4_
C_5H_5	$C_{25}H_{22}P_2$	–	CO	I	162, **15**, _1_
C_5H_5	$C_{25}H_{22}P_2$	C_2H_3N	CO	–	167, **16**, _17_
C_5H_5	$C_{25}H_{22}P_2$	$C_{12}H_{10}P$	CO	–	85, **5**, _15;_ 88, **6**, _9_
C_5H_5	$C_{26}H_{20}P_2$	–	CO	–	166, **16**, _9/12_
C_5H_5	$C_{26}H_{20}P_2$	–	CO	Br	163, **15**, _8_
C_5H_5	$C_{26}H_{20}P_2$	–	CO	Cl	163, **15**, _7_
C_5H_5	$C_{26}H_{20}P_2$	–	CO	I	163, **15**, _9_
C_5H_5	$C_{26}H_{20}P_2$	–	CO	$SnCl_2$	163, **15**, _10_
C_5H_5	$C_{26}H_{24}P_2$	–	CO	–	166, **15**, _5/7_
C_5H_5	$C_{26}H_{24}P_2$	–	CO	Br	162, **15**, _3_
C_5H_5	$C_{26}H_{24}P_2$	–	CO	I	162, **15**, _4_
C_5H_5	$C_{26}H_{24}P_2$	–	–	N_2	103
C_5H_5	$C_{26}H_{24}P_2$	C_2H_3N	CO	–	167, **16**, _18_
C_5H_5	$C_{26}H_{24}P_2$	C_2H_3O	CO	–	162, **15**, _5_
C_5H_5	$C_{26}H_{24}P_2$	$C_{12}H_{10}$	CO	–	84, **5**, _10_
C_5H_5	$C_{27}H_{26}P_2$	–	CO	–	166, **16**, _8_
C_5H_5	$C_{27}H_{26}P_2$	–	CO	I	163, **15**, _6_
C_5H_5	$C_{27}H_{26}P_2$	C_2H_3N	CO	–	167, **16**, _19_
C_5H_5	$C_{28}H_{24}P_2$	–	CO	–	167, **16**, _13_
C_5H_5	$C_{28}H_{24}P_2$	–	CO	Br	163, **15**, _11_
C_5H_5	$C_{36}H_{30}NiP_2$	–	CO	–	149
C_5H_5	$C_{36}H_{30}P_2Pt$	–	CO	–	149
C_5H_5	$C_{52}H_{48}Cl_2\text{-}CoP_4$	–	CO	–	167, **16**, _15_
C_5H_5	$C_{54}H_{52}Cl_2\text{-}CoP_4$	–	CO	–	167, **16**, _16_
C_5H_6					
$CH_3CH{=}C{=}C{=}CH_2$	–	–	CO	–	13, **2**, _2_

$-C(CH_3)=CH-CH=CH--$	–	–	CO	–	28, **3**, *2*
$-CH=C(CH_3)-CH=CH--$	–	–	CO	–	29, **3**, *3*
C_5H_6O	–	–	CO	–	39, **3**, *77*
$C_5H_6O_2$	–	–	CO	–	33, **3**, *34*
$C_5H_9O_3P$	–	–	CO	$SnCl_2$	141, **11**, *10;* 142, **11**, *13*
C_6F_6	–	–	CO	–	3, **1**, *12*
C_6H_5ClSn	C_5H_5	–	CO	–	126, **9**, *28*
C_6H_5S	C_5H_5	–	CO	–	82
C_6H_6	–	–	CO	–	67, **4**, *21*
C_6H_7	–	–	CO	Br	155, **13**, *15, 16*
C_6H_7	–	–	CO	Cl	155, **13**, *14*
C_6H_7	–	–	CO	I	152; 155, **13**, *17, 18*
C_6H_7	–	–	CO	$SnBr_2$	127, **9**, *35*
C_6H_7	–	–	CO	$SnCl_2$	127, **9**, *34*
C_6H_7	–	–	CO	SnF_2	127, **9**, *33*
C_6H_7	–	–	CO	SnI_2	127, **9**, *36*
C_6H_7	C_5H_5	–	CO	$SnBr_2$	127, **9**, *42*
C_6H_7	C_5H_5	–	CO	$SnCl_2$	127, **9**, *41*
C_6H_7	C_5H_5	–	CO	SnI_2	127, **9**, *43*
C_6H_7	C_5H_5	$C_{15}H_{27}ClCo-O_3PSn$	CO	–	127, **9**, *44*
C_6H_7	$C_{15}H_{27}ClCo-O_3PSn$	C_5H_5	CO	–	127, **9**, *44*
C_6H_8					
$H_2C=CHCH=CH-CH=CH_2$	–	–	CO	–	1, **1**, *2*
$CH_3CH=C=C=CHCH_3$	–	–	CO	–	13, **2**, *4*
$(CH_3)_2C-C-C-CH_2$	–	–	CO	–	14, **2**, *5*
$-C(CH_3)-C(CH_3)-CH=CH-$	–	–	CO	–	29, **3**, *4*
$-C(CH_3)=CH-CH=C(CH_3)-$	–	–	CO	–	29, **3**, *5*
$-C(CH_3)=CH-C(CH_3)=CH-$	–	–	CO	–	29, **3**, *7*
$-CH=C(CH_3)-C(CH_3)=CH-$	–	–	CO	–	29, **3**, *6*
$-C(C_2H_5)=CH-CH=CH-$	–	–	CO	–	39, **3**, *75*
$C_6H_8O_2$					
$-C(OCH_3)=CH-CH=C(OCH_3)-$	–	–	CO	–	34, **3**, *42*
$-C(OH)=CCH_3-CCH_3=C(OH)-$	–	–	CO	–	34, **3**, *39*
$-C(OH)=CC_2H_5-CH=C(OH)-$	–	–	CO	–	33, **3**, *35*
$C_4H_2(OCH_3)_2$	–	–	CO	SO_2	61
$C_6H_{10}Ge$	C_5H_5	–	CO	–	117, **8**, *21*
$C_6H_{10}Sn$	C_5H_5	–	CO	–	126, **9**, *22*

$C_6H_{15}O_3P$	C_5H_5	–	CO	$SnCl_2$	140, **11**, *6;* 141, **11**, *12*
$C_6H_{15}P$	C_5H_5	–	CO	$SnCl_2$	140, **11**, *1*
$C_6H_{16}P_2$	C_5H_5	–	CO	I	162, **15**, *2*
$C_6H_{16}P_2$	C_5H_5	–	–	N_2	103
$C_6H_{16}Si_2$	C_5H_5	–	CO	–	115, **8**, *9*
$C_6H_{18}Si_3$	C_5H_5	–	CO	–	115, **8**, *8*
$C_7H_5BrMnO_2Sb$	C_5H_5	–	CO	–	110, **7**, *8*
$C_7H_6S_2Sn$	C_5H_5	–	CO	–	124, **9**, *10*
C_7H_7D	–	–	CO	–	96
C_7H_7S	C_5H_5	–	–	S_2	99
C_7H_8					
Spiro[2.4]-hepta–4,6–diene	–	–	CO	–	2, **1**, *6*
Bicyclo[2.2.1]-hepta–2,5–diene	–	CH_3S	CO	–	78
Cyclohexa–2,4–diene–1–methylen-diyl–	–	–	CO	–	94
C_7H_8O	–	–	CO	–	64, **4**, *1;* 65, **4**, *2*
$C_7H_8O_3$	–	–	CO	–	35, **3**, *51*
C_7H_{10}	–	–	CO	–	15, **2**, *8*
$C_8H_4N_2O_4$	–	–	CO	–	3, **1**, *14*
$C_8H_5ClMoO_3Sn$	C_5H_5	–	CO	–	126, **9**, *29*
C_8H_6	–	–	CO	–	38, **3**, *70, 71*
C_8H_8					
2 a, 2 b, 4 a, 4 b, tetra-hydro[cd]pentalene	–	–	CO	–	2, **1**, *10*
Tricyclo[4.2.0.0^{2,5}]-octa–3,7–diene	–	–	CO	–	2, **1**, *4, 5*
$C_8H_8Cl_4$	–	–	CO	–	32, **3**, *23*
$C_8H_8O_4$	–	–	–	–	
$C_4H_2(OOCCH_3)_2$	–	–	–	–	34, **3**, *43*
$C_4H_2(COOCH_3)_2$	–	–	CO	–	30, **3**, *13*
C_8H_{10}	–	–	CO	–	37, **3**, *62*
$C_8H_{10}N_2O_6$	–	–	CO	–	3, **1**, *13*
$C_8H_{10}O$	–	–	CO	–	91/3
$C_8H_{10}O_4$	–	–	CO	–	75/6
C_8H_{12}					
$(CH_3)_2C=C=C=C(CH_3)_2$	–	–	CO	–	15, **2**, *9*
cycloocta–1,5–diene	–	–	CO	–	2, **1**, *3*
$C_4(CH_3)_4$	–	–	CO	–	30, **3**, *15*
$C_8H_{12}As_2CrO_6$	C_5H_5	–	CO	–	109, **7**, *1*
$C_8H_{12}As_2MoO_4$	C_5H_5	–	CO	–	109, **7**, *2*
$C_8H_{12}As_2O_4$	C_5H_5	–	CO	–	109, **7**, *3*
$C_8H_{12}O_2$					
–C(OH)=C(C$_4$H$_9$–n)–CH=C(OH)–	–	–	CO	–	33, **3**, *36*

-C(OH)=C(C$_4$H$_9$-t)- CH=C(OH)	–	–	CO	–	33, **3**, *37*
-C(OH)=C(C$_2$H$_5$)- C(C$_2$H$_5$)=C(OH)-	–	–	CO	–	34, **3**, *40*
C$_8$H$_{12}$O$_4$	–	–	CO	–	35, **3**, *48*
C$_8$H$_{14}$N$_2$	–	–	CO	–	36, **3**, *60*
C$_8$H$_{16}$MgO$_2$	C$_5$H$_5$	–	CO	–	143/4
C$_8$H$_{18}$Ge	C$_5$H$_5$	–	CO	–	117, **8**, *22*
C$_8$H$_{18}$Sn	C$_5$H$_5$	–	CO	–	126, **9**, *20, 21*
C$_8$H$_{20}$Ge$_2$	C$_5$H$_5$	–	CO	–	117, **8**, *26*
C$_8$H$_{20}$O$_5$P$_2$	C$_5$H$_5$	–	CO	–	167, **16**, *14*
C$_8$H$_{20}$O$_5$P$_2$	C$_5$H$_5$	–	CO	I	163, **15**, *12*
C$_8$Mn$_2$O$_8$S$_2$	C$_5$H$_5$	–	CO	–	107/8
C$_9$H$_7$	C$_5$H$_5$	–	CO	I	152
C$_9$H$_7$	–	–	CO	SnBr$_2$	127, **9**, *39*
C$_9$H$_7$	–	–	CO	SnCl$_2$	127, **9**, *38*
C$_9$H$_7$	–	–	CO	SnF$_2$	127, **9**, *37*
C$_9$H$_7$	–	–	CO	SnI$_2$	127, **9**, *40*
C$_9$H$_7$	C$_5$H$_5$	–	CO	SnBr$_2$	128, **9**, *46*
C$_9$H$_7$	C$_5$H$_5$	–	CO	SnCl$_2$	128, **9**, *45*
C$_9$H$_7$	C$_5$H$_5$	–	CO	SnI$_2$	128, **9**, *47*
C$_9$H$_8$N$_2$O	–	–	CO	–	39, **3**, *74*
C$_9$H$_8$O$_5$	–	–	CO	–	66, **4**, *10*
C$_9$H$_9$ClO$_4$	–	–	CO	–	35, **3**, *52*
C$_9$H$_{10}$O	–	–	CO	–	74/5
C$_9$H$_{10}$O$_4$	–	–	CO	–	34, **3**, *46*
C$_9$H$_{12}$O	–	–	CO	–	65, **4**, *5*
C$_9$H$_{12}$O$_5$	–	–	CO	–	65, **4**, *7*
C$_9$H$_{14}$O$_2$	–	–	CO	–	2, **1**, *7*; 35, **3**, *53*
C$_9$H$_{21}$O$_3$P	C$_5$H$_5$	–	CO	–	141, **11**, *7*
C$_9$H$_{21}$O$_3$P	C$_5$H$_5$	C$_{12}$H$_{10}$P	CO	–	85, **5**, *13*; 88, **6**, *7*
C$_9$H$_{21}$O$_3$P	C$_{12}$H$_{10}$P	C$_5$H$_5$	CO		85, **5**, *13*; 88, **6**, *7*
C$_9$H$_{21}$P	C$_5$H$_5$	–	CO	–	140, **11**, *2, 3*
C$_9$H$_{21}$P	C$_5$H$_5$	C$_{12}$H$_{10}$P	CO	–	84, **5**, *8*
C$_9$H$_{21}$P	C$_{12}$H$_{10}$	C$_5$H$_5$	CO		84, **5**, *8*
C$_{10}$H$_8$	–	–	CO	–	13, **2**, *3*
C$_{10}$H$_8$O$_2$	–	–	CO	–	33, **3**, *38*
C$_{10}$H$_{10}$	–	–	CO	–	96
C$_{10}$H$_{10}$Ge	C$_5$H$_5$	–	CO	–	117, **8**, *23*
C$_{10}$H$_{10}$MgN$_2$	C$_5$H$_5$	–	CO	–	143/4
C$_{10}$H$_{10}$Sn	C$_5$H$_5$	–	CO	–	126, **9**, *23*
C$_{10}$H$_{12}$	–	–	CO	–	17, **2**, *20*
C$_{10}$H$_{12}$O$_4$	–	–	CO	–	30, **3**, *14*
C$_{10}$H$_{14}$	–	–	CO	–	37, **3**, *63*
C$_{10}$H$_{14}$O	–	–	CO	–	92/3
C$_{10}$H$_{16}$O$_2$	–	–	CO	–	35, **3**, *47*

$C_{10}H_{17}P$	–	–	CO	–	77/8
$C_{10}H_{20}O_4Si_2$	–	–	CO	–	37, **3**, *61*
$C_{12}H_4F_{12}O_2$	–	–	CO	–	34, **3**, *45*
$C_{12}H_{10}As$	C_5H_5	–	CO	–	84, **5**, *5*
$C_{12}H_{10}Ge$	C_5H_5	–	CO	–	117, **8**, *24*
$C_{12}H_{10}O_4S_2Sn$	C_5H_5	–	CO	–	124, **9**, *11*
$C_{12}H_{10}P$	$C_3H_9O_3P$	C_5H_5	CO	–	84, **5**, *7;* 85, **5**, *12;* 88, **6**, *5*
$C_{12}H_{10}P$	C_5H_5	–	CO	–	84, **5**, *3;* 88, **6**, *2;* 158, **14**, *8*
$C_{12}H_{10}P$	C_5H_5	–	CO	H	169
$C_{12}H_{10}P$	C_5H_5	$C_3H_9O_3P$	CO	–	84, **5**, *7;* 85, **5**, *12;* 88, **6**, *5*
$C_{12}H_{10}P$	C_5H_5	$C_6H_{15}O_3P$	CO	–	88, **6**, *6*
$C_{12}H_{10}P$	C_5H_5	$C_6H_{15}P$	CO	–	84, **5**, *6;* 85, **5**, *11;* 88, **6**, *3*
$C_{12}H_{10}P$	C_5H_5	$C_9H_{21}O_3P$	CO	–	85, **5**, *13;* 88, **6**, *7*
$C_{12}H_{10}P$	C_5H_5	$C_9H_{21}O_3P$	CO	–	84, **5**, *8*
$C_{12}H_{10}P$	C_5H_5	$C_{18}H_{15}O_3P$	CO	–	84, **5**, *9;* 85, **5**, *14;* 88, **6**, *8*
$C_{12}H_{10}P$	C_5H_5	$C_{18}H_{15}P$	CO	–	88, **6**, *4*
$C_{12}H_{10}P$	C_5H_5	$C_{25}H_{22}P_2$	CO	–	85, **5**, *15;* 88, **6**, *9*
$C_{12}H_{10}P$	C_5H_5	$C_{26}H_{24}P_2$	CO	–	84, **5**, *10*
$C_{12}H_{10}P$	$C_6H_{15}O_3P$	C_5H_5	CO	–	88, **6**, *6*
$C_{12}H_{10}P$	$C_6H_{15}P$	C_5H_5	CO	–	84, **5**, *6;* 85, **5**, *11;* 88, **6**, *3*
$C_{12}H_{10}P$	$C_9H_{21}O_3P$	C_5H_5	CO	–	85, **5**, *13;* 88, **6**, *7*
$C_{12}H_{10}P$	$C_9H_{21}P$	C_5H_5	CO	–	84, **5**, *8*
$C_{12}H_{10}P$	$C_{18}H_{15}O_3P$	C_5H_5	CO	–	84, **5**, *9;* 85, **5**, *14;* 88, **6**, *8*
$C_{12}H_{10}P$	$C_{18}H_{15}P$	C_5H_5	CO	–	88, **6**, *4*
$C_{12}H_{10}P$	$C_{25}H_{22}P_2$	C_5H_5	CO	–	85, **5**, *15;* 88, **6**, *9*
$C_{12}H_{10}P$	$C_{26}H_{24}P_2$	C_5H_5	CO	–	84, **5**, *10*
$C_{12}H_{10}S_2Sn$	C_5H_5	–	CO	–	124, **9**, *8*
$C_{12}H_{10}Sb$	C_5H_5	–	CO	–	160, **14**, *24*
$C_{12}H_{10}Si$	C_5H_5	–	CO	–	115, **8**, *6*
$C_{12}H_{10}Sn$	C_5H_5	–	CO	–	126, **9**, *25*
$C_{12}H_{11}P$	C_5H_5	$C_{24}H_{20}P_2$	–	–	103
$C_{12}H_{11}P$	$C_{24}H_{20}P_2$	C_5H_5	–	–	103

$C_{12}H_{12}$	–	–	CO	–	32, **3**, *24/6*
$C_{12}H_{12}O_8$	–	–	CO	–	33, **3**, *30*
$C_{12}H_{13}As$	–	–	CO	–	9
$C_{12}H_{13}P$	–	–	CO	–	8
$C_{12}H_{14}Sn$	C_5H_5	–	CO	–	126, **9**, *24*
$C_{12}H_{16}$	–	–	CO	–	37, **3**, *64*
$C_{12}H_{20}$					
$-C(C_4H_9-t)=CH-CH=$ $C(C_4H_9-t)-$	–	–	CO	–	29, **3**, *8*
$-C(C_4H_9-t)=CH-$ $C(C_4H_9-t)=CH-$	–	–	CO	–	30, **3**, *9*
$-C_4(C_2H_5)_4-$	–	–	CO	–	31, **3**, *16*
$C_{12}H_{24}N_4O_4Sn$	–	–	CO	–	9
$C_{12}H_{27}O_3P$	C_5H_5	–	CO	$SnCl_2$	141, **11**, *8*
$C_{12}H_{27}P$	C_4H_4	–	CO	–	22, **2a**, *1;* 23
$C_{12}H_{27}P$	C_5H_5	–	CO	$SnCl_2$	140, **11**, *4*
$C_{12}H_{27}P$	$C_{28}H_{16}$	–	CO	–	22, **2a**, *7;* 23
$C_{12}H_{27}P$	$C_{28}H_{20}$	–	CO	–	22, **2a**, *4;* 23; 60
$C_{13}H_{20}O$					
$-C_4H_2(C_4H_9-t)_2CO-$	–	–	CO	–	65, **4**, *3, 4*
$-C_4(C_2H_5)_4CO-$	–	–	CO	–	65, **4**, *6*
$C_{14}H_8Cl_2$	–	–	CO	–	38, **3**, *73*
$C_{14}H_{10}$	–	–	CO	–	38, **3**, *72*
$C_{14}H_{20}$	–	–	CO	–	16, **2**, *18*
$C_{14}H_{24}$	–	–	CO	–	15, **2**, *10*
$C_{15}H_{16}$	–	–	CO	–	16, **2**, *17*
$C_{15}H_{27}ClCoO_3PSn$	C_5H_5	C_6H_7	CO	–	127, **9**, *44*
$C_{15}H_{27}ClCoO_3PSn$	C_6H_7	C_5H_5	CO	–	127, **9**, *44*
$C_{16}H_{10}Br_2$	–	–	CO	–	36, **3**, *57*
$C_{16}H_{10}Mo_2O_6Sn$	C_5H_5	–	CO	–	126, **9**, *26*
$C_{16}H_{12}$					
$C_6H_5CH=C=C=C=$ CHC_6H_5	–	–	CO	–	14, **2**, *6*
$(C_6H_5)_2C=C=C=CH_2$	–	–	CO	–	14, **2**, *7*
$-CH=C(C_6H_5)-$ $C(C_6H_5)=CH-$	–	–	CO	–	30, **3**, *10*
$-C(C_6H_5)=CH-CH=$ $C(C_6H_5)-$	–	–	CO	–	30, **3**, *11*
$-C(C_6H_5)=CH-$ $C(C_6H_5)=CH-$	–	–	CO	–	30, **3**, *12*
$C_{16}H_{12}O_2$	–	–	CO	–	34, **3**, *41*
$C_{16}H_{20}$	–	–	CO	–	33, **3**, *32*
$C_{16}H_{20}O_8$	–	–	CO	–	33, **3**, *31*
$C_{16}H_{36}O_4Si_4$	–	–	CO	–	35, **3**, *50*

$C_{17}H_{12}O$	–	–	CO	–	66, **4**, *12*
$C_{17}H_{20}O$	–	–	CO	–	66, **4**, *16*
$C_{17}H_{28}O$	–	–	CO	–	66, **4**, *13*
$C_{18}H_{12}Cl_2$	–	–	CO	–	2, **1**, *9*
$C_{18}H_{12}O_4$	–	–	CO	–	34, **3**, *44*
$C_{18}H_{14}$	–	–	CO	–	2, **1**, *8*
$C_{18}H_{15}ClPPt$	C_5H_5	–	CO	–	149
$C_{18}H_{15}O_3P$	C_4H_4	–	CO	–	22, **2a**, *3*
$C_{18}H_{15}O_3P$	C_5H_5	–	CO	SnCl$_2^\bullet$	141, **11**, *9*
$C_{18}H_{15}O_3P$	C_5H_5	$C_{12}H_{10}$	CO	–	84, **5**, *9;* 85, **5**, *14;* 88, **6**, *8*
$C_{18}H_{15}O_3P$	$C_{12}H_{10}$	C_5H_5	CO	–	84, **5**, *9;* 85, **5**, *14;* 88, **6**, *8*
$C_{18}H_{15}O_3P$	$C_{28}H_{16}$	–	CO	–	22, **2a**, *9*
$C_{18}H_{15}O_3P$	$C_{28}H_{20}$	–	CO	–	22, **2a**, *6*
$C_{18}H_{15}P$	C_4H_4	–	CO	–	22, **2a**, *2;* 60
$C_{18}H_{15}P$	C_5H_5	$C_{12}H_{10}P$	CO	–	88, **6**, *4*
$C_{18}H_{15}P$	$C_{12}H_{10}P$	C_5H_5	CO	–	88, **6**, *4*
$C_{18}H_{15}P$	$C_{28}H_{16}$	–	CO	–	22, **2a**, *8*
$C_{18}H_{15}P$	$C_{28}H_{20}$	–	CO	–	22, **2a**, *5;* 60/1
$C_{18}H_{16}$	–	–	CO	–	31, **3**, *17, 18*
$C_{18}H_{21}P$	–	–	CO	–	8/9
$C_{18}H_{28}O_4$	–	–	CO	–	35, **3**, *54*
$C_{19}H_{16}O$	–	–	CO	–	66, **4**, *11;* 67, **4**, *17, 18*
$C_{19}H_{18}O_2$	–	–	CO	–	35, **3**, *49*
$C_{19}H_{28}O_5S$	–	–	CO	–	36, **3**, *55*
$C_{20}H_{16}O_4$	–	–	CO	–	36, **3**, *56*
$C_{20}H_{20}$	–	–	CO	–	31, **3**, *19*
$C_{20}H_{20}MgN_4$	C_5H_5	–	CO	–	144
$C_{21}H_{16}O_5$	–	–	CO	–	66, **4**, *15*
$C_{21}H_{20}O$	–	–	CO	–	66, **4**, *14*
$C_{22}H_{28}Si_2$	–	–	CO	–	36, **3**, *58*
$C_{22}H_{36}$	–	–	CO	–	17, **2**, *21*
$C_{23}H_{19}F_6OP$	–	–	CO	–	76
$C_{23}H_{28}OSi_2$	–	–	CO	–	67, **4**, *19*
$C_{24}H_{15}Br_3$	–	–	CO	–	67, **4**, *22*
$C_{24}H_{20}P_2$	C_5H_5	$C_{12}H_{11}P$	–	–	103
$C_{24}H_{20}P_2$	$C_{12}H_{11}P$	C_5H_5	–	–	103
$C_{24}H_{28}$					
$-C_4(C_6H_5)_2-$	–	–	CO	–	36, **3**, *59*
$(C_4H_9-t)_2-$					

cyclo-$C_4(C_6H_5)_2$-$(C_4H_9-t)_2$	–	–	CO	–	79/80
t-$C_4H_9(C_6H_5)$-C=C=C=C($C_6H_5)C_4H_9$-t	–	–	CO	–	15, **2**, *11*
$C_{24}H_{32}$	–	–	CO	–	37, **3**, *65*
$C_{25}H_{22}P_2$	C_2H_3N	C_5H_5	CO	–	167, **16**, *17*
$C_{25}H_{22}P_2$	C_5H_5	–	CO	–	166, **16**, *3, 4*
$C_{25}H_{22}P_2$	C_5H_5	–	CO	I	162, **15**, *1*
$C_{25}H_{22}P_2$	C_5H_5	C_2H_3N	CO	–	167, **16**, *17*
$C_{25}H_{22}P_2$	C_5H_5	$C_{12}H_{10}P$	CO	–	85, **5**, *15;* 88, **6**, *9*
$C_{25}H_{22}P$	$C_{12}H_{10}P$	C_5H_5	CO	–	85, **5**, *15;* 88, **6**, *9*
$C_{25}H_{29}OP$	–	–	CO	–	76
$C_{26}H_{20}P_2$	C_5H_5	–	CO	–	166, **16**, *9/12*
$C_{26}H_{20}P_2$	C_5H_5	–	CO	Br	163, **15**, *8*
$C_{26}H_{20}P_2$	C_5H_5	–	CO	Cl	163, **15**, *7*
$C_{26}H_{20}P_2$	C_5H_5	–	CO	I	163, **15**, *9*
$C_{26}H_{20}P_2$	C_5H_5	–	CO	$SnCl_3$	163, **15**, *10*
$C_{26}H_{24}$	–	–	CO	–	15, **2**, *12*
$C_{26}H_{24}P_2$	C_2H_3N	C_5H_5	CO	–	167, **16**, *18*
$C_{26}H_{24}P_2$	C_2H_3O	C_5H_5	CO	–	162, **15**, *5*
$C_{26}H_{24}P_2$	C_5H_5	–	CO	–	166, **16**, *5/7*
$C_{26}H_{24}P_2$	C_5H_5	–	CO	Br	162, **15**, *3*
$C_{26}H_{24}P_2$	C_5H_5	–	CO	I	162, **15**, *4*
$C_{26}H_{24}P_2$	C_5H_5	–	–	N_2	103
$C_{26}H_{24}P_2$	C_5H_5	C_2H_3N	CO	–	167, **16**, *18*
$C_{26}H_{24}P_2$	C_5H_5	C_2H_3O	CO	–	162, **15**, *5*
$C_{26}H_{24}P_2$	C_5H_5	$C_{12}H_{10}P$	CO	–	84, **5**, *10*
$C_{26}H_{24}P_2$	$C_{12}H_{10}P$	C_5H_5	CO	–	84, **5**, *10*
$C_{26}H_{36}$	–	–	CO	–	37, **3**, *66*
$C_{27}H_{26}P_2$	C_5H_5	–	CO	–	166, **16**, *8;* 167, **16**, *19*
$C_{27}H_{26}P_2$	C_5H_5	–	CO	I	163, **15**, *6*
$C_{27}H_{29}O_5P$	–	–	CO	–	76/7
$C_{28}F_{20}$	–	–	CO	–	31, **3**, *22*
$C_{28}H_{16}$	–	–	CO	–	17, **2**, *19*
$C_{28}H_{16}$	$C_{12}H_{27}P$	–	CO	–	22, **2a**, *7;* 23
$C_{28}H_{16}$	$C_{18}H_{15}O_3P$	–	CO	–	22, **2a**, *9*
$C_{28}H_{16}$	$C_{18}H_{15}P$	–	CO	–	22, **2a**, *8*
$C_{28}H_{16}Cl_4$	–	–	CO	–	31, **3**, *21*
$C_{28}H_{20}$					
$(C_6H_4)_2C=C=C=$$C(C_6H_5)_2$	–	–	CO	–	16, **2**, *13*
$(C_6H_5)_2C=C=C=$$C(C_6H_5)_2$	$C_{12}H_{27}P$	–	CO	–	22, **2a**, *4;* 23

$(C_6H_5)_2C=C=C=$ $C(C_6H_5)_2$	$C_{18}H_{15}O_3P$	–	CO	–	22, **2a**, *6*
$(C_6H_5)_2C=C=C=$ $C(C_6H_5)_2$	$C_{18}H_{15}P$	–	CO	–	22, **2a**, *5*
$-C_4(C_6H_5)_4-$	–	–	CO	–	31, **3**, *20*
Tetraphenylcyclo-butadiene	–	–	CO	–	80
Tetraphenylcyclo-butadiene	$C_{12}H_{27}P$	–	CO	–	60
Tetraphenylcyclo-butadiene	$C_{18}H_{15}P$	–	CO	–	60/1
$C_{28}H_{24}P_2$	C_5H_5	–	CO		167, **16**, *13*
$C_{28}H_{24}P_2$	C_5H_5	–	CO	Br	163, **15**, *11*
$C_{28}H_{40}$	–	–	CO	–	37, **3**, *67, 68*
$C_{29}H_{16}Cl_4O$	–	–	CO	–	66, **4**, *9*
$C_{29}H_{20}O$	–	–	CO	–	65, **4**, *8*
$C_{30}H_{24}$	–	–	CO	–	16, **2**, *14*
$C_{30}H_{24}O_2$	–	–	CO	–	16, **2**, *15*
$C_{30}H_{29}O_3P$	–	–	CO	–	76
$C_{30}H_{44}$	–	–	CO	–	38, **3**, *69*
$C_{32}H_{20}$	–	–	CO	–	32, **3**, *27/9*
$C_{33}H_{20}O$	–	–	CO	–	67, **4**, *20*
$C_{33}H_{29}OP$	–	–	CO	–	76
$C_{34}H_{25}P$	–	–	CO	–	78
$C_{35}H_{25}$	–	–	CO	Hg	145, **12**, *4*
$C_{36}H_{24}$	–	–	CO	–	16, **2**, *16*
$C_{36}H_{30}NiP_2$	C_5H_5	–	CO	–	149
$C_{36}H_{30}P_2Pt$	C_5H_5	–	CO	–	149
$C_{52}H_{48}Cl_2CoP_4$	C_5H_5	–	CO	–	167, **16**, *15*
$C_{54}H_{52}Cl_2CoP_4$	C_5H_5	–	CO	–	167, **16**, *16*
$C_{100}H_{70}Cu_4Ir_2P_2$	–	–	CO		81
$C_{108}H_{86}Cu_4Ir_2P_2$	–	–	CO	–	81

Table of Conversion Factors

Force	N	dyn	kg
1 N (Newton)	1	10^5	0.1019716
1 dyn	10^{-5}	1	1.019716×10^{-6}
1 kg	9.80665	9.80665×10^5	1

Pressure	Pa	bar	kg/m²	at	atm	Torr	lb/in²
1 Pa (Pascal) = 1 N/m²	1	10^{-5}	1.019716×10^{-1}	1.019716×10^{-5}	0.986923×10^{-5}	0.750062×10^{-2}	145.038×10^{-6}
1 bar = 10^6 dyn/cm²	10^5	1	10.19716×10^3	1.019716	0.986923	750.062	14.5038
1 kg/m² = 1 mm H_2O	9.80665	0.980665×10^{-4}	1	10^{-4}	0.967841×10^{-4}	0.735559×10^{-1}	1.42233×10^{-3}
1 at = 1 kg/cm²	0.980665×10^5	0.980665	10^4	1	0.967841	735.559	14.2233
1 atm = 760 Torr	101325	1.01325	1.033227×10^4	1.033227	1	760	14.69595
1 Torr = 1 mm Hg	133.3224	1.333224×10^{-3}	13.59510	1.359510×10^{-3}	1.315789×10^{-3}	1	19.3368×10^{-3}
1 lb/in² = 1 psi	6.89476×10^3	68.9476×10^{-3}	703.070	70.3070×10^{-3}	68.0460×10^{-3}	51.7128	1

Work, Energy, Heat	J	kWh	kcal	Btu	MeV
1 J (Joule) = 1 Ws = 1 Nm = 10^7 erg	1	2.778×10^{-7}	2.388×10^{-4}	9.478×10^{-4}	6.242×10^{12}
1 kWh	3.6×10^6	1	859.845	3412.14	2.247×10^{19}
1 kcal	4186.8	1.163×10^{-3}	1	3.96832	2.614×10^{16}
1 Btu (British thermal unit)	1055.06	2.931×10^{-4}	0.251996	1	6.586×10^{15}
1 MeV	1.602×10^{-13}	4.45×10^{-20}	3.82×10^{-17}	1.518×10^{-15}	1

Power	kW	PS	kg m/s	kcal/s
1 kW = 10^{10} erg/s	1	1.35962	101.9716	0.238846
1 PS	0.735499	1	75	0.1757
1 kg m/s	9.807×10^{-3}	0.0133333	1	2.342×10^{-3}
1 kcal/s	4.1868	5.692	426.939	1

according to: Kraftwerk Union Information. Technical and Economic Data on Power Engineering. Mühlheim (Ruhr) 1978.

References:

1) International Union of Pure and Applied Chemistry. Manual of Symbols and Terminology for Physicochemical Quantities and Units. Butterworth, London 1970.
2) The International System of Units (SI). National Bureau of Standards Specl. Publ. 330. 1972 Edition.
3) H. Ebert, Physikalisches Taschenbuch. 5th Ed., Vieweg, Wiesbaden 1976.
4) F.W. Küster, A. Thiel, K. Fischbeck, Logarithmische Rechentafeln. 101st Ed., W. de Gruyter, Berlin 1972.
5) E. Padelt, H. Laporte, Einheiten und Größenarten der Naturwissenschaften. 3rd Ed., VEB Fachbuchverlag, Leipzig 1976.
6) H.J. Gray, A. Isaacs, A New Dictionary of Physics. 2nd Ed., Longman, London 1975, p. 587/98.
7) Balser, Kayser, Das internationale System der Einheiten. Umrechnungsfaktoren aller englischen und deutschen Maßeinheiten in das SI. Verlag Heisler, Stuttgart 1967.
8) J.F. Cordes, Das neue internationale Einheitensystem, Naturwissenschaften **59** [1972] 177/82.

Reihenfolge (Systemnummern) der im Gesamtwerk behandelten Elemente

Gmelin System of Elements and Compounds

System-Nr.	Symbol	Element	System-Nr.	Symbol	Element
1		Edelgase	37	In	Indium
2	H	Wasserstoff	38	Tl	Thallium
3	O	Sauerstoff	39	Sc	Scandium
4	N	Stickstoff		Y	Yttrium
5	F	Fluor		La	Lanthan
6	Cl	Chlor		Ce–Lu	Lanthanide
7	Br	Brom	40	Ac	Actinium
8	J	Jod	41	Ti	Titan
	At	Astat	42	Zr	Zirkonium
9	S	Schwefel	43	Hf	Hafnium
10	Se	Selen	44	Th	Thorium
11	Te	Tellur	45	Ge	Germanium
12	Po	Polonium	46	Sn	Zinn
13	B	Bor	47	Pb	Blei
14	C	Kohlenstoff	48	V	Vanadium
15	Si	Silicium	49	Nb	Niob
16	P	Phosphor	50	Ta	Tantal
17	As	Arsen	51	Pa	Protactinium
18	Sb	Antimon	52	Cr	Chrom
19	Bi	Wismut	53	Mo	Molybdän
20	Li	Lithium	54	W	Wolfram
21	Na	Natrium	55	U	Uran
22	K	Kalium	56	Mn	Mangan
23	NH_4	Ammonium	57	Ni	Nickel
24	Rb	Rubidium	58	Co	Kobalt
25	Cs	Caesium	59	Fe	Eisen
	Fr	Francium	60	Cu	Kupfer
26	Be	Beryllium	61	Ag	Silber
27	Mg	Magnesium	62	Au	Gold
28	Ca	Calcium	63	Ru	Ruthenium
29	Sr	Strontium	64	Rh	Rhodium
30	Ba	Barium	65	Pd	Palladium
31	Ra	Radium	66	Os	Osmium
32	Zn	Zink	67	Ir	Iridium
33	Cd	Cadmium	68	Pt	Platin
34	Hg	Quecksilber	69	Tc	Technetium[1])
35	Al	Aluminium	70	Re	Rhenium
36	Ga	Gallium	71	Np,Pu...	Transurane[2])

HCl

$CrCl_2$

$ZnCrO_4$

$ZnCl_2$

Dem einzelnen Element werden alle Verbindungen mit denjenigen Elementen zugeordnet, die im Gmelin-System vor diesem Element stehen. Bei dem Element Zink mit der System-Nr. 32 stehen z.B. alle Verbindungen mit den Elementen der System-Nr. 1 bis 31.

The material under each element number contains all information on the element itself as well as on all compounds with other elements which precede this element in the Gmelin System.
For example, zinc (system number 32) as well as all zinc compounds with elements numbered from 1 to 31 are classified under number 32.

---

[1]) Diese System-Nr. ist im Jahre 1941 unter der Bezeichnung „Masurium" erschienen.
[2]) Bearbeitung erfolgt im Rahmen des Ergänzungswerkes zur 8. Auflage.

Periodensystem der Elemente mit Gmelin Systemnummern siehe Innenseite des vorderen Deckels